The World Food Economy

The World Food Economy

**Douglas Southgate,
Douglas H. Graham, and
Luther Tweeten**

Department of Agricultural, Environmental,
and Development Economics
The Ohio State University
Columbus, OH, USA

John Wiley & Sons, Inc.

VP and Publisher	George Hoffman
Acquisition Editor	Lacey Vitetta
Senior Editorial Assistant	Emily McGee
Assistant Marketing Manager	Diane Mars
Marketing Assistant	Laura Finley
Production Manager	Janis Soo
Assistant Production Editor	Elaine S. Chew
Cover Designer	Seng Ping Ngieng
Cover Photo	© SuperStock

This book was set in 10/12.5 Point Palatino Roman by Thomson Digital, Noida, India and printed and bound by Courier Westford, Inc. The cover was printed by Courier Westford, Inc.

This book is printed on acid free paper. ⊗

Founded in 1807, John Wiley & Sons, Inc. has been a valued source of knowledge and understanding for more than 200 years, helping people around the world meet their needs and fulfill their aspirations. Our company is built on a foundation of principles that include responsibility to the communities we serve and where we live and work. In 2008, we launched a Corporate Citizenship Initiative, a global effort to address the environmental, social, economic, and ethical challenges we face in our business. Among the issues we are addressing are carbon impact, paper specifications and procurement, ethical conduct within our business and among our vendors, and community and charitable support. For more information, please visit our website: www.wiley.com/go/citizenship.

Copyright © 2011 John Wiley & Sons, Inc. All rights reserved. No part of this publication may be reproduced, stored in a retrieval system or transmitted in any form or by any means, electronic, mechanical, photocopying, recording, scanning or otherwise, except as permitted under Sections 107 or 108 of the 1976 United States Copyright Act, without either the prior written permission of the Publisher, or authorization through payment of the appropriate per-copy fee to the Copyright Clearance Center, Inc. 222 Rosewood Drive, Danvers, MA 01923, website www.copyright.com. Requests to the Publisher for permission should be addressed to the Permissions Department, John Wiley & Sons, Inc., 111 River Street, Hoboken, NJ 07030-5774, (201)748-6011, fax (201)748-6008, website http://www.wiley.com/go/permissions.

Evaluation copies are provided to qualified academics and professionals for review purposes only, for use in their courses during the next academic year. These copies are licensed and may not be sold or transferred to a third party. Upon completion of the review period, please return the evaluation copy to Wiley. Return instructions and a free of charge return shipping label are available at www.wiley.com/returnlabel. Outside of the United States, please contact your local representative.

of Congress Cataloging-in-Publication Data

, Douglas DeWitt, 1952-
  world food economy / Douglas Southgate, Douglas H. Graham, and Luther Tweeten.—2nd ed.
    p. cm.
  s bibliographical references and index.
  8-0-470-59362-2 (pbk.)
    —Economic aspects. 2. Food supply. I. Graham, Douglas H. II. Tweeten, Luther G.

4 2010

2010043303

tates of America

Contents

Preface to the Second Edition — ix
Acknowledgments — xi

1 **Introduction** — 1
   1.1 Our Focus — 4
   1.2 Chapter Outline — 7
   Key Words and Terms — 10
   Study Questions — 10

2 **The Demand Side: How Population Growth and Higher Incomes Affect Food Consumption** — 11
   2.1 Classic Malthusianism, Its Modern Variants, and Its Critiques — 12
   2.2 Demographic Transition — 20
   2.3 Trends in Human Numbers, Past and Present — 31
   2.4 Food Consumption and Income — 34
   2.5 Demand Trends and Projections — 38
   2.6 Summary and Conclusions — 41
   Key Words and Terms — 41
   Study Questions — 42
   Appendix: The Fundamental Economics of Demand — 43

3 **The Supply Side: Agricultural Production and Its Determinants** — 49
   3.1 The Nature of Agriculture — 50
   3.2 Increases in Agricultural Supply — 59
   3.3 Has Intensification Run Its Course? — 75
   3.4 The Food Economy Beyond the Farm Gate — 82
   3.5 Trends in Per Capita Production — 85
   Key Words and Terms — 87
   Study Questions — 88
   Appendix: The Fundamental Economics of Supply — 89

4 Aligning the Consumption and Production of Food over Time — 95
  4.1 The Desirability of Competitive Equilibrium — 97
  4.2 Public Policy and Markets for Farm Products — 100
  4.3 Historical Trends in the Scarcity of Agricultural Products — 106
  4.4 Outlook for the Twenty-First Century — 111
  Key Words and Terms — 113
  Study Questions — 114
  Appendix: The Coordination of Decentralized Decision Making — 114

5 Agriculture and the Environment — 122
  5.1 Diagnosing and Correcting Environmental Market Failure — 124
  5.2 Agriculture and Climate Change — 127
  5.3 Farmland Degradation — 131
  5.4 Agriculture and Deforestation — 136
  5.5 Agricultural Development and the Environment — 140
  Key Words and Terms — 144
  Study Questions — 144
  Appendix: Market Failure and Its Remedies — 145

6 Globalization and Agriculture — 151
  6.1 The Theory of Comparative Advantage — 152
  6.2 Trade Distortions and the Economic Impacts — 154
  6.3 The Debate over Globalization — 158
  6.4 Potential Gains from Agricultural Trade Liberalization — 161
  6.5 Multilateral Trade Negotiations and Agriculture — 165
  6.6 The Case for Free Trade Still Stands — 170
  Key Words and Terms — 171
  Study Questions — 172
  Appendix: Two-Country Illustrations of Comparative Advantage — 172

7 Agriculture and Economic Development — 178
  7.1 Economic Expansion and Structural Transformation — 180
  7.2 Agriculture's Role in Economic Development — 187
  7.3 Trying to Develop at Agriculture's Expense — 189
  7.4 Agricultural Development for the Sake of Economic Growth and Diversification — 193
  7.5 Summary and Conclusions — 196
  Key Words and Terms — 197
  Study Questions — 197

8	**Striving for Food Security**	**199**
	8.1 Who and Where Are the Food-Insecure?	200
	8.2 Achieving Food Security	204
	8.3 The Food Security Synthesis and Economic Development	208
	8.4 The Standard Model and Communitarian Values	217
	Key Words and Terms	220
	Study Questions	220
9	**Interregional Differences and Similarities**	**221**
	9.1 Economic Growth and Income Distribution	221
	9.2 Population Dynamics	226
	9.3 Agriculture's Response to Demand Growth	234
	9.4 Summary	238
	Key Words and Terms	240
	Study Questions	240
10	**Affluent Nations**	**241**
	10.1 Standards of Living	242
	10.2 Population Dynamics	245
	10.3 The Food Economy	249
	10.4 Dietary Change and Consumption Trends	259
	10.5 Summary	264
	Key Words and Terms	266
	Study Questions	266
11	**Asia**	**267**
	11.1 Trends in GDP per Capita	268
	11.2 Population Dynamics	274
	11.3 Agricultural Development	279
	11.4 Dietary Change, Consumption Trends, and Food Security	285
	11.5 Summary	290
	Key Words and Terms	291
	Study Questions	291
12	**Latin America and the Caribbean**	**292**
	12.1 Trends in GDP per Capita	293
	12.2 Population Dynamics	298
	12.3 Agricultural Development	302
	12.4 Dietary Change, Consumption Trends, and Food Security	311
	12.5 Summary	315
	Key Words and Terms	318
	Study Questions	318

13	**The Middle East and North Africa**	**319**
	13.1 Political Realities and Economic Trends	320
	13.2 Population Dynamics	324
	13.3 Agricultural Development	328
	13.4 Dietary Change, Consumption Trends, and Food Security	335
	13.5 Summary	339
	Key Words and Terms	340
	Study Questions	340
14	**Eastern Europe and the Former Soviet Union**	**341**
	14.1 Economic Trends since the Fall of Communism	343
	14.2 Demographic Trends	349
	14.3 The Agricultural Sector	353
	14.4 Dietary Change, Consumption Trends, and Food Security	362
	14.5 Summary	370
	Key Words and Terms	371
	Study Questions	371
15	**Sub-Saharan Africa**	**372**
	15.1 Trends in GDP per Capita	375
	15.2 Demographic Trends	380
	15.3 Agricultural Development	386
	15.4 Consumption Trends and Food Security	395
	15.5 Summary	402
	Key Words and Terms	403
	Study Questions	403
16	**The Global Food Economy in the Twenty-First Century**	**404**
	16.1 Victims of Our Own Success?	407
	16.2 The New Food Economy	410
	16.3 The Changing Role of Government	412
	16.4 Back to the Future Food Economy?	414
	Key Words and Terms	418
	Study Questions	418
Abbreviations and Acronyms		*419*
Map Annex		*421*
References		*427*
Index		*441*

Preface to the Second Edition

Since the first edition of this book was released, in October 2006, the global food economy has grabbed headlines regularly. Spiking prices of dietary staples in late 2007 and the first half of 2008 shook the public's complacency about the availability of affordable food. Moreover, policies that make farm products expensive, such as trade restrictions and biofuel subsidies, were singled out for criticism.

We have revised *The World Food Economy* in response to recent developments and also have filled gaps in the original edition of the book. For example, the discussion of long-term improvements in human well-being in Chapter 2, which addresses the demand for food, now draws on the insights of economist Robert Fogel, who won a Nobel Prize in part for documenting the profound and varied benefits of better nutrition. A new section on energy and agriculture and another on agribusinesses have been added to Chapter 3, which is about the supply side of the food economy. Also, we now examine recent price movements in Chapter 4 (markets and public policy), what might be done about climate change in Chapter 5 (agriculture and the environment), recent multilateral trade negotiations in Chapter 6 (agriculture and globalization), and the continuing struggle against hunger in Chapter 8 (food security).

In the closing chapter, we also ponder issues highlighted in popular writings and films about the food system in places like the United States. Some of this media coverage causes us to wonder whether agriculture, which now produces commodities in enormous volumes with a tiny share of the workforce, might have become a victim of its own success. This possibility is worth considering in light of proposals to go back to the future in various ways in the food economy.

Throughout the book, information about the world food economy has been updated. Also, details of economic analysis formerly contained in the main text have been moved to appendices in Chapters 5 and 6—exactly as is done in Chapters 2, 3, and 4. We have adopted this approach for the sake of readers desiring a thorough treatment of our subject matter, but with minimal reference to economic theory. Also new are the highlighting and listing of key words and terms. We hope this new feature enhances the value of *The World Food Economy* as a textbook.

Like the first edition, this new version of the book provides a broad introduction to the economics of food and agriculture. It does not really substitute for more specialized literature. In Chapter 2, for example, trends in consumption are examined by drawing on the bare essentials of demand theory. Likewise, fundamental economic ideas about production are harnessed in Chapter 3 to interpret the broad course of agricultural development. The same approach is followed in the rest of the book. No doubt, readers who have taken an economics course or two will grasp some material a little quicker than those with no prior exposure to the "dismal science." However, this book has been written with a general audience in mind.

Like practically all economists, the three authors are convinced that goods and services are best allocated in free markets. Although we are not shy about pointing out the advantages of markets in this book, the value of a functional government is not neglected. For one thing, the state is in a unique position to satisfy the institutional prerequisites of a market economy, such as reliable enforcement of contracts and property rights. Moreover, governments play a central role in providing public goods, such as transportation infrastructure and research and development. The authors highlight the importance of governmental contributions such as these to a successful, market-based economy.

Finally, PowerPoint slides containing tables, figures, and maps are available on the Instructor's Companion Site, at www.wiley.com/college/southgate. Answers to end-of-chapter study questions are provided there as well. In addition, the authors are happy to respond to queries about the book; just contact Douglas Southgate, at southgate.1@osu.edu.

Acknowledgments

As we have worked on the two editions of this book, we have benefited enormously from advice and insights shared by numerous colleagues and students. Various faculty members at The Ohio State University have enlightened us on key facets of the global food economy, suggested useful readings, or both. Others have reviewed draft chapters or served as a sympathetic sounding board. We are grateful to all of them: Dale Adams, Trevor Arscott, Joyce Chen, Wen Chern, William Flinn, D. Lynn Forster, Claudio González-Vega, Richard Hamilton, Eugene Jones, David Kraybill, Rattan Lal, Allan Lines, Richard Meyer, Norman Rask, Ian Sheldon, Brent Sohngen, Donald Thomas, Cameron Thraen, Stanley Thompson, and Carl Zulauf.

Similar assistance has been provided by colleagues at other institutions: Dennis Avery, who directs the Center for Global Food Issues; Ian Coxhead of the University of Wisconsin; Robert Gitter of Ohio Wesleyan University; Donald Hertzmark, an economic consultant based in Washington; John Sanbrailo, Executive Director of the Pan American Development Foundation; Sarah Scherr of the University of Maryland; and Robert Thompson of the University of Illinois.

Sarah Lowder, formerly with the U.N. Economics and Social Commission for Asia and the Pacific and now working in the U.N. Food and Agriculture Organization (FAO), has provided detailed information about food aid, which we have used in both editions of this book. Ms. Lowder earned her doctorate at Ohio State, as did another individual who has been generous with her advice: Maria Pagura, also of the FAO. Information and other help have been provided as well by Caroline Boin, Julian Morris, and Kendra Okonski of the International Policy Network in London.

Draft chapters of this edition were examined by Philipp Aerni (Swiss Federal Institute of Technology), Kyosti Aravuori (University of Helsinki), Thomas DeGregori (University of Houston), Drew L.

Kershen (University of Oklahoma), T. E. Kleynhans (Stellenbosch University), Sharon May (Maryville College), Dianne Meredith (University of California at Davis), Julia Paxton (Ohio University), David Schodt (St. Olaf College), David Stead (University College Dublin), Alex Winter-Nelson (University of Illinois), and one anonymous reviewer. In addition, Donald Balmer, emeritus professor of political science at Lewis and Clark College, and Mark Elwood, a highly respected teacher of history from Worthington, Ohio, provided useful feedback on Chapters 1 through 4.

Data contained in tables in Chapters 9 through 15 were originally compiled by a pair of individuals who were studying at Ohio State at the time: Elizabeth Myre and Myriam Elizabeth Southgate. Ms. Southgate began the process of updating these tables for the second edition. Ariana González, a 2010 graduate of Northwestern University, continued this work. Cristina Graham-Nagy then completed the tables and gave everything a thorough editing.

With only a couple of exceptions, all figures in the book were prepared by Janice DiCarolis, who works with us in Ohio State's Department of Agricultural, Environmental, and Development Economics. Two other members of the departmental staff have logged many hours on this book: Judy Luke and Susan Sheller. Another staff member in the department, Mathew Gonzales, helped in the preparation of various tables.

We would be remiss if we did not thank students from Ohio State and a number of other institutions in the United States and elsewhere who have contacted us, occasionally to point out a confusing passage but also to express appreciation for a better understanding of the global food economy.

The first edition of this book was written thanks to the interest and encouragement of Seth Ditchik, formerly an editor at Basil Blackwell. Likewise, crucial guidance for the second edition has been provided by George Lobell, our editor at Wiley-Blackwell until early 2010, as well as his successor, Lacey Vitetta. We are grateful to them as well as their current and former colleagues: Constance Adler, Elaine Chew, Laura Finley, Emily McGee, Laura Stearn, Elizabeth Wald, and Stephanie Zicko.

Finally, we have received steadfast support from our wives—Myriam, Jane, and Eloyce. Sincerely grateful for their patience, we dedicate this new edition of *The World Food Economy* to them.

1

Introduction

As recently as 2006, inexpensive food seemed to be the right and natural order of things. Corrected for inflation, **prices** of farm products had been at historic lows for a couple of decades. As a result, hundreds of millions of poor people who otherwise would have gone hungry could afford an adequate diet. Cheap food also allowed everyone to increase his or her purchases of other goods and services, thereby stimulating the economic diversification that is an essential part of **economic development**. At the same time, savings increased. This in turn spurred investment and **economic growth**, as development requires.

To be sure, not everyone eats well even when food prices are low. Civil war and ethnic cleansing can easily lead to mass starvation. Just as alarming as scenes from refugee camps in places like the Horn of Africa, however, are reports issued periodically by the U.N. **Food and Agriculture Organization (FAO)** that hundreds of millions of human beings do not eat enough for a healthy and productive life.

But unsolved problems in the late twentieth century and the early part of this century were much less severe than those of the 1960s and 1970s, when warnings of imminent famine were rife. Four decades ago, environmentalist Paul Ehrlich insisted that many parts of the world were beyond hope. Where there was no reasonable chance of feeding a country's population from domestic production and imports, he called for the immediate cessation of all donations (Ehrlich, 1968, pp. 141–149). Ehrlich's views were not unique by any means. According to an influential report issued nearly 40 years ago, the world economy would collapse not long after the turn of the twenty-first century barring immediate curbs on human reproduction and economic growth (Meadows et al., 1972).

As we now know, alarm of this sort was not justified by later events. People have not bred themselves to oblivion and the world is not running short of edible output. Nevertheless, popular thinking about the race between food demand and food availability continues to be swayed by exponents of gloom and doom from times past. For those still influenced by Ehrlich and others like him, a magnificent human accomplishment remains obscured. In the face of demographic expansion of unprecedented proportions, food supplies have not just kept pace, but consistently have outstripped increases in demand.

Yet cheap nourishment cannot be taken for granted and, in case a reminder was needed, the food economy commanded everyone's attention again a few years ago. As recently as July 2007, a bushel of wheat changed hands for $6 and soybeans were bought and sold for less than $10/bushel. Six months later, the prices of wheat and soybeans had shot past $10/bushel and $13/bushel, respectively, and were still rising. After June 2008, markets softened and prices fell most of (though not all) the way back to their former levels. Reflecting on this experience, we are obliged to study why food availability increased faster than food demand for many years, precisely so that the race between demand and supply is not lost in the future.

The first thing to understand about the food economy is that production did not go up during the second half of the twentieth century because farmers were responding to higher prices, for the simple reason that food is less expensive now than it used to be. Did supplies increase because the agricultural workforce was growing? No. In many countries, that workforce actually shrank. This was true in the United States, where crop and livestock output has reached unprecedented levels and yet where the farm population today is less than one-sixth what it was in 1940 (Gardner, 2002, p. 94). In many other parts of the world, the farm population did not decline, although output growth greatly exceeded increases in agricultural employment.

The expansion of farmed area explains no more than a small share of the growth in food output since the middle of the twentieth century. Instead, the main reason for this growth has been **technological improvement**, which has boosted **yields** (i.e., output from a given area of farmland). Thanks to this improvement, much more production is obtained today from a given amount of inputs—or, equivalently, the same output is now produced with far fewer inputs.

Another reason for increased abundance is that **specialization and trade** have proliferated in the food economy. Where commercial exchange is minimal and self-sufficiency is the norm, as is the case in the poorest parts of the world, most of the working population labors mightily to grow paltry amounts of food. All available land that lends itself to agriculture is cultivated, as are many fields that are poorly suited to farming. Relative to everyone's meager earnings, food is expensive; for a typical family, the cost of a minimally adequate diet represents a serious financial burden.

Precisely the opposite circumstances are observed where specialization and trade flourish, as happens in a robust **market economy**. Most people work outside of agriculture. As a rule, farming is confined to settings well suited to crop production and non-agricultural uses of land, including nature reserves, are extensive. Rather than being acutely scarce, food is so available that its expense is well within the reach of the vast majority of households.

Food economies of the kind that exist in the United States do not emerge spontaneously. To unleash market forces, the state needs to establish an enabling institutional environment—an environment expressed by the term **rule of law** and characterized by transparent and even-handed enforcement of contracts and property rights. Government also facilitates commercial exchange by underwriting the **infrastructure** needed for transportation and communications. Without this infrastructure, trade is impeded and rational specialization is difficult. Furthermore, gains from specialization are enhanced by productivity-enhancing investments in **human capital** and new technology—investments often made possible in one way or another by government.

As market forces are unleashed, specialization reaches an advanced stage within the food economy. Rather than being self-sufficient, farmers purchase agricultural inputs and marketing and processing services from **agribusinesses**. The ultimate beneficiaries of the cost savings created by this exchange are consumers. In the United States, food expenditures, including payments for restaurant meals, amount to 10 percent of total household income and the food ingredients supplied by farms make up no more than one-fifth of these expenditures, or 2 percent of consumers' earnings (Gardner, 2002, p. 155). The other four-fifths of what Americans spend on nourishment are payments for the inputs and services provided, with great efficiency, by agribusinesses.

Agribusinesses are also a source of technological innovation. Over the years, advances in machinery and seeds have come from the private sector. Today, major corporations are dominant players in **biotechnology**. These enterprises, then, are no less important than farmers and the government. Indeed, it is useful to think of the food economy as a tripod, with each leg supporting and being supported by the other two legs. Just as this economy could not exist without crop and livestock producers, its efficiency would be seriously impaired if either government or agribusinesses were absent.

As the food economy develops, the impacts reverberate far and wide. Aside from improved nutrition and diminished hunger, accelerated economic diversification, and investment and growth, people start living their lives in profoundly different ways. For example, they stop having children merely to augment the family's stock of (unskilled) labor. Instead, more consideration is given to the rewards of educating and training offspring. In particular, the number of children in each household falls so that parents can invest more in the skills and capabilities of each child. Family size also declines because women enjoy better opportunities for education and employment. Among the consequences of the decline in family size caused by rising prosperity and female economic empowerment is, obviously, decelerating growth in the demand for food.

Compared with pessimistic assessments of the adequacy of food supplies that grabbed headlines in the late 1960s and early 1970s, today's concerns about the global food economy are less urgent. More than 80 percent of humankind is adequately fed and will continue to be so for the foreseeable future (Chapter 8). Moreover, easing the suffering of the hungry minority of the world population would not be prohibitively expensive, but rather is well within our reach. If the specter of global famine truly loomed, any and all measures to boost agricultural production would be universally applauded. The very strength of current opposition to some of these measures—including the biotechnological process of modifying one agricultural species by inserting a gene from an entirely different part of the plant or animal kingdom—is evidence of the progress made to date in supplying people with enough of the food they desire at prices they can afford.

While pessimism about the adequacy of food supplies is not warranted, neither is complacency. Much of the world invests too little in agriculture and its improvement. Moreover, governmental interference with market forces diminishes incentives for food production in many places. These are some of the reasons why hundreds of millions of people eat too little and the food economy is not contributing as much as it could to everyone's well-being.

1.1 Our Focus

Given the stakes involved, the world's food economy merits careful examination. In this book, trends in demand and supply are analyzed, those during the recent past as well as those anticipated in the years to come. While properly acknowledging the decline in food **scarcity** that occurred during the second half of the twentieth century, we emphasize that further declines are not guaranteed and that fully alleviating hunger remains a challenge.

To assess what the future holds for producers and consumers of food, one may suppose that trends from the recent past will continue into the future. However, simple extrapolation of this sort would be misleading. During the last 100 years, the number of mouths to be fed skyrocketed, from a little more than 1.5 billion in 1900 to a little less than 2.5 billion in 1950 and ultimately surpassing 6.0 billion by 2000. But even before the turn of the century, there were clear signs that population growth was decelerating and, hence, that food demand will not increase as quickly during the next few decades as it did in the 1900s. Likewise, the pace of supply growth is changing. In particular, the rise in crop yields that has coincided with gains in agricultural productivity may slacken. It is possible that demand growth, slower though it is sure to be, will match increments in supply during the first part of the twenty-first century. If so, the trend toward diminished food scarcity that was experienced during the second half of the twentieth century will stall. Scarcity, and thus food prices, might even go up for a while.

For much of the human race, a departure from recent trends would hardly be noticeable. As already mentioned, the value of unprocessed agricultural commodities is roughly 2 percent of total household income in the United States. Even a 50 percent increase in the former value, which is more pessimistic than most available forecasts, would only oblige the typical American family to reallocate 1 percent of its budget. But for hundreds of millions of people in South Asia, Sub-Saharan Africa, and other impoverished regions, any increase in scarcity would create real hardship. For example, spiking prices in 2007 and 2008 pushed as many as 155 million people into the sort of absolute poverty that leads to nutritional deprivation (World Bank, 2008).

Increased food scarcity has impacts aside from the suffering that poor people endure if the things they eat become more expensive. Some of these impacts are environmental. For example, rising prices of crops and livestock cause agricultural land use to expand. As this happens, natural habitats and the benefits they provide, such as the conservation of biological diversity, are lost. Other consequences of rising food prices are more narrowly economic. It bears repeating that diminished food scarcity has contributed to growth and development by increasing savings and investment and accelerating economic diversification. If food becomes scarcer in a place where a sizable portion of household earnings is spent on edible goods, then the economy diversifies at a slower pace and investment and growth diminish. Also, low-income nations are not the only losers if they must struggle harder to escape hunger and poverty.

Their trading partners, actual as well as potential, forego the benefits of the additional commerce that materialize whenever inexpensive food adds to purchasing power anywhere and everywhere.

To understand why food becomes more or less scarce as well as the various consequences of such a change, the conceptual framework that economics provides has no substitute. As emphasized at the beginning of any introductory textbook or course on the subject, economics examines how people respond to scarcity, which comes about whenever supply is limited relative to demand. Economists give special attention to specialization and trade as well as investment and technological improvement as ways to deal with scarcity.

The global food economy being as complex as it is, each and every response to scarcity cannot be analyzed exhaustively in a single volume. On the supply side, this book focuses more on **production agriculture** than on agribusiness. The former is a distinctive sort of enterprise, not least because of its being inseparable from the natural environment. In contrast, agribusinesses have a lot in common with firms outside the food economy. In this book, overall trends in the availability of farm commodities are examined. We also analyze the impacts of changes in land use and other inputs as well as improvements in productivity. Certain elements of supply—in particular, the cereal grains consumed directly by people or fed to livestock—are examined more than others.

On the demand side, the focus is likewise on overall trends, with special attention paid to population growth and improved living standards as primary drivers of changes over time in food consumption. Given this emphasis, food safety, public acceptance of **genetically modified organisms (GMOs)**, and related topics are not examined in great detail. Also touched on lightly are some important aspects of human nutrition. One of these is education aimed at informing people about what is good to eat and what is not. Another aspect has to do with improvements in water supplies and sanitation needed to eradicate intestinal parasites, which prevent hundreds of millions of people from receiving full nourishment from the food they consume. When it comes to economic development, our overriding concern is to clarify linkages between the agricultural sector and the rest of the economy and to analyze how these linkages are altered as an economy grows and diversifies, as opposed to exhausting the topic of economic development.

In this book, we do not duplicate the content of more specialized contributions to the economics literature. In a comprehensive study of food consumption, one that addresses all aspects of human nutrition and food safety, consumer behavior is modeled in sophisticated ways. Likewise, texts on agricultural economics contain a thorough analysis of such topics as farmers' choices among inputs and production methods, their response to risks created by variable weather and economic fluctuations, and the performance of markets in which they purchase inputs and sell outputs. By the same token, books on the economics of development provide a comprehensive treatment of topics in that field.

Recognizing the availability of specialized volumes, we have written this book precisely so that the overlap among food consumption, agricultural production, and development can be better understood. The pages that follow contain more observations about agricultural production and development economics than a book focused on consumption alone. Similarly, we look more carefully at population growth and its impact on food demand than the authors of agricultural economics texts normally do. In addition, while agriculture is relegated to little more than a single chapter in many books on development economics, linkages between agriculture and the rest of the economy are, to repeat, one of our major interests.

Aside from addressing the overlap among three interrelated topics, we try to derive maximum insights from minimal economic theory. Fundamental economic concepts, which are reviewed in the appendices of various chapters, are used to make sense of changing demands for food, trends in supply, as well as changes in the markets in which the things we eat are traded. The same fundamental concepts can be put to good use in an analysis of the economy-wide impacts of a rise or fall in food scarcity.

A particular benefit of not trying to duplicate volumes that use more advanced theory and have a narrower topical focus is that we are able to examine the availability of food not just at a global level, but at the regional and even national levels as well. This allows differences in the food economy from one country or part of the world to the next to be highlighted.

1.2 Chapter Outline

Our examination of the world's food economy begins, in Chapter 2, on the demand side. During the twentieth century, death rates plummeted in poor countries without simultaneous reductions in birth rates, so human numbers swelled. However, there is clear evidence that death and birth rates are now converging, particularly in Asia and Latin America. The global population could well peak midway through this century, perhaps within the lifetimes of students assigned to read this book in a college class. As demographic expansion slackens, other factors influencing the demand for food, most notably higher standards of living, will become more important.

Responses to demand growth are analyzed in Chapter 3. Attention is paid to production increases during the twentieth century as well as how output growth is likely to be achieved during the decades to come. The role of expanded agricultural land use is documented, as are its costs. So is the role of increased use of energy, water, agricultural chemicals, and other non-land inputs. Emphasis is placed on the impacts of productivity growth and the contributions that government and agribusinesses have made to this growth. We stress that, for food production to keep pace with demand in the future, investment that raises productivity must continue.

How markets bring consumption into line with production, thereby enhancing economic well-being, is discussed in Chapter 4. We describe the changes observed in food markets as demand goes up because of growth in population, income, and other variables, supply increases because of technological innovation and other reasons, or both. As demonstrated by events in 2007 and 2008, commodity prices are affected by government policies in various ways. In addition, we examine the downward trend in inflation-adjusted commodity values during the twentieth century as well as the prospects for continuation of this trend, which to repeat is not guaranteed.

For the most part, commercial exchange that is competitive and unregulated is enormously beneficial. However, the marketplace often leaves environmental problems unresolved. Some problems have to do with governmental interference with market forces: keeping water prices low, for instance. Others, perhaps including changes in the global climate, are a consequence of what economists refer to as market failure, a standard example being output that is too high and prices that are too low because producers escape paying for the harm they do to the environment. In Chapter 5, we survey the damage done in the food economy to natural resources because of market failure. Also examined is the environmental cost of misguided governmental intervention in the marketplace. Of special concern is the habitat loss that corresponds to an expansion of agricultural land use. Similarly important is land degradation, resulting from erosion and other processes. This problem is especially severe in Sub-Saharan Africa, not in spite of but because that region has benefited less than other parts of the world from advances in agricultural technology.

As reported in Chapter 6, a sizable part of the world's food supply is traded internationally. Also recapitulated is the case for specialization and commercial exchange across national borders—a case dating back to the early years of modern economics. Reviewed as well in the chapter is the inefficiency that has resulted as agricultural trade has been suppressed—inefficiency that harms poor countries that can produce crops inexpensively for international markets.

Chapter 7 focuses on relations between agriculture and the rest of the economy. The literature on the contributions that the former sector makes to general development is surveyed. Also assessed is the performance of different countries in which governmental treatment of agriculture varied markedly during the twentieth century. A lesson we stress is that governmental investment in education, science, and information services to raise agricultural productivity has had much higher economic pay-offs than governmental manipulation of prices and incomes. Also highlighted are the synergies between productivity growth in agriculture and dynamism in other sectors.

Lest a review of broad trends in demand, supply, and commodity values causes us to forget the many people who still go hungry, we acquaint the reader in Chapter 8 with the dimensions of food insecurity around the world. Nearly 7 billion people are alive today, of whom as many as 1 billion lack economic access to the food required for a healthy and productive life. This deplorable situation has less to do with scarce supplies than with the penurious incomes of those suffering hunger. Economic development of the kind that lifts people out of poverty, then, must be the centerpiece of a strategy for alleviating food insecurity.

The rest of the book is regionally organized. Food economies differ in terms of climate and natural resources, history and culture, and recent economic development. Interregional differences are summarized in Chapter 9. Each of the six chapters that follow focuses on a single part of the world: the world's affluent places (Western Europe, North America, Japan, South Korea, and Oceania); Asia; Latin America and the Caribbean; the Middle East and North Africa; Eastern Europe and the Former Soviet Union; and Sub-Saharan Africa. Data on principal features of the food economy at the national level are presented in each of these regional chapters, in which patterns of economic growth, population dynamics, agricultural development, and food consumption and security are analyzed.

This book's survey of the world food economy yields a number of conclusions, which are summarized in Chapter 16. One is that, if policies are put in place that encourage the proper utilization of natural resources and current and emerging technologies, production of food and other goods and services should continue to increase, as has happened in recent decades. Moreover, the benefits of this expansion can be widely distributed and the food-insecure portion of the human population can keep on declining. As the specter of hunger recedes, other concerns about the food economy will grow more prominent. Among these are the environmental impacts that might result as new technologies for crop and livestock production are developed and applied, thereby making food less scarce. Also of mounting concern is the poor health that results from over-eating—a phenomenon observed not just in the United States and other wealthy countries, but also in places where famine threatened just a generation or two ago. We expect that age-old challenges will persist in the food economy during the twenty-first century as novel problems emerge.

Key Words and Terms

agribusinesses	market economy
biotechnology	price
economic development	production agriculture
economic growth	rule of law
Food and Agriculture Organization (FAO)	scarcity
genetically modified organism (GMO)	specialization and trade
human capital	technological improvement
infrastructure	yield

Study Questions

1. In the late 1960s and early 1970s, Paul Ehrlich and others forecast imminent famine throughout the world. Describe what has happened and what has not happened in the food economy to avert this disaster.
2. What are the three main elements of the food economy and how are these elements interrelated?
3. Describe the economy-wide impacts of agricultural development.
4. Compare and contrast the impacts of scarcer agricultural commodities in a poor country and in an affluent place, such as the United States.
5. Compare and contrast this volume with textbooks on agricultural economics and development economics.

2

The Demand Side

How Population Growth and Higher Incomes Affect Food Consumption

---

Trends in food consumption used to be easy to understand. Little attention needed to be paid to living standards since the vast majority of people barely eked out a living and since variations in earnings were negligible, not just from year to year but from generation to generation. So growth in human numbers was the only reason why food demand increased. If the population was going up by x percent, then food consumption was rising by x percent as well.

Estimates of demand growth such as these were positively alarming whenever it was supposed that human numbers were on an exponential trajectory. If the population compounded steadily, one could identify its **doubling time** by applying the **Rule of 72**: that is, dividing 72 by the annual growth rate. For example, Ecuador's population was increasing by 2.8 percent per annum during the early 1980s (World Bank, 2009b), which implies a doubling time of a little more than 25 years (72 ÷ 2.8 = 25.7). Had that rate not diminished, as indeed it has, then the country's population, which was a little under 8 million in 1980, would have reached 16 million in 2005, 32 million in 2030, and so forth. It would not have taken long for Ecuador's capacity to feed itself from domestic production and imports to be overwhelmed.

However, no exponential trend lasts indefinitely. One possible brake on human numbers—the purported tendency of food supplies always to increase slower than the population—was emphasized 200 years ago by Thomas Malthus, the author of a classic treatise on demography. Still influential, his observations comprise a suitable point of departure for this chapter's discussion of trends in food demand. But as we shall see, profound demographic changes have occurred around the world since the late 1700s—changes having much to do with economic progress and the empowerment of women, which happen to go hand in hand.

Much of this chapter is devoted to analyzing recent trends in population, trends that could well lead to a stabilization of human numbers in just a few decades. We also stress that changes in living standards are bound to affect food demand more as population growth wanes.

2.1 Classic Malthusianism, Its Modern Variants, and Its Critiques

An ordained clergyman in the Church of England, Thomas Malthus was also the first professor of what was then called political economy (Pullen, 1987). Born in 1766, he witnessed the dawn of the Industrial Revolution, when the steam engine and other technology were first harnessed for the mass production of textiles and other goods. But agriculture was still an important part of the British economy, accounting for a large share of total employment (Cipolla, 1965, p. 67). The sector therefore commanded the attention of Malthus and other practitioners of **classical economics** (as we now call it) of the late eighteenth and early nineteenth centuries—including David Ricardo, who shaped our thinking about specialization and trade in a market economy (Chapter 6).

Before Malthus was born, demographic expansion accelerated, both in Great Britain and on the European continent. A sensible question for him and others to consider, then, was how long this expansion could continue (Pullen, 1987). In addressing this question, Malthus (1963) formulated his **principle of population**, which rests on a simple characterization of human behavior. Strikingly at variance with modern views, to say the least, this characterization is best expressed by two words: vice and misery.

Vice, by which Malthus meant profligate (and procreative) sex, was posited to arise whenever food supplies exceed what is required for bare subsistence. That is, any positive difference between what people eat and minimum dietary requirements would not only keep hunger and illness at bay, but would also cause people to reproduce fast enough for the population to grow exponentially (e.g., from 1,000 to 1,100, to 1,210, to 1,331, etc.). But if consumption was below bare subsistence, because people had reproduced beyond the number that can be adequately fed, then misery would be experienced, with deaths caused by starvation and disease exceeding births. Malthus contended that the only way to avoid either exponential growth or demographic contraction is for human diets to be barely adequate, nothing more and nothing less. Otherwise, vice would set in or its opposite, misery.

Malthus conceded that population growth is possible over the long term, yet constrained by trends in food availability. Furthermore, he contended that, unlike human numbers, agricultural output rises linearly (e.g., from 1,000 to 1,100, to 1,200, to 1,300, etc.) over time, not exponentially. This view is reasonable if technological advances in agriculture are insignificant.

Of course, an exponential trajectory for population cannot be maintained indefinitely if food availability is following a linear path. Furthermore, the collision between the two trends can be wrenching if rapid reproduction continues even as the threshold of bare subsistence is being passed. This possibility is illustrated in Figure 2.1, in which an upward-sloping line depicts the trend in agricultural output and the other curve represents population (which varies nonlinearly as the years pass) multiplied by a minimally adequate diet.

At an initial date, 0, food supplies exceed subsistence requirements. According to Malthus's principle of population, this abundance induces vice and an exponential increase in the number of people and the amount of food they must eat. But by year t^*, basic dietary needs have drawn even with food availability, so an overshoot is imminent. As human numbers continue growing exponentially, average consumption falls

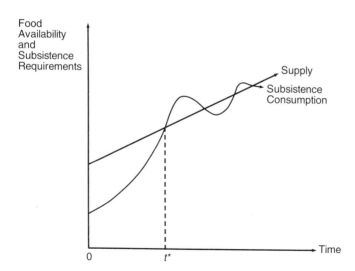

Figure 2.1 Malthusian trends in food demand and supply

below the subsistence threshold. Soon afterwards, the population contracts because of disease and starvation. As shown in Figure 2.1, the amplitude of the alternating cycle of overshoot and collapse may diminish over time, with consumption levels eventually settling in at bare subsistence.

Overshoot and collapse can happen even if the principle of population, as originally conceived by Malthus, is not really operative. In particular, people might have too many children if their individual choices about reproduction are not influenced by the wider environmental consequences. To explain this sort of decision making, Hardin (1968) coined the term, **tragedy of the commons** (Box 2.1). But regardless of whether the principle of population is realistic, there is a tragedy of the commons, or a combination of the two, humankind's lot is to experience periodic episodes when the population contracts due to food shortages. As depicted in Figure 2.1, even the good times between these episodes set the stage for subsequent demographic collapse. In light of findings of this sort, is it any wonder that the scholarly discipline founded by Malthus and other classical economists often has been referred to as the **dismal science**?

Box 2.1   Overpopulation as a tragedy of the commons

At a time when concerns about overpopulation and famine were reaching a crescendo, Hardin (1968) did not blame these problems on human ignorance—a failure to take note of dwindling per capita food supplies, for example. Instead, his explanation focused on the discrepancy between the interests of individual households and those of society as a whole.

To understand excessive reproduction as a tragedy of the commons, bear in mind that a typical household stands to gain from bringing another child into the world—in terms of the net contributions he or she makes to household earnings, for example. But while parents can be counted on to assess how the well-being of their household is affected by additional offspring, they neglect other impacts of population growth, such as diminished per capita food supplies for other people. In other words, the costs of reproduction are largely shared, rather than being shouldered entirely by individual households. As a result, reproduction is excessive.

Neo-Malthusianism

To recognize that the implications of Malthus's ideas are grim does not constitute a refutation, of course. As some see it, these ideas have not been disproved at all. Indeed, the classic **Malthusian** model has proved resilient, its variants cropping up regularly through the years.

In the early 1970s, for example, the **Club of Rome**—an influential group of civic, governmental, and business leaders from around the world—commissioned a team of natural scientists and systems engineers to study the long-term consequences of demographic and economic expansion (Meadows et al., 1972). The model these experts developed features a series of feedback loops, both positive and negative. For example, economic growth leads to investment, which in turn stimulates growth. Also, growth lowers environmental quality, which increases mortality and dampens growth. But the model's core is Malthusian. Trends in population, industrial output, pollution, and so on are all exponential. If allowed to run unchecked, these trends inevitably overshoot limits imposed by the world's fixed endowments of fossil fuels, farmland, and other natural resources.

Meadows et al. (1972) predicted that, in the absence of measures such as comprehensive birth control and tight limits on resource use, a few decades of exponential growth in the human population and its economy would exhaust available resources. The ensuing collapse would be permanent. Rather than foreseeing oscillations of the population around a linear, though positively sloped, trend in food availability, as depicted in Figure 2.1, the Club of Rome's experts expected that the current age of rapid growth and industrialization would be succeeded by conditions similar or even inferior to what preceded it. As shown in Figure 2.2, the world's population would experience a precipitous and irreversible decline. Indeed, per capita supplies of food and industrial products during the twenty-first century would end up lower than the levels in 1900, when accelerated growth was barely under way.

The alarming predictions made by Meadows et al. (1972) have much to do with their model's not incorporating an intrinsic feature of economic life—namely, markets. Among other things, markets constitute an important feedback mechanism for dealing with resource scarcity. Consider, for example, the various ways that producers and consumers of petroleum respond to its growing more scarce, due to demographic or economic expansion. Sellers can be counted on to raise prices. Higher prices, in turn, accelerate the search for new supplies as well as the development

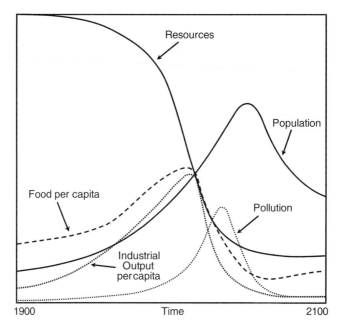

Figure 2.2 Standard run of Club of Rome model
Source: Meadows et al. (1972), p. 124.

of alternative energy sources and conservation technology. Even if exploration proves fruitless and technological progress does not happen, higher prices induce customers to cut back their purchases, thereby postponing the date when oil deposits are exhausted.

The omission of markets by Meadows et al. (1972) can be rationalized by observing that markets do not exist for various environmental resources, including the air we breathe and others of vital importance. As will be explained in Chapter 5, absent markets have to do with the absence of property rights, which are a prerequisite for commerce. But granted that various natural resources are not owned by anyone, the complete neglect of markets by Meadows et al. (1972) causes them to ignore all the substitution, conservation, and development of new resource supplies that take place as economically motivated firms and households respond to rising environmental scarcity.

Something else neglected in the classic Malthusian model and its later variants is the simple human desire for self-improvement and the results observed when people are individually empowered to act on that desire. Lasting improvements in living standards cannot be reconciled with people thriving or perishing pretty much as animals do, reproducing with a will whenever there is temporary respite from bare subsistence and plagued by misery at other times. By the same token, individual desires for a better life are all but impossible to factor into the Club of Rome model since it lacks the best setting for acting on these desires, which is the marketplace.

By going to the trouble to write a book, Meadows et al. (1972) implicitly conceded that at least some human beings can recognize an unsustainable trend when presented with convincing evidence and also can be counted on to support measures needed to avoid a catastrophic collapse. But individual adaptation to mounting scarcity in a market setting is not really a part of their model. Instead, coercive, regulatory action, presumably to be carried out by a central authority that is benevolent as well as capable, is emphasized.[1]

Is the World Dismally Malthusian?

Interestingly, Malthus himself did not entirely discount the possibility of sustained improvements in human well-being. In the second edition of his book, published five years after the first edition, the world's first professor of economics suggested that humankind's prospects were not entirely dismal.

> "(Though) our future prospects respecting the mitigation of the evils arising from the principle of population may not be so bright as we could wish, yet they are far from entirely disheartening and by no means preclude gradual and progressive improvement in human society" (quoted in Johnson, 2000, pp. 1–2).

Since these words were written, Malthus's optimism, guarded as it was, has been vindicated, to a degree and in ways that no one living 200 years ago could have imagined.

The reasons for the "progressive improvement in human society" that Malthus pondered and that has accelerated during the past two centuries have been investigated by a pair of economists at the University of Chicago: the late D. Gale Johnson and Nobel-laureate Robert Fogel. Both these economists have focused on the consequences of changing agricultural practices, such as mechanization and crop rotation during and since the 1700s, that have increased the amount of food raised by each farmer.

---

[1] Ehrlich (1969) referred to an extreme proposal in *The Population Bomb*. "One plan often mentioned involves the addition of temporary sterilants to water supplies or staple food. Doses of the antidote would be carefully rationed by the government to produce the desired population size" (p. 122).

Increases in the productivity of farm labor created a **second agricultural revolution** (Fogel, 2004a, p. 21) that has unleashed a process that Fogel and Costa (1997) call **technophysio evolution**: "a synergism between technological advances and physiological improvements that has produced a form of human evolution that is biological but not genetic, rapid, culturally transmitted, and not necessarily stable" (Fogel, 2004b, pp. 645–646). As explained in Box 2.2, technophysio evolution has had a profound impact on human well-being, including the adequacy of our diets and how long we live.

Box 2.2 The second agricultural revolution and technophysio evolution

The second agricultural revolution and the resulting gains in food output have created myriad benefits. For instance, the pauper class, which often was too weak to work due to inadequate diets and which comprised more than a tenth of the British population as recently as the middle 1800s, largely disappeared during the second half of the nineteenth century thanks to expanded food supplies (Fogel, 2004b). Moreover, the second agricultural revolution has set in motion technophysio evolution.

In simple terms, technophysio evolution has worked like this. Once output-per-farmer started to go up, thanks to advances such as the seed-planting drill (invented by Jethro Tull in 1701) and the winnowing machine (invented by James Sharp in 1777), people with better diets grew taller. Even more significantly, they also became more robust—as reflected in a rise in the average **body mass index (BMI)**. Too many people in the United States and other countries weigh too much these days (Chapters 10 and 16). But when the second agricultural revolution was getting under way and for a long time afterward, obesity was practically unheard of and a sizable portion of the population, including though not limited to the pauper class, had BMIs that were undesirably low. Under these circumstances, better nutrition and ensuing increases in **physiological capital**, as measured by average BMIs, made the population more resistant to infectious (or communicable) diseases, such as influenza and smallpox, which were the main cause of mortality not so long ago.

Better nutrition is the most important reason why physiological capital has accumulated in modern times, although broader access

> to clean water and sanitation services and, after World War II, the widespread use of antibiotics and other pharmaceuticals also have had an impact. Together, all these developments have allowed the human race to make an "escape from premature death," to quote the title of a recent book by Fogel (2004a).

Technophysio evolution has greatly increased labor productivity, in part because our physical stamina has improved. In addition, growth in output-per-farmer has allowed for the release of rural labor to nonagricultural sectors, such as industry. Furthermore, productivity gains have allowed for an expansion of research, development, and related activities, which in turn have created additional productivity gains. Over time, more output has been produced with fewer inputs of labor and other factors. As this has happened, technophysio evolution has continued.

Johnson (2000) has documented the profound benefits of these developments—benefits already enjoyed in or within reach of every part of the world, he emphasizes. For one thing, improvements in agricultural productivity have driven down the cost of food. Human nutrition has improved, and not just by a small amount and not just in a few places. Furthermore, people are considerably healthier today than they were 200 years ago, as indicated by increases in **life expectancy at birth**. Finally, people around the world have responded to these changes not by stepping up reproduction, but rather by doing the opposite. On average, women today have many fewer children than their not-too-distant female ancestors did.

Contrary to what Malthus and his contemporaries expected to happen, rapid population growth has coincided with significant improvements in the average person's diet. Human numbers did not rise above 1 billion until 1820 or so. Since then, an increase of more than 500 percent has occurred, the global population reaching 6 billion at the turn of the twenty-first century. Malthusian analysis would suggest that this increase should have been accompanied by a drop in per capita consumption of food. However, precisely the opposite has happened. During the 1600s, long before the principle of population was proposed and before the second agricultural revolution began, the typical European ate no more than 1,700 calories per day—about the same as average daily consumption throughout the world in the middle of the twentieth century. By the early 1960s, the global average had risen to 1,940 calories per day. During the 1990s, the global daily average was approaching 2,600 calories and less than 10 percent of the world's people lacked reliable access to 2,200 calories per day of food (Johnson, 2000).

Health improvements have been at least as significant. In 1900, **infant mortality rates** in Europe ranged from 88 per thousand live births in Norway to 221 per thousand in Austria (Cipolla, 1965, p. 83). China's rate 60 years later was in this same range: 157 per thousand. Since then, it has gone down to 62 per thousand (Johnson, 2000). Thanks mainly to improved survival of infants and young children, human longevity has increased. Two hundred years ago, English and French newborns had a life expectancy at birth of 30 years. By 1900, this number was approaching 50 in industrialized nations (Cipolla, 1965, p. 86). Life spans have shot up faster in India during the past century—from 23 years in 1900, to 32 during the 1940s, to 43 in 1960, to 62 years in 1996 (Johnson, 2000).

Living standards also have gotten better. Between 1500 and 1820, growth in income per capita was barely noticeable, amounting to just 0.04 percent per annum. From 1820 to 1992, the yearly rate of increase picked up to 1.21 percent. During the nineteenth century, growth in average incomes was concentrated in Europe and other places Europeans had colonized. But during the past 100 years, growth has become more universal, leaving few parts of the world unaffected, and if anything has accelerated in recent decades (Maddison, 1995, pp. 20–21).

According to Malthus's model, better diets, improved health, and higher incomes ought to have caused a skyrocketing of human fertility. However, his principle of population has proved to be a poor guide to recent trends. In some places, most notably China, strict controls on family size, of the sort advocated by Meadows et al. (1972) and other modern Malthusians, have been applied. But many other parts of the world have experienced declines in fertility comparable to China's without resorting to coercive measures. The latter result can be explained only in terms of people, especially women, deciding that their individual interests are best served by having fewer children and also being able to act accordingly.

2.2 Demographic Transition

The desire for fewer children has arisen and been acted on only after the beginning of the **demographic transition**, which historically got under way thanks to a sustained and unprecedented decline in mortality. With reference to Figure 2.3A, in which the transition is depicted abstractly,

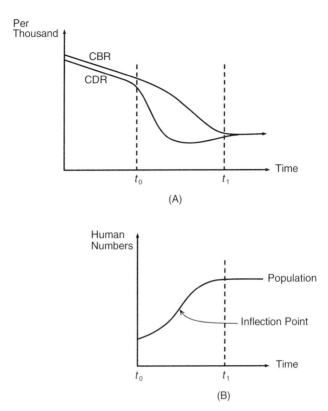

Figure 2.3 The demographic transition
A. Changes over time in birth and death rates
B. Changes over time in population

this decline occurs after date t_0. The entire process has run full course once the age composition of the population has stabilized, with old people comprising a larger share of total numbers than in times past (Lee, 2003). While each and every part of the world has experienced mortality decline and while fertility has fallen in all but a handful of nations, there is no country where age composition has reached equilibrium. Thus, demographic transition is still happening everywhere. However, quite a few places have reached the point represented in Figure 2.3A by t_1, at which human numbers have stopped increasing because the birth rate has fallen into line with the death rate.

By definition, a population is in **demographic equilibrium**—defined in terms of equivalent birth and death rates as well as a stable age composition—both before and after the transition. However, the two equilibria, pre- and post-transition, differ sharply. Characteristic of the impoverished, rural communities in which practically all humankind dwelt for millennia, the pre-transition state features annual **crude death rates (CDRs)** that are extremely high by the standards of the modern world: 40 per thousand or higher.[2] If there is no net immigration, then a community with a CDR in this range must have a **crude birth rate (CBR)** that is correspondingly high in order to avoid population contraction. Suppose a village loses 4.5 percent of its residents to illness every year. This means that there must be 45 live births annually per thousand villagers if demographic equilibrium is to be maintained. With males comprising about half the population, 9 percent of the female population must deliver a child each year. Of course, the percentage for those females who are of childbearing age is higher still. Prior to the demographic transition, then, the **total fertility rate (TFR)**, which is the median number of children that a woman delivers during her lifetime, is elevated. For example, TFRs 30 years ago in Honduras, Kenya, and a number of other countries where the transition was not far along exceeded six children per woman (World Bank, 2009b).

CDRs, CBRs, and TFRs all being high before the demographic transition begins—again, at t_0—these fundamental indicators decline to much lower levels as the transition proceeds. Conceivably, the transition could occur with no **natural increase**, which comprises the difference between the CBR and the CDR. All that would be needed is for birth and death rates to decline at the same pace. However, this never occurs. Instead, a positive gap opens up after t_0 because mortality drops off first. This reflects people's enthusiasm for availing themselves of food supplies, health care, and public sanitation as all these improve, thanks to development. In contrast, **contraception** is adopted more gradually during the early years of the demographic transition, which prevents the TFR from falling as rapidly as the CDR.

Even after family size starts shrinking, in response to factors such as better educational and job opportunities for women (see below), the CBR may not change much and almost always stays well above the CDR. This is partly a delayed consequence of better infant and child survival. Compared to what happened before the demographic transition began, many more of the babies born at date t_0 or soon afterward, when the TFR is close to its pre-transition level, live past their fifth birthday. Hence, large numbers of

---

[2]The convention among demographers is to state most rates as numbers per thousand, rather than as percentages. For example, a CDR of 18 per thousand is equivalent to 1.8 percent annual mortality.

females are reaching sexual maturity 15 to 20 years after the transition is set in motion by a drop in the CDR. Each of these women would have to deliver far fewer children than her mother or grandmother did for natural increase to be avoided. However, this never occurs, so natural increase is an intrinsic feature of the demographic transition.

As population specialists put it, the sizable female cohort reaching sexual maturity 15 to 20 years into the demographic transition creates **demographic momentum**. In addition to resulting from better infant survival, this momentum is sustained as long as TFRs remain above the **replacement level** of fertility, which is approximately 2.1 births per woman. But the replacement level is eventually reached, almost always beyond the inflection point in the time-path of total human numbers (Figure 2.3B). During this part of the demographic transition, which ends at date t_1 in Figure 2.3, the CBR falls toward the CBR.

One economist has offered a succinct explanation for the natural increase observed once the demographic transition has begun: "Rapid population growth commenced not because human beings suddenly started breeding like rabbits but rather because they stopped dying like flies" (Eberstadt, 1995, p. 21). People respond to the welcome achievement of diminished mortality, especially for the young, by cutting back on family size. However, holding the grim reaper at bay creates demographic momentum, as large numbers of females reach childbearing age. Accordingly, natural increase stays positive. As this momentum dissipates, which always happens eventually, birth and death rates come together at a low level. As shown in Figure 2.3B, human numbers are much higher once this equivalence is achieved than what they were before the demographic transition began.

The Modern Transition in the Developing World

Regardless of where and when it has happened, some basic features of the demographic transition are always observed, such as the lag between CDR declines and falling birth rates. However, specific trends in CBRs and CDRs have varied considerably from time to time and from place to place. By today's standards, the transition in northern and western Europe, which got under way 250 years ago or more, proceeded at a leisurely pace, with natural increase never exceeding 15 per thousand (equal to 1.5 percent) annually. In contrast, natural increase rose above 30 per thousand in a number of developing countries during the second half of the twentieth century, thanks mainly to precipitous drops in mortality. Although CBRs have fallen as well, the wide gaps that have opened up between birth and death rates in Africa, Asia, and Latin America explain two fundamental demographic facts. One is that human numbers have gone up faster since the Second World War than at any other time in history. The other is that most of the increase has occurred in developing countries.

These facts can be appreciated by examining historical trends in Great Britain and Scandinavia (Table 2.1). By the 1750s, CDRs in these places had fallen to or below 30 per thousand, while CBRs were around 35 per thousand—except for Finland, where the birth rate was 45 per thousand (column 1). During the next two centuries, the two rates went down slowly. Indeed, the decline in mortality halted at times. The CDR in England and Wales remained at 23 per thousand from the early to middle 1800s (columns 2 and 3). One reason for this is that people were moving to cities, where industrial employment was increasing and yet where living conditions were insalubrious—especially in crowded neighborhoods inhabited by the poor. Kremer (2002) points out that achieving lower mortality during this period, long before the discovery of antibiotics and other pharmaceutical products for combating disease, required a sizable investment in potable water systems and sanitary infrastructure. Progress was slow, as reflected in the data reported in Table 2.1. But after the middle of the nineteenth century, much of this infrastructure was in place and diets were improving (see above), which allowed for an appreciable drop in CDRs. Not coincidentally, natural increase peaked in England and Wales and Scandinavia during the late 1800s and early 1900s (column 4). CDRs did not stabilize at 9 to 12 per thousand until the middle 1900s, more than 200 years after the decline began, and CBRs reached this level even more recently (column 5).

If the demographic transition were proceeding this slowly today in Africa, Asia, and Latin America, the prospects for a stable population this century would be remote and we would be facing a global catastrophe of the sort predicted by the Club of Rome. However, the transition has been much faster in the developing world than it was in places like Great Britain and the Nordic nations. Consider mortality trends. Different from what occurred in Europe and North America, other parts of the world have not had to discover factors contributing to the spread of mortal illness on their own before dealing with these factors. Instead, Africa, Asia, and Latin America have been able to apply knowledge and technology originally developed elsewhere. Using pharmaceuticals to contain mortality, which was not an option in Europe during the 1800s, has turned out to be cheap and quick. This is one reason why CDRs have declined rapidly (Kremer, 2002). Due partly to **acquired immune deficiency syndrome (AIDS)**, which is caused by the spread of the **human**

Table 2.1 Crude birth and death rates in Great Britain and Scandinavia from 1750s to 2000

Country	1751–1755 (1)		1801–1805 (2)		1851–1855 (3)		1905–1909 (4)		1950 (5)		2000 (6)	
	CBR	CDR	CBR	CDR	CBR	CDR	CBR	CDR	CBR	CDR	CBR	CDR
England and Wales*	35	30	34	23	34	23	27	15	16	12	11	10
Finland	45	29	38	25	36	28	31	18	25	10	11	10
Norway	34	25	28	24	33	17	27	14	19	9	13	10
Sweden	37	26	31	24	32	22	26	15	16	10	10	10

*2000 data are for the United Kingdom as a whole.
Sources: Cipolla (1965, p. 81) for 1750s through 1950; World Bank (2009b) for 2000.

Table 2.2 Crude birth and death rates in selected countries from 1960 to 2007

Country	1960–1965		1970–1975		1985–1990		2005–2007	
	CBR	CDR	CBR	CDR	CBR	CDR	CBR	CDR
	(1)		(2)		(3)		(4)	
Tanzania	49	22	50	18	48	13	40	13
Bangladesh	47	21	48	20	40	14	25	8
Senegal	49	24	48	20	44	13	36	9
Indonesia	44	22	38	16	27	9	19	6
Syria	48	16	47	11	40	6	27	3
Thailand	41	10	33	8	22	7	15	9
Ecuador	45	13	39	10	32	7	21	5
Chile	34	11	25	8	23	6	15	5
Poland	17	7	19	9	16	10	10	10
Sweden	16	10	13	11	13	12	11	10

Source: World Bank (2009b).

immunodeficiency virus (HIV), Tanzania's CDR is still high. But in Bangladesh, Senegal, and many other nations where death rates 50 years ago (Table 2.2, column 1) were comparable to European rates in the early 1800s (Table 2.1, column 2), little more than three decades were required for CDRs to fall to the low single digits. The same adjustment took a century or more in Great Britain and Scandinavia.

With CDRs falling a lot during the twentieth century, natural increase has accelerated. Adding to the gap between birth and death rates was that CBRs were high in Africa, Asia, and Latin America before the demographic transition began. In half the countries listed in Table 2.2, birth rates in 1960 (column 1) were at least as high as Finland's rate in 1750. However, CBRs have dropped significantly in these countries, just as their CDRs have not taken long to decline. Once again, Tanzania is an exception. Although human fertility went down there during the late twentieth century, demographic momentum has kept the birth rate high. But elsewhere, the falloff has been dramatic. Whereas 200 years were required for Finland's CBR to drop from the middle 40s to the middle 20s, Bangladesh, Indonesia, and Syria are among the many developing countries that accomplished the same feat in about four decades (Table 2.2, column 4).

The schematic representation of the adjustment from high CBRs and CDRs to low rates contained in Figure 2.3A corresponds to the demographic transition that began in most of the developing world during the twentieth century. This transition has started with a sudden plunge in CDRs, which in many countries have stayed below the level observed at

t_1 (i.e., around 10 per thousand) for a number of years. An annual rate of 4 or 5 per thousand reflects the low average age of the population, which results from diminished mortality of infants and young children and which indicates the strength of demographic momentum both now and in the near future. For some time after t_0, the CBR does not change much, because the TFR remains elevated and because the effects of diminishing fertility are largely offset by demographic momentum. Obviously, natural increase accelerates as long as declines in the CBR do not come close to matching those in the death rate. However, this part of the transition is over once human fertility is sustained at or near the replacement level. Once this happens, natural increase is being driven almost entirely by demographic momentum. This momentum is not spent until t_1 in Figure 2.3, when as mentioned above human numbers reach a peak.

The Revolution in Human Fertility

While declining mortality catalyzes the demographic transition, the transition's duration is governed by how quickly human fertility falls from its pre-transition level. The numbers reported in columns 1 and 2 of Table 2.3 show that much of humanity truly experienced a revolution in human fertility in recent decades. In a single generation, TFRs in places like Bangladesh, Indonesia, and Syria have fallen by more than half. The median number of children for female Ecuadorians was five in 1980; for their daughters, a quarter century later, the number was not much above the replacement level. In Chile and Thailand, women now bear fewer than two children each. Human fertility even has gone down in Tanzania.

Family size rarely declines without the use of contraception, and the inverse relationship is strong between TFRs and the prevalence of birth control (Table 2.3, column 3). This prevalence is low in Tanzania and Senegal, where human fertility remains elevated. In contrast, widespread use of contraceptives coincides with women bearing fewer children on average, even where most people belong to religious faiths led by men who oppose birth control—for instance in Chile and Ecuador, where the majority is Roman Catholic, and in Bangladesh, Indonesia, and Syria, which are Muslim nations. The World Bank (2009b) does not report contraceptive prevalence for Poland and Sweden. But these are countries where women routinely do what is required to avoid having children, so the data are not worth collecting.

That birth control is widely practiced in places with low TFRs does not prove that the former is a fundamental, as opposed to a proximate, cause of the latter. Instead, contraception is something used by women who

Table 2.3 Fertility decline between 1980 and 2007 and causal factors in 2007 or recent years

Country	1980 TFR (births per woman) (1)	2007 TFR (births per woman) (2)	Percentage of Females 15 to 49 Years Old that Use Contraception (3)	Urbanized Percentage of Population (4)	GDP per capita (PPP $) (5)	Illiterate Percentage of Adult Females (6)	Infant Mortality Rate (deaths per 1000 births) (7)
Tanzania	6.7	5.2	26 (in 2005)	25	1,184	34	73
Bangladesh	6.1	2.8	56	27	1,247	52	47
Senegal	7.0	5.1	12 (in 2005)	42	1,737	67 (in 2006)	59
Indonesia	4.4	2.2	61 (in 2006)	50	3,711	11 (in 2006)	25
Syria	7.3	3.1	58 (in 2006)	54	4,276	23	15
Thailand	3.5	1.8	77 (in 2006)	33	7,394	7	6
Ecuador	5.0	2.6	73 (in 2004)	65	7,437	18	20
Chile	2.8	1.9	58 (in 2006)	88	13,858	3	8
Poland	2.3	1.3	Unavailable	61	16,089	1	6
Sweden	1.7	1.8	Unavailable	85	36,712	Unavailable	3

Source: World Bank (2009b).

have decided to limit fertility. This decision is influenced by various factors, some of which (e.g., religious affiliation) are noneconomic. However, diminished childbearing also reflects economic choices made by women and their partners—choices influenced by interrelated factors such as **urbanization**, living standards, female literacy, and infant mortality.

There is an inverse relationship between fertility and urbanization (Table 2.3, column 4). The reason why rural families tend to be larger, in the developing world and elsewhere, is that an extra pair of hands can always be employed fetching water or firewood, tending livestock, and so forth. Incentives are different in urban settings, where fewer people have large families merely out of a desire to have a large pool of family labor. Of course, access to contraception is better in urban settings than in the countryside. The same is true of opportunities for education and employment.

In the ten-country sample, Senegal is clearly a place where low urbanization helps to explain a high TFR. In contrast, two out of three Ecuadorians live in cities, which is one reason why the median number of children per woman has fallen below three. Bangladesh and Thailand seem to be outliers. But numbers for the former country, where two-thirds of the population are rural, are a little misleading. As humorist P. J. O'Rourke has gone to the trouble to find out, average population density in Bangladesh, where approximately 160 million people live (World Bank, 2009b), exceeds 1,000 inhabitants per square kilometer. This happens to be greater than the population density of Fremont, California—a suburban community located along the southeastern shore of the San Francisco Bay. In other words, much of the Bangladeshi countryside is more crowded than metropolitan areas in affluent parts of the world (O'Rourke, 1994, p. 53), which has an effect on human fertility. Thailand is more interesting: a place where other factors driving down the TFR more than offset the impacts of limited urbanization.

As is true of urbanization, living standards (Table 2.3, column 5), which are commonly gauged by **gross domestic product (GDP)** per capita with a correction for **purchasing power parity (PPP)**,[3] rise as development takes place and as a rule are inversely related to human fertility. Earnings constrain a household's "investment" in offspring—an investment that can be quantitative (i.e., having more children) or qualitative (e.g., spending more on the next generation's education), as Becker and

---

[3] A country's GDP equals the value of all goods and services produced within its borders. PPP adjustments reflect the degree to which prices in any given country, especially for nontraded services, are higher or lower than prices in the United States (Perkins, Radelet, and Lindauer, 2006, pp. 36–39).

Lewis (1976) stress. One expects most households to respond to income growth by increasing this investment. However, there are two reasons for TFRs to fall as GDP per capita increases. First, improved average earnings translate into a higher opportunity cost of time, including the time spent raising children. Second, the desire for smaller families is reinforced as rising affluence inclines parents toward qualitative investments in their offspring, rather than the quantitative alternative.

The general tendency of human fertility to go down as living standards go up is illustrated by a pair of Southeast Asian nations. As reported in Table 2.3, Indonesia's GDP per capita (column 5) is appreciably below Thailand's and its TFR is noticeably higher. But there are some interesting exceptions to this general relationship, including in prosperous nations that have experienced rising affluence along with increased fertility in recent years (Chapter 10).

No factor affects human fertility more than **female empowerment**, as often measured in terms of educational attainment.[4] Since women with more years of schooling have better employment options, the opportunity cost of time they spend raising children rises. For example, two-thirds of the women in Senegal can neither read nor write, which is why most of them have many offspring; but in Chile and Thailand, female illiteracy practically does not exist (Table 2.3, column 6), which facilitates women's participation in the labor force. Thailand is especially interesting because its TFR is comparable to China's, even though it never has used any of the draconian policies that the world's most populous nation has implemented to reduce fertility.

To recapitulate, women who are empowered, because they are educated and have access to jobs outside the home, generally have fewer children than women with neither schooling nor jobs. Decisions at the individual level depend on where a woman lives (i.e., in a city or in the countryside), how much she and other members of the household earn, and other economic circumstances. Also having an effect is the frequency with which young children die. In fact, the inverse relationship between the infant mortality rate (Table 2.3, column 7) and the TFR comprises clear proof that childbearing is, indeed, a conscious decision, at least for empowered women. As the chances that a newborn will live past his or her fifth birthday go up, such women stop going through an extra pregnancy or two merely so that they will end up with the desired number of

---

[4] Another indicator of female empowerment is women who work (outside the home) as a portion of the labor force. This indicator is referred to in this book's regionally focused chapters 10 through 15.

children. There is no better evidence that, once impediments to female empowerment are overcome, human fertility is driven by choices made by people desiring a better life, not Malthus's principle of population.

2.3 Trends in Human Numbers, Past and Present

As the data presented in Tables 2.1 and 2.2 indicate, the course of demographic transition varies considerably from country to country. Specific observations about population growth and its determinants in Africa, Latin America, and other parts of the globe are offered later in the book. The paragraphs that follow focus on long-term trends for the world as a whole.

The most obvious thing to notice about these broad trends is how quickly human numbers have shot up in recent centuries. Approximately 12,000 years ago, our ancestors, who numbered no more than 10 million, were taking advantage of the end of the last Ice Age to start raising crops (Deevey, 1960, cited in Matras, 1977, p. 34). By Roman times, ten millennia later, about 250 million human beings were alive (Clark, 1967, cited in Matras, 1977, p. 36). During the next 1,500 years, Asia, Europe, and other parts of the world suffered occasionally from devastating plagues. The population grew slowly, only reaching 427 million in 1500 (Table 2.4, column 7). The only place to experience demographic contraction since then has been Latin America, where most of the indigenous population succumbed quickly to smallpox, measles, and other communicable diseases that were unknown

Table 2.4   Estimated world population by region from 1500 to 2000 (millions)

Date	Africa (1)	Asia* (2)	Latin America (3)	Europe* (4)	North America (5)	Oceania (6)	World (7)
1500	85	225	40	74	1	2	427
1750	106	498	16	167	2	2	791
1800	107	630	24	208	7	2	978
1850	111	801	38	284	26	2	1262
1900	133	925	74	430	82	6	1650
1950	217	1335	162	572	166	13	2465
1975	400	2262	318	822	243	21	4066
2000	780	3460	513	960	314	30	6057

*Russia's entire population is included in European totals, not Asia's.
Sources: Matras (1977, p. 36) for 1500 through 1950; UNDP (2002b, pp. 162–165) for 1975 and 2000.

in the region before Europeans arrived (column 3). Otherwise, unprecedented growth has occurred during the last half millennium, with human numbers increasing 14-fold between 1500 and 2000 (column 7).

In addition to multiplying, the population has changed its distribution around the planet since 1500. When Columbus landed in the Caribbean, 18 percent of all human beings lived in Europe (including Asiatic Russia), North America, and Oceania (Table 2.4, columns 4, 5, and 6). More than four-fifths of the population was in Africa, Asia (excluding Russia), and Latin America (columns 1, 2, and 3). The latter share declined slightly during the next 250 years or so and then fell sharply, as Europe and other parts of the world with predominantly European populations experienced demographic transition. Annual growth of the combined populations of these places, which was slightly below 0.50 percent in the second half of the eighteenth century, accelerated to nearly 0.75 percent between 1800 and 1850. Even faster growth, 1.00 percent per annum, occurred during the second half of the nineteenth century. By the early 1900s, more than three in every ten people lived in Europe, Russia, Canada, the United States, Australia, and New Zealand (Table 2.4).

With two world wars and the genocides and mass murders perpetrated by Hitler and Stalin, increases in population in Europe and its "offshoots" slowed, though did not turn negative, during the first half of the twentieth century. After the Second World War, growth picked back up, although by this time natural increase was higher in the developing world. This pattern has held ever since, which means that Africans, Asians, and Latin Americans now comprise a larger share of the total population. Equal to 73 percent in 1975, that share is approaching 80 percent, which is about what it was 250 years ago. Even though TFRs are now recovering a little in Europe and its offshoots (Myrskylä, Kohler, and Billari, 2009), more than four-fifths of the human race will soon be living in other parts of the world, as was the case before the demographic transition of modern times began.

As emphasized in this chapter, natural increase is not about people reproducing more. Rather, it is mainly a consequence of avoiding premature mortality. Asian, Latin American, and now African women have responded to this change by having fewer children. Even with demographic momentum, the revolution in human fertility has pulled down CBRs, dramatically so during and since the waning years of the twentieth century (Table 2.2).

People whose knowledge of demographic facts is dated—to be specific, anyone who has not examined CBRs and CDRs during the last two decades—are bound to be surprised by how close human numbers are to stabilizing in places like Chile and Thailand. In such countries, demographic momentum is now the only reason why there is still a gap, albeit a modest one, between birth and death rates. Likewise, many are unaware that Bangladesh, Ecuador, Syria, and many other countries, where natural increase equaled or exceeded 2.5 percent as recently as the late 1980s (Table 2.2), will soon be in the same position that Chile and Thailand are in now.

Popular misunderstanding of current population trends cannot be faulted too severely since those trends—especially the recent slide in TFRs and CBRs—have caught professional demographers a little by surprise. Regardless, most of them now agree that, for the human race as a whole, the inflection point in the S-shaped curve that tracks population during the demographic transition (Figure 2.3B) has been surpassed. Demographers now concern themselves mainly with predicting how soon, not whether, the human population will stop growing.

The standard source of demographic projections is the **United Nations Population Division (UNPD)**. During the early 1990s, its medium forecast of human numbers in 2025 was 8.47 billion (UNPD, 1993, pp. 284–285); since then, this forecast has been revised downward—to 8.01 billion, which exceeds the current population of 6.83 billion by 17 percent (UNPD, 2009, p. 38). The latest estimates of global population in the middle of this century are as follows: 7.96 billion (low), 9.15 billion (medium), and 10.46 billion (high), as shown in Figure 2.4 (UNPD, 2009, p. 1). Something important to note here is that the difference between the medium projection for 2050 and the 6.06 billion people alive in 2000 is smaller than population growth during the second half of the twentieth century (Figure 2.4)—not just in relative terms but in absolute terms as well.

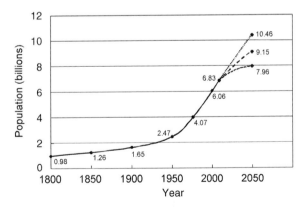

Figure 2.4 World population from 1800 to 2050
Sources: Table 2.4 and UNPD (2009).

Many centuries have passed since the last decline in world population. If UNDP forecasts are to be believed, a contraction will occur again in just a few decades, although for an entirely new reason. Rather than resulting from an upswing in mortality, as occurred during the fourteenth century when a succession of plagues swept across the Eastern Hemisphere, human numbers are now on course to stop growing and then fall a little around 2050 or soon afterward because the median TFR for the world as a whole is fast approaching the replacement level. As highlighted recently on the cover of a leading news magazine, the "population problem is solving itself." This is not a consequence of a pruning back of the human race, as envisioned by Malthus. Neither is it because Chinese policies to control fertility have been adopted by other countries, as Malthus's recent adherents have advocated. Instead, human numbers are on course to stabilize thanks to voluntary reductions in human fertility (*The Economist*, 2009a).

2.4 Food Consumption and Income

As observed at the beginning of this chapter, predicting the future consumption of food used to be a straightforward exercise. Standards of living improved little over time, so forecasting food demand involved little more than extrapolating demographic trends.

With human numbers growing more slowly, trends in food consumption are being driven more and more by other factors. Among these are cultural influences, such as Islam's prohibition on consuming pork and Hindus' aversion to livestock products. Another example of cultural influences on diet is the apparent disinclination of Chinese toward sweets. Per capita consumption of sugar is low in China—much lower, for example, than in India, where earnings are not as high.

Other factors influencing food demand are demographic. As the demographic transition proceeds, fewer women are pregnant at any given time, which cuts down on food requirements. Adults, who eat more than children do, comprise an ever larger portion of the population as fertility declines, which leads to a rise in per capita consumption (Foster and Leathers, 1999, pp. 173–174). On the other hand, people of retirement age account for a larger segment of the population in places where natural increase is dwindling or turning negative (Beck, 2009). As a rule, the elderly eat less on average than adolescents and young and middle-aged adults. Patterns of food consumption in Japan, for instance, reflect the

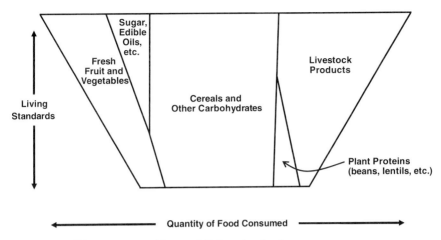

Figure 2.5 Dietary composition and living standards

fact that seniors comprise a large part of the country's population, as they do wherever the demographic transition is far advanced.

Economic factors also have gained importance as a driver of consumption trends. Agricultural development puts downward pressure on commodity prices (Chapters 4 and 7). As food becomes cheaper relative to other goods and services, household budgets are reallocated, with more food consumed though with food purchases claiming a smaller share of total earnings. In addition, changes in income matter.

The relationship between living standards and diet is depicted in Figure 2.5. The poorest people in the world, whose earnings do not cover the cost of a minimally adequate diet, subsist almost entirely on starchy carbohydrates. In addition, they often consume food (and water) that is contaminated with microbes and other harmful things, although this is not represented in Figure 2.5. As living standards rise, per capita consumption of staple carbohydrates increases. Furthermore, people opt for diets that are safer and more varied. In affluent settings, diets are diverse, with a wide range of fruits and vegetables, livestock products, and edible oils being eaten. In addition, aversion to products that may threaten human health is acute, even when the threat is miniscule.

Income growth does not cause purchases of each and every food item to rise. Starchy carbohydrates that are cheap, such as cassava in many settings, are **inferior goods** in the sense that consumers' purchases of these items decline as incomes rise. In particular, the poor respond to income growth by switching to cereal products (e.g., rice, noodles, bread, etc.), which are more palatable and which also are easy to prepare (Timmer, Falcon, and Pearson, 1983, pp. 22–35). This switch largely explains why a 10 percent increase in income causes the demand for grain to go up by 6 percent or so as long as PPP-adjusted GDP per capita is below $3,000. At the opposite end of the earnings spectrum, grain consumption isn't affected much by income growth (World Bank, 2009a, p. 65). Also, some affluent individuals refuse to eat products that are genetically modified or may contain minute pesticide residues. In light of the negligible health risks posed by these products (Ames and Gold, 1989), choices such as these reflect just how strong the demand for food safety is in wealthy settings.

After the switch from inferior carbohydrates to cereal products is made, income growth in the developing world almost always coincides with increased consumption of milk, eggs, and other livestock products. These products account for just one-third (in cereal equivalents)[5] of the diet in Bangladesh, which is destitute. The same dietary share in Thailand, which has experienced much more development, is nearly two-thirds. In South Korea, which was poor right after the Second World War and is now wealthy, livestock products account for more than 70 percent of all cereal equivalents consumed (Chapter 11).

There are other reasons why aggregate food expenditures exhibit the fundamental property of a **normal good**, which is that these expenditures rise with earnings. One is that, as income goes up, the opportunity cost of time increases. This causes food preparation at home to be replaced by food preparation elsewhere. In-home consumption of frozen and microwaveable meals grows. Of course, demand for these products is linked to refrigerators, microwaves, and other appliances, which also fall in the category of normal goods. Also in that category is entertainment, which arguably is part of what one enjoys (besides the convenience of staying out of the kitchen) at many restaurants. As people experiencing better living standards demand more food preparation and other services, the value of raw ingredients as a portion of total food expenditures diminishes. In the United States, this portion has fallen below 20 percent (Gardner, 2002, p. 155). Moreover, livestock products and food services of various sorts are all normal goods.

---

[5] The cereal equivalents of livestock products are found by multiplying the calorie content of such products by their respective conversion rates—that is, the calorie content of plant products required to produce one calorie's worth of livestock. The conversion rates for eggs, poultry, and pork are 4 to 1. For milk, the rate is 8 to 1. Eleven calories' worth of plant products are required to produce 1 calorie of beef or mutton (Foster and Leathers, 1999, p. 177).

The sensitivity of food purchases to a rise or fall in earnings is expressed by the **income elasticity** of food demand. This measure is analogous to **own-price elasticity**, which is discussed in the appendix to this chapter. Just as the latter compares relative change in consumption of a particular good to relative change in that good's price, income elasticity equals relative change in consumption divided by relative change in earnings. For example, the income elasticity of food demand is 0.40 if a 10 percent change in income causes purchases of food to go up by 4 percent.

Unlike own-price elasticity, which is negative because higher prices lead to diminished consumption, income elasticity for a normal good is positive. Furthermore, the elasticity of overall food expenditures with respect to income is generally below 1.00. True, there are individual goods—for example, gourmet meals consumed in expensive restaurants—for which income elasticity of demand is above this threshold. Also, very poor families spending most but not all they have on food might decide to buy food with any incremental money that comes their way, in which case income elasticity is greater than 1.00.[6] However, these are exceptional instances. As a rule, the proportion of income spent on food declines as income rises. This relationship, called **Engel's Law**, implies that the income elasticity of food demand is generally less than 1.00 and falls as income per capita increases. Where poverty is rampant, a 10 percent increase in earnings might cause food purchases to rise by 7 percent or more, which implies an income elasticity of 0.70 or higher.[7] In contrast, this elasticity is 0.40 or lower in wealthy places, such as the United States.

International patterns of income elasticity of demand are entirely consistent with Engel's Law. ERS-USDA (2003) reports estimates of this variable for 117 countries around the world, including the ten listed in Tables 2.2 and 2.3. Reported in Table 2.5 are income elasticities of food demand (column 1) for these countries: elasticities that are positive though inversely related to GDP per capita, of course. An extreme case is Tanzania, where GDP per capita is only $1,208 and income elasticity is 0.80. At the other extreme is Sweden, which is much more prosperous and where the income elasticity of food demand is 0.36.

---

[6] Define the incremental income, which is spent entirely on food, as Δ. If $F < I$, then $[\Delta \div \frac{1}{2}(F + \Delta + F)] \div [\Delta \div \frac{1}{2}(I + \Delta + I)] > 1.00$.

[7] Among poor households, income elasticity and own-price elasticity of the dietary staple (e.g., cassava or rice) that dominates the diet have similar absolute values. If there are no close substitutes for the staple and if expenditures on it comprise a large share of earnings, the effects of a price change on poor households' purchasing power and their consumption of the staple are comparable to the effects of a change in their incomes.

Table 2.5 Annual growth in population, average income, and food demand in selected countries

Country	Population Growth (%) in 2007 (1)	Growth in Per Capita GDP (%) in 2007 (2)	Income Elasticity of Food Demand (3)	Demand Growth (%)* (4)
Tanzania	2.44	4.52	0.80	6.06
Bangladesh	1.64	4.70	0.73	5.07
Senegal	2.77	1.93	0.74	4.20
Indonesia	1.15	5.10	0.69	4.67
Syria	2.46	4.04	0.68	5.21
Thailand	0.61	4.12	0.65	3.29
Ecuador	1.04	1.59	0.71	2.17
Chile	0.98	4.07	0.59	3.38
Poland	−0.05	6.70	0.58	3.84
Sweden	0.74	1.97	0.36	1.45

*The formula for this calculation is: Δ demand = Δ population + (elasticity x Δ income per capita), where Δ denotes percentage growth. As indicated in footnote 8, this definition does not contain a tiny interaction term, Δ population x (elasticity x Δ income per capita).
Sources: World Bank (2009b) for growth in population and GDP per capita; ERS-USDA (2003) for income elasticities.

International variations, of the sort on display in Table 2.5, are matched within many individual countries, especially where disparities between the rich and poor are acute. In places with great inequality, one finds a small minority living and eating about as well as the average Swede. Meanwhile, large numbers are no better off than the typical Tanzanian. Being an average measure for an entire nation, GDP per capita does not capture this sort of difference. Examination of the distribution of income (Chapter 7) is also required.

2.5 Demand Trends and Projections

In this chapter, we have examined population growth, past and present. Linkages between income and food consumption also have been considered. We now turn our attention to current trends in food demand. Presented as well in this section are estimates of what global consumption will be later in the twenty-first century.

A simple measure of demand growth is obtained by adding population growth to the increase in per capita consumption, which is found by multiplying the income elasticity of food demand by growth in GDP per capita. This measure omits an interaction term, which registers the additional food consumed by a population's new "recruits" and which invariably is tiny.[8] Reported in Table 2.5 is food-demand growth (column 4) for our sample of ten countries, along with underlying trends in GDP per capita and population.

The most interesting thing to observe about Table 2.5 is that, in eight of the ten countries, gains in per capita consumption are influencing overall food demand more than population growth. One of the two exceptions is Senegal, where the economy has not expanded much and which has one of the fastest rates of population growth in the world. The other exception among the ten is Sweden, where the income elasticity of food demand has fallen to a low level—as is consistent with Engel's Law. In contrast, growth in food consumption per capita has more of an impact than demographic expansion in Ecuador, where GDP per capita went up by just 1.59 percent in 2007, not to mention the seven other nations.

Moving from measures of current trends in food demand to estimates of demand at a later date is a challenge. One reason for this is that projecting population in the future is problematic, mainly due to declines in human fertility that are difficult to anticipate. Similarly difficult to forecast are living standards in the future. Yet another reason why pinpointing future demand is not easy is that changes in income affect income elasticities.

In an analysis of global food demand during the twenty-first century, Tweeten (1998) considers the interaction between GDP per capita and income elasticity. Conceding that the latter is being driven down in various parts of the world by income growth, he also points out that ever larger portions of the increase in human numbers are occurring in poor countries,

---

[8]Compare and contrast, for example, the simple measure of demand growth with a measure that includes the interaction term for a place experiencing a 4 percent increase in per capita consumption (which, as noted above, is found by multiplying income elasticity by growth in GDP per capita) and demographic expansion of 2 percent. Demand growth, including the interaction term, is:

$$[(1.04 \times 1.02) - 1.00] \times 1.00 = 6.08 \text{ percent.}$$

In contrast, the simple measure of growth, which to repeat leaves out interaction, is:

$$4 \text{ percent} + 2 \text{ percent} = 6 \text{ percent.}$$

Even in this hypothetical example, which features rapid increases in per capita consumption as well as human numbers, the interaction term is just 0.08 percent (equal to 4 percent multiplied by 2 percent), which is negligible.

Table 2.6 Growth in food demand from 2009 to 2050

	Relative Growth (%)
Low UNPD Population Forecast*	
0.2% yearly growth in consumption per capita	27
0.3% yearly growth in consumption per capita	31
Medium UNPD Population Forecast*	
0.2% yearly growth in consumption per capita	45
0.3% yearly growth in consumption per capita	52
High UNPD Population Forecast*	
0.2% yearly growth in consumption per capita	66
0.3% yearly growth in consumption per capita	73

*Figure 2.4.

where income elasticities remain elevated, as natural increase tapers off in more affluent places. Due to this latter reality, per capita consumption could continue growing at the same annual rate, 0.3 percent. However, the possibility of slower growth—0.2 percent per annum, to be specific—also merits consideration. This is largely because the income elasticity of grain demand has reached low levels in many emerging economies, including Brazil, China, and Russia (World Bank, 2009a, p. 72).

Precise estimates of future population, living standards, and annual growth in per capita consumption of food are not available, so multiple projections of food demand in 2050 are provided in Table 2.6. Every single figure corresponds to the combined impacts of population growth between 2009 and the middle of this century as well as compounded increases in consumption per capita.[9] Given the UNPD's medium demographic prediction (Figure 2.4) as well as the assumption that per capita consumption will grow at an annual rate of 0.3 percent, we predict that demand in 2050 will be 52 percent above the level in 2009. Note that this increase exceeds 34 percent, which is the relative difference between the UNPD's medium forecast for 2050 (9.15 billion) and the current population (6.83 billion). The gap has to do with increases in average consumption, of course.

To be sure, food demand will go up more if human numbers reach 10.46 billion in the middle of this century, as the UNPD anticipates if

---

[9] Different from the simple estimates of annual demand growth reported in Table 2.5, Table 2.6's forecasts do not leave out interactions between demographic expansion and increases in per capita consumption. As mentioned above, these interactions are of minor importance in any given year. However, they grow sizable if they are compounded over a longer period, between 2009 and 2050, for example.

fertility declines more slowly. But it is also possible for demand growth to decelerate, because there is less demographic expansion, annual improvements in per capita consumption slow to 0.2 percent, or a combination of the two. As reported in Table 2.6, food demand ought to increase between 2009 and 2050 by at least 27 percent and as much as 73 percent.

2.6 Summary and Conclusions

Accustomed as many people are to thinking about trends in human numbers in Malthusian terms, the nature and consequences of the demographic transition now happening in Latin America, Asia, and Africa have yet to be fully appreciated. Rather than increasing because TFRs are going up, the population has risen in the past mainly because death rates have fallen. People have not responded to this happy change by cutting back instantaneously on family size. However, there has been a revolution in human fertility, with women around the developing world having far fewer children on average than their mothers and grandmothers. Because of this revolution, growth is decelerating and there is an excellent chance that human numbers will peak well before the end of this century, perhaps around 2050.

This is not to say that demand for farm products will stabilize any time soon. Depending on demographic and economic expansion, overall food consumption could increase by three-fourths during the next four decades (Table 2.6). Demand for agricultural output might increase even faster, if more sugar, corn, and other commodities are converted into substitutes for fossil fuels (Chapter 4). Clearly, agricultural development will still be a major challenge for many more years.

Key Words and Terms

acquired immune deficiency syndrome (AIDS)
body mass index (BMI)
ceteris paribus
classical economics
Club of Rome

contraception
crude birth rate (CBR)
crude death rate (CDR)
demographic equilibrium
demographic momentum
demographic transition

dismal science
doubling time
Engel's Law
female empowerment
gross domestic product (GDP)
human immunodeficiency
    virus (HIV)
income elasticity
infant mortality rate
inferior goods
life expectancy at birth
Malthusian
marginal utility (MU)
natural increase
normal good
own-price elasticity
physiological capital
principle of population
purchasing power parity (PPP)
replacement level (of fertility)
Rule of 72
second agricultural revolution
technophysio evolution
total fertility rate (TFR)
tragedy of the commons
U.N. Population Division
    (UNPD)
urbanization

Study Questions

1. Describe Thomas Malthus's principle of population, including the view of human behavior on which this principle rests.
2. Compare and contrast overshoot and collapse in Malthus's model of trends in food demand and supply and overshoot and collapse in the systems-engineering model developed for the Club of Rome.
3. Identify critical missing elements of the Club of Rome model and assess the implications of these omissions.
4. Describe the second agricultural revolution and the technophysio evolution it catalyzed.
5. What happens to death and birth rates at the beginning of the demographic transition? When is the transition at an end?
6. Why does natural increase always occur during the demographic transition?
7. Explain why mortality has declined faster in the developing world during the twentieth century than it did in places like Great Britain and Scandinavia during the 1800s.
8. What factors have caused human fertility to decline throughout the developing world in recent decades?
9. Does an improvement in human well-being always lead to diminished TFRs?
10. Describe demographic momentum and its contribution to natural increase.

11. To improve your understanding of the demographic transition, look up and analyze data on death and birth rates and TFRs since 1960 or so for a few countries, including one where GDP per capita is elevated, another with a low average income, and one or two more that are in between.
12. Contrast the income elasticity of food demand in countries where PPP-adjusted GDP per capita is $3,000 or less with the income elasticity of food demand in nations with much higher standards of living.
13. Describe likely trends in global food demand during the first half of the twenty-first century.

Appendix: The Fundamental Economics of Demand

Economic analysis of changes in food consumption is a complicated task given the host of factors that need to be considered. To make the undertaking tractable, some abstraction from complex reality is desirable. With respect to consumption of any particular commodity, it is convenient to suppose that everything influencing consumers' purchases is unchanging—everything, that is, aside from the commodity's price. This assumption—referred to with a Latin phrase, *ceteris paribus* ("other things being equal")—allows for a general consumption function, in which all factors affecting people's purchases are represented, to be transformed into the simple abstraction of a demand curve, which relates consumption of a good solely to its own price. With such a curve identified, adjustments over time in demand can be examined. This growth is caused by changes in precisely those things, including population size and household incomes, that must be held constant so that a demand curve can be traced out in the first place.

Measuring price on the vertical axis and tracking yearly consumption on the horizontal axis, we can plot out a demand curve for corn in the United States (Figure 2.6). The curve has a downward, or negative, slope, which implies an inverse relationship between price and quantity. For example, raising the price from $2.00 to $2.50 per bushel causes yearly consumption to fall from 9.20 billion to 8.80 billion bushels, *ceteris paribus*. In contrast, quantity demanded increases if the price declines, again barring any change in other variables affecting consumption.

That the demand curve slopes downward reflects a fundamental characteristic of consumers' preferences. While people are generally happy to

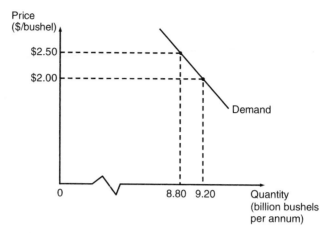

Figure 2.6 U.S. corn demand

receive more of all sorts of things, the satisfaction (or utility) they derive from another unit of any particular item grows smaller as their consumption of that item increases. This characteristic of consumers' preferences, called diminishing **marginal utility (MU)**, explains why the marginal value of a good or service—that is, the maximum amount of money that is offered for the last, or marginal, unit—falls as the total quantity of the good or service rises. The demand curve's negative slope is a manifestation of this declining marginal value. Diminishing MU also explains why, from the perspective of suppliers, prices must be cut to get consumers to increase their purchases.

Own-Price Elasticity

While it can be taken for granted that a demand curve slopes downward, the sensitivity of quantity demanded to price changes varies considerably from product to product and from consumer to consumer. This sensitivity, called own-price elasticity, is found by dividing relative change in consumption by relative change in price.

As long as quantity and price move in opposite directions, as is generally the case, own-price elasticity is negative. If dividing the relative change in quantity by the relative change in price yields an absolute value less than one, then the elasticity of demand is low. To say the same thing, demand is inelastic, which means that consumption is not responsive to price changes. Suppose, for example, that a 50 percent decline in price causes consumption to go up by 10 percent. Dividing the latter change by the former yields an own-price elasticity of −0.20, which is smaller in absolute magnitude than −1.00 (i.e., unitary elasticity). In contrast, the relative change in quantity may exceed the relative change in price, in which case the elasticity of demand is high. For example, a 20 percent increase in price might cause consumption to go down by 40 percent. With own-price elasticity equal to −2.00, which is obviously greater in absolute magnitude than the unitary benchmark of −1.00, demand is elastic. Consumption changes quite a lot in response to lower or higher prices.

As long as two points on a single demand curve have been identified, own-price elasticity can be estimated. Let P and Q_D represent one combination of price and quantity consumed on a particular curve. If the price rises to P', then there will be movement to another point on that curve, with consumption falling to $Q_{D'}$. Own-price elasticity is:

$$\begin{aligned} &\text{relative quantity change} \div \text{relative price change} \\ &= [\Delta Q \div \text{average of } Q_{D'} \text{ and } Q_D] \div [\Delta P \div \text{average of } P' \text{ and } P] \\ &= [(Q_{D'} - Q_D) \div \tfrac{1}{2}(Q_{D'} + Q_D)] \div [(P' - P) \div \tfrac{1}{2}(P' + P)]. \end{aligned} \quad (2.1)$$

In Figure 2.6, let P and Q_D be $2.00 per bushel and 9.20 billion bushels per annum, respectively, and P' and Q_D' be $2.50 per bushel and 8.80 billion bushels per annum, respectively. The change in prices, ΔP, is positive, $0.50, while ΔQ_D is negative, −0.40 billion bushels. Since P and Q_D move in opposite directions, own-price elasticity of demand is less than zero:

$$\begin{aligned} &[-0.40 \text{ billion} \div \tfrac{1}{2}(8.80 \text{ billion} + 9.20 \text{ billion})] \\ &\quad \div [\$0.50 \div \tfrac{1}{2}(\$2.50 + \$2.00)] \\ &= [-0.40 \text{ billion} \div 9.00 \text{ billion}] \\ &\quad \div [\$0.50 \div \$2.25] = -0.20. \end{aligned}$$

Since the absolute value of elasticity is less than the unitary cutoff, one concludes that U.S. demand for corn is inelastic.

Two demand curves, with markedly different own-price elasticities, are shown in Figure 2.7. In the case of Jimmy's demand for jam (part A), own-price elasticity happens to be low, which is manifested by his curve's steep slope. In contrast, Betty's consumption of peanuts is highly sensitive to the price of the good, which is reflected by the gentle slope of her demand curve (part B). When the own-price elasticity of Jimmy's

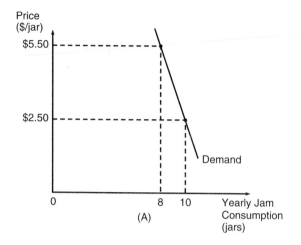

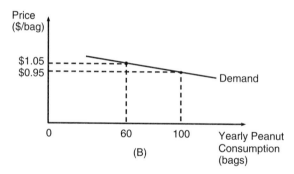

Figure 2.7 Demand elasticity
A. Jimmy's demand for jam
B. Betty's demand for peanuts

demand for jam is calculated, the number we come up with has a low absolute value:

$$[-2 \div \tfrac{1}{2}(8+10)] \div [3 \div \tfrac{1}{2}(5\tfrac{1}{2}+2\tfrac{1}{2})] = [-2 \div 9] \div [3 \div 4] \approx -0.30.$$

With Betty and peanuts, the absolute value of own-price elasticity is high:

$$[40 \div \tfrac{1}{2}(60+100)] \div [-0.10 \div \tfrac{1}{2}(1.05+0.95)]$$
$$= [40 \div 80] \div [-0.10 \div 1.00] = -5.$$

Comparing absolute values of these estimates to unitary elasticity, one concludes that Jimmy's demand for jam is inelastic and Betty's demand for peanuts is elastic.

Changes in Demand

As long as there is no variation at all in the number of consumers in a market, their incomes and tastes, and the prices of other goods and services, then consumption of the good being exchanged in that market changes only because that good's price rises or falls. However, any adjustment in the former set of factors brings about an entirely new relationship between consumption and price. In other words, there is a new demand curve, one that is either closer to or farther from the vertical axis.

As shown in Figure 2.8, a reduction in demand is represented by displacement of the demand curve toward the vertical axis, which means that less is consumed at any given price. Demand can fall for various reasons. A contraction in the population of consumers is one reason. Another is lower prices of substitute goods. For example, demand for beef declines if chicken becomes cheaper. There are also complementary goods. If these grow more expensive, then less will be consumed at any given price. A case in point is hot dogs and buns; if the price of one of these goods goes up, then demand for the other falls. In addition, income influences demand. If consumers have diminished earnings, then demand for most goods falls. Finally, economists recognize that tastes—a catch-all category that encompasses everything from whims influenced by passing fashion to dietary requirements related to age and other demographic variables—influence demand. Diminished food consumption caused by an aging of

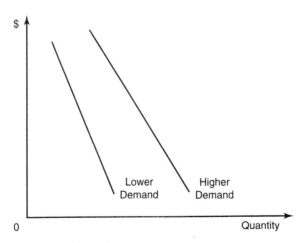

Figure 2.8 Changes in demand

the population would be regarded by economists as a fall in demand caused by changing tastes.

Just as it can decline, demand can grow. Such an increase is represented by a movement of the demand curve away from the vertical axis, which means that more is consumed at any given price. The causes of this sort of shift are a mirror image of the causes of falling demand. One of these is population growth, which of course is a major focus of this chapter. Demand also increases because of higher prices of substitute goods or lower prices of complementary goods. It can also go up because of changing tastes. Likewise, income growth causes consumption of normal goods to be higher at any given price.

Just as own-price elasticity indicates the sensitivity of consumption to price changes, *ceteris paribus*, income elasticity indicates how much demand for some good changes in response to higher or lower earnings, all else remaining the same. Suppose, for example, a 5 percent rise in earnings leads the Smith family to increase its spending on food by 2 percent. Under these circumstances, its income elasticity of food demand is 0.40 (2 percent divided by 5 percent).

If spending on food is known at a pair of earnings levels, then income elasticity can be estimated. Let F and Y represent one combination of food expenditures and income, respectively, and F' and Y' represent another combination. Income elasticity is:

$$\text{relative change in food purchases} \div \text{relative income change}$$
$$= [\Delta F \div \text{average of } F' \text{ and } F] \div [\Delta Y \div \text{average of } Y' \text{ and } Y] \quad (2.2)$$
$$= [(F' - F) \div \tfrac{1}{2}(F' + F)] \div [(Y' - Y) \div \tfrac{1}{2}(Y' + Y)].$$

For the sake of illustration, consider a family of six in Bangladesh, where GDP per capita was $1,241 in 2007. Its earnings increase, from $7,600 to $8,400. This causes its food expenditures to rise from $4,815 to $5,185. Its income elasticity of demand is:

$$[(5,185 - 4,815) \div \tfrac{1}{2}(5,185 + 4,815)] \div [(8,400 - 7,600)$$
$$\div \tfrac{1}{2}(8,400 + 7,600)]$$
$$= [370 \div 5,000] \div [800 \div 8,000]$$
$$= [7.4\%] \div [10.0\%] = 0.74.$$

3

The Supply Side

Agricultural Production and Its Determinants

With human numbers and living standards continuing to rise, the amount of food people eat is nowhere near cresting. At a minimum, production needs to go up by one-fourth between 2009 and 2050 if demand growth (Table 2.6) is not to outstrip increases in supply. In fact, output may have to rise 75 percent during the next four decades if food is not to become scarcer.

One way to raise production is to make wider use of farming's main natural resource, which is land. An alternative is to increase the application of other inputs, such as labor, machinery and fuel, fertilizers and pesticides, and water. Also possible are changes in the technology that farmers use to convert inputs into edible output. Each of these responses to demand growth has been an important feature of agricultural progress during the last century. Likewise, increases in food production will result both from **extensification** (i.e., using more land for crop and livestock production) and **intensification** (resulting from increased application of non-land inputs, technological change, or a combination of the two) during the next 100 years.

This chapter begins with some broad observations about farming: its sensitivity to environmental conditions as well as its close resemblance to the economic ideal of perfect competition. Special attention is given to the second half of the twentieth century, when increases in demand were unprecedented yet exceeded by supply growth. Continuation of this trend is certainly possible, though by no means inevitable.

In the pages that follow, we assess prospects for the further extensification of agriculture as well as the sector's continued intensification, especially the increases in crop yields resulting from technological improvement. Also examined is energy use in the food economy, the rising importance of agribusiness, and the trade-offs resulting from the conversion of crops into **biofuels**.

3.1 The Nature of Agriculture

For most people in affluent places and a good many in the developing world, farming is a vaguely old-fashioned activity, something engaged in long ago by grandparents or their forebears. Yet agriculture represents something quite new in human experience, a development dating back no more than 12,000 years. For innumerable millennia before people planted crops and cared for livestock, our distant ancestors fed themselves exclusively by hunting and gathering.

To be sure, hunter-gatherers came up with countless incremental improvements. Before the **agricultural revolution**,[1] people undoubtedly figured out that removing weeds would promote the growth of food-bearing plants. Simple observation also would have revealed that vegetation is more robust where some manure or a discarded fish has been left to decay. But from practices based on observations like these to agriculture, with crops and livestock raised in fields tended by farmers, a great leap was required.

Until that leap occurred, human existence itself was tenuous. Even in the most luxuriant setting, such as a rainforest, extensive tracts of land were (and are) required to support a small nomadic group of hunter-gatherers. Food storage was rare, and survival was jeopardized by drought, floods, insect plagues, frost, and disease, not to mention the predations of rival human bands. Few of these threats were predictable and the total population was miniscule. No more than 10 million people were alive 12,000 years ago (Chapter 2). Moreover, urban culture and all its arts and industry were inconceivable as long as a few hunter-gatherers, each using a substantial amount of natural resources, were scattered across the face of the earth.

The problem our distant ancestors faced was that they exercised almost no control over the biological processes that sustained them. Today, these same processes remain the underlying source of sustenance, although farmers now manipulate biological nature intensively. As a result, far fewer resources are needed to feed one person, which has allowed human numbers to rise. By the standards of the age, there was a population explosion in various parts of the Western Hemisphere as corn farming,

---

[1] This was actually the first agricultural revolution, which is distinct from the second such revolution of modern times (Fogel, 2004a, p. 21).

which originated in southern Mexico and Central America, spread northward and southward. In the Western Hemisphere and elsewhere, the existence of an agricultural population that could produce more food than it required for its own nourishment also made incipient urbanization possible (Diamond, 1997, pp. 86–92). It is significant, for example, that the oldest towns excavated in the Middle East—such as ancient Jericho, which is 9,000 years old—were founded after the agricultural revolution (Cipolla, 1965, pp. 18–25).

Farmers' manipulation of biological processes takes myriad forms. Domesticated animals are protected from parasites and pathogens and their reproduction is controlled. The soil in which the seeds of crops are planted is managed, both to prevent its being eroded by wind and rain and to maintain its fertility. **Fertilization** traditionally has meant the application of manure, which contains the three macronutrients required for plant growth: nitrogen, phosphorus, and potassium. These days, it is more common to apply chemical fertilizers, which are industrial and mining products, to replace the macronutrients used up as crops grow. Insecticides are another manufactured product. Ever a concern, **weed control** has long been accomplished by hoeing and other forms of cultivation. Here again, many farmers now use an industrial input instead: herbicides, to be specific. Finally, just as fertilization replenishes soil fertility, rainfall can be augmented by **irrigation**, the additional water applied to farm fields being extracted from underground **aquifers** or diverted from rivers, lakes, or streams.

Although nature is managed in various ways on farms, crop and livestock production remains essentially biological, which has a number of ramifications. One of these is that large volumes of organic residuals are produced along with edible commodities. For example, the rice or corn we eat comprises a tiny portion of a rice or corn plant. Domesticated animals may be fed some of the stalks and leaves. But much of this material is left in the field to decompose, thereby adding to the organic content of soil. This is largely beneficial since organically rich soils hold water and nutrients, which crops utilize. Beneficial use is also made of the organic residuals of livestock production. As already mentioned, livestock wastes contain macronutrients needed by plants. However, these macronutrients are not utilized exclusively by the crops farmers raise. Whenever the rate at which manure (or fertilizer) is applied to fields exceeds plant uptake, there is a good chance that nutrients attached to eroded soil or dissolved in runoff water will find their way into waterways or aquifers, which has various consequences. For example, algae and other aquatic organisms proliferate, as they feed on nutrients transported into waterways from farm fields. This proliferation deprives fish of dissolved oxygen, which cuts into their numbers.

The water pollution resulting from excessive concentrations of waste by-products is an impediment to the industrialization of livestock production, which involves confining many animals to small spaces.[2] As a matter of fact, wide expanses of rangeland continue to be grazed by cattle and other domesticated species. Crop farming is similarly extensive, although selected horticultural products are raised intensively in greenhouses and other controlled settings.

Agriculture, Soils, and Climate

Much more than other economic sectors, agriculture is exposed to the elements. A sustained downpour and associated flooding can wash away crops. In addition, extreme heat is a problem, as is excessive cold. Far from the equator, in places like northern Russia and Canada, the growing season can be curtailed by heavy frosts in the late spring or early autumn. This problem is less serious farther from the polar regions. Indeed, **temperate-zone** winters are ideal since these are strong enough to kill off insects and other pests. Something else that benefits agriculture in the middle latitudes is the availability of soils that are fertile and not too susceptible to erosion and other forms of degradation (Sánchez and Logan, 1992). Also favorable are the many hours of daylight during the growing season, which promote photosynthesis and therefore rapid plant growth. It is no mystery, then, that temperate settings with adequate hydrologic resources and fertile soils are important sources of staple grains, such as corn and wheat (which was originally cultivated in the Middle East). Among these areas are the American Midwest, the Argentine *pampas*, and the Ukraine.

Farming in the **tropics and subtropics** is more challenging, in part because of poor land quality (Box 3.1). Another problem is that solar radiation is intense near the equator, which accelerates evaporation and transpiration. Conditions in many tropical and subtropical settings are arid or semiarid, which means that irrigation is a prerequisite for farming. One dry place with irrigated agriculture is the Punjab, which straddles the Indian-Pakistani border and where the Green Revolution (discussed later in this chapter) has had a substantial impact.

---

[2] Also arousing concern are the conditions in which confined livestock live (Pollan, 2002).

> **Box 3.1 Tropical and subtropical soils**
>
> There are settings in the low latitudes with excellent farmland. Soils of recent volcanic origin, in Central America and East Africa for example, are well suited to crop production. Also, river valleys in the Middle East, South Asia, and elsewhere are well known for their fertile soils.
>
> But by and large, soil quality is not good close to the equator. Stewart, Lal, and El-Swaify (1991) estimate that approximately 57 percent of tropical and subtropical soils are Alfisols, Oxisols, and Ultisols, each of which tends to suffer from chemical limitations (e.g., low fertility, aluminum toxicity, etc.), has undesirable physical properties (e.g., high susceptibility to erosion or compaction), or both. In contrast, Entisols, Inceptisols, and Mollisols, which lend themselves well to farming, comprise 32 percent of all soils within $23^1/_2$ degrees of the equator, but 46 percent in the temperate latitudes (Sánchez and Logan, 1992).

Heavy rains occur elsewhere in the low latitudes. However, precipitation tends to occur in heavy storms, which wash away soil and nutrients rapidly. Also, organisms that harm people or the species they consume thrive in places that are always warm and humid. Crude death rates 50 years ago in places like Indonesia and Senegal (Table 2.2) were comparable to Scandinavian rates in the early 1800s (Table 2.1). No doubt, winters that were long and hard used to cut back on human longevity in far-northern settings. However, the lack of a season when subfreezing temperatures kill off mosquitoes and other vectors that transmit parasites and pathogens is a clear disadvantage. To this day, diseases such as malaria are responsible for widespread morbidity and mortality in the tropics and subtropics. Likewise, Black Sigatoka, which attacks bananas, is an example of a plant pathogen capable of decimating harvests. Furthermore, weed growth is a constant headache for farmers in places that are permanently warm and moist.

Two agricultural species of major importance are typical of the tropics and subtropics. One is sugarcane, which grows so rapidly that it usually outcompetes weedy species in the fields where it is planted. Domesticated in Asia, this crop gradually migrated westward. Sugarcane's introduction in the Western Hemisphere, where the indigenous population collapsed during the 1500s (Chapter 2), coincided with the beginnings of the trans-Atlantic slave trade. The other species is rice, which has long been the mainstay of Asian diets. Production of this grain is water-intensive, especially if rice seedlings are transplanted to flooded fields (called paddies) as a measure for weed control and irrigation. This transplanting also makes rice production labor-intensive. The traditional cultures of rural Asia were shaped in no small way by the need to organize the hydrologic and human inputs of rice farming.

The most pleasant environments in the world are found in another tropical and subtropical setting: hills and mountain valleys. These places have permanently springlike climates, with warm afternoons and cool nights, and have fewer insects and other pests than at lower elevations. Conditions are ideal for the production of coffee and tea, which are the traditional cash crops of hilly areas in the low latitudes. Also, horticultural products, including cut flowers, are now finding their way to international markets from upland farms in Mesoamerica and the Andes, Southeast Asia, and East Africa.

Specialization and Diversity in the Food Economy

Due to variations in temperature, rainfall, and soil fertility, agriculture is heterogeneous. However, differences in how edible goods are produced do not reflect environmental differences, alone. From place to place, the economic forces and public policies that influence the level and mix of farm inputs and outputs vary considerably.

Consider, for example, how the same crop, corn, is grown in two strikingly different places. One of these is Tanzania, which is one of the world's poorest countries, and the other is the United States, which of course is one of the richest. A typical Tanzanian farm occupies a few hectares[3] of unirrigated ground. Manufactured **capital** (i.e., economists' synonym for productive capacity) is similarly meager, only comprising hoes, sickles, and other equipment worth $50 or so. It is also exceptional to have draft animals (e.g., oxen), which are costly for a poor household to maintain, so agricultural equipment is wielded with muscle power provided by farmers themselves. Human muscle power being as limited as it is, a typical household never cultivates more than a couple of hectares at any time. The harvest is barely enough to satisfy the nutritional needs of the farming family, and oftentimes less than that. Without a marketable surplus, there is no cash for

---

[3] There are 2.471 acres in one hectare, which has 10,000 square meters. To visualize the dimensions of a hectare, Americans should think about a square area with 100-meter sides (about the length of a U.S. football field plus one end zone).

buying improved seeds and fertilizers, which could be used to raise yields. Even if there were a surplus, commerce is preempted by the poor state of roads and other infrastructure.

The situation could hardly be more different in the American Midwest, where a typical farming family possesses 500 hectares or more as well as $500,000 worth of tractors, combines, and so forth. Fossil fuels and electricity (not draft animals, obviously) power all this equipment and machinery, thus enabling a household to work its entire holding without hired labor. Practically all output is sold and the use of improved seeds, fertilizer, and other purchased inputs is routine. Production technology is up to date, with inputs converted into grain with great efficiency. Transportation and communications infrastructure is similarly advanced, which allows enormous amounts of inputs and output to move from sellers to buyers at a low cost.

Disparities of comparable magnitude also exist between irrigated rice production in the Mississippi Delta of the United States and upland rice farming in West Africa. The former is characterized by large, capital-intensive operations that take advantage of the latest technology, rely heavily on purchased inputs, and sell all output. In contrast, subsistence farmers in Africa tend to practice **shifting cultivation**.[4] This process begins with the clearing of bush forest with saws, axes, and cutlasses (machetes) and the burning of residue, which releases nutrients into the soil. A crop or two of rice is grown without additional fertilization, after which vegetables, root crops, or tubers are produced for a couple of years. To restore soil fertility after this cycle, a field must be left undisturbed for 15 to 20 years. Thus, shifting cultivation requires extensive tracts of land per farmer to be sustainable, since 15 to 20 hectares must lie fallow for every cultivated hectare. However, this system, which makes no use of chemical inputs, becomes unsustainable if the fallow period is curtailed, as frequently occurs because of population growth.

Environmental variation creates differences among small impoverished farms in the developing world. Yet these farms have much in common with one another. In particular, they are self-sufficient to a high degree, which is not coincidental at all. Economic progress requires specialization and trade, with every economic actor (be it a household or an entire nation) concentrating on the activity in which it has a **comparative advantage** (Chapter 6) and making exchanges with other actors with different comparative advantages. In poor, rural areas, little distinguishes one household from another, which

---

[4] Another term for bush-fallow farming of this sort is swidden agriculture. Nearly synonymous as well is slash-and-burn farming—a term that can have pejorative connotations.

means that bartering or commercial transactions among them yield modest gains. Also, economic interaction is impeded by infrastructural deficiencies. Thus, much of the African, Asian, and Latin American countryside is mired in economic isolation, poverty, and hunger.

Specialization and trade are salient features of the food economy in affluent parts of the world. Farms being diverse, those with rich soils produce crops while livestock graze where land quality is low and water is scarce. A rice farm in Japan, which may encompass as little as one hectare, is highly mechanized so the operator can spend time earning nonagricultural income. In contrast, a sheep station in inland Australia can be 1,000 to 100,000 times larger and, due to its being located far from any urban area, is operated by a full-time rancher. Of greater importance is that an evolution has occurred from a food economy comprising self-reliant operations to one made up of farms focused on crop and livestock production proper as well as a wide array of firms specializing in the production of inputs and services used on and by those farms.

In times past, on-farm processing of food (e.g., pickling, canning, and smoking) was routine. Likewise, it was not unusual for an individual operator to produce various crops and livestock and sell directly to consumers, in public markets for example. Nowadays, many farmers raise just one or two commodities. Rather than producing inputs like seeds and fertilizer on their own, they buy these from agribusinesses, which enjoy **economies of scale or size**[5] and other advantages in the marketplace. Farmers also rely on specialized firms for **marketing** services, even though self-reliance for storage and transportation is an option. This reflects the growing predominance of services throughout the economy (Chapters 7 and 16).

Regardless of the inputs and services they provide, agribusinesses have been distrusted since the nineteenth century. During the 1870s and 1880s, for instance, local monopolization of rail transport was the rule in the farm country west of Chicago; as a result, freight rates were high and service was limited, much to the detriment of grain producers (Cochrane, 1979, pp. 93–94). Likewise, sanitation was poor in a number of meatpacking plants, as Upton Sinclair and other muckraking journalists let the U.S. public know in the early 1900s. A legal and regulatory framework now exists to curb such excesses, thanks to political action by populist farmers and progressive reformers around the turn of the twentieth century. Beyond the United States, the misdeeds (some exaggerated or imaginary,

---

[5] Any economic activity in which per-unit (or average) costs are minimized only at a high level of output is said to be characterized by economies of scale or size.

others not) of the United Fruit Company, which dominated the banana business in the Western Hemisphere for several decades, continue to be chronicled (Chapman, 2007).

As indicated later in this chapter, there is little evidence today of monopolization and other abuses, in part because of **antitrust** legislation, pure-food laws, and other such arrangements. Nevertheless, agribusinesses continue to be a focus of populist suspicion. Critics pay little heed to the rural poverty that persists, in Sub-Saharan Africa and other places, where there are no specialized firms to provide inputs and marketing services to farmers (Collier, 2008). By the same token, they do not concede that farms in places like Europe and North America are as prosperous as they are precisely because agribusiness is robust.

Specialization and trade comprise much of the basis for rural prosperity in affluent nations. However, commercial interaction among myriad farms and agribusinesses could not happen in the absence of a sizable public investment. Governments construct and maintain the roads and navigable waterways along which trade flows. Likewise, commerce would be hard to conduct without courts and related institutions that enforce contracts and property rights—institutions required for the very existence of markets. In addition, public funding has been provided for agricultural **research and development**, which are the source of new technology for crop and livestock production, as well as **extension** (or outreach), which is about transferring technology to producers. Publicly funded education creates the foundation for the application of science to food production, insofar as educated farmers are better able to learn about and apply technological advances. Also, the spread of irrigated crop production has been enabled by dams and canals built, maintained, and operated by the public sector.

Some governmental involvement in the food economy is driven by a desire to subsidize farmers, keep food cheap, or a combination of the two. Much of the public support for irrigated agriculture is a case in point. But there is another reason why government often takes the lead in providing infrastructure, establishing legal institutions, and supporting research, extension, and education, which is that many of the services yielded by these investments are best categorized as **public goods**. Two textbook (and nonagricultural) illustrations are military defense and environmental quality. Both of these goods have a critical characteristic in common, which is the impossibility of excluding nonpaying consumers. If a paying subset of the population were to be defended against foreign attack, for example, then denying the same protection to those who do not pay would be impossible. Likewise, expenditures that a segment of the population might make to reduce pollution benefit not just them, but everyone else as well.

The impossibility of excluding nonpaying beneficiaries makes private firms, which are motivated by profits, unwilling to supply public goods. To make sure these goods are produced, then, governmental intervention is necessary. Sometimes the intervention is direct, consisting of public funding for defense as well as rural roads and bridges, agricultural research and development, and other activities that yield public goods utilized by the food economy. Another option is to create legal arrangements, such as **intellectual property rights (IPRs)**, which make the private production of public goods rewarding. For example, a patent on a genetically modified organism (GMO) created thanks to agricultural biotechnology gives the patent holder exclusive legal control over that organism. Due to this control, anyone wishing to use the GMO must make a payment to the patent holder. Without IPRs, the high cost of excluding nonpaying users of the products of innovation would discourage private research and development.

To summarize, a well-developed food economy is made up of three elements. One of these, of course, is production agriculture, which comprises farms and ranches where crops and livestock are raised. A second element is agribusiness, the emergence of which reflects the increased specialization and trade that create prosperity. The third is government, which furnishes the public goods used by production agriculture and agribusiness. Each of these three elements is essential and, if any one were to disappear, the remaining two would suffer.[6]

Notwithstanding the importance of different actors in the food economy, emphasis is placed in this chapter and throughout this book on production agriculture. Government as a source of public goods, as well as laws and regulations that influence market performance, is an important topic. But while we compare the benefits and costs of, say, technological advances originating in the public sector, we do not address the motivation of governmental investment or policymaking. Neither does this book focus on agribusiness, for the simple reason that little distinguishes the enterprises that supply agricultural inputs and marketing services from firms in other parts of the economy.

---

[6] An up-to-date example of symbiosis in the food economy has to do with technological innovation aimed not at finding ways to produce standard commodities more efficiently, but rather at the development of new foods, with better nutritional properties for instance. Agribusinesses profit by succeeding in this activity. There are also benefits for farmers contracted by these agribusinesses to grow new varieties. Furthermore, government contributes by underwriting the basic science (a public good) that makes the creation of new foods possible.

Some enterprises engaged in agricultural production differ little from firms outside the food economy. In a laying hen enterprise, for example, trucks deliver feed at one end of a building complex housing 5,000 birds or more. This feed is distributed on conveyor belts to hens that spend their lives in cages with tilted floors. The eggs they lay roll down these floors onto another network of conveyor belts that lead to a part of the facility where mechanized washing and packaging occur. Clean, packaged eggs are then loaded on trucks to be transported to market. Pointing to **concentrated animal feeding operations (CAFOs)** of this sort, some critics decry "factory farming." They have a point since these businesses are organized much like any other industrial enterprise, with ownership by public corporations, routine use of debt financing, and assembly-line production.

However, factory farms remain exceptional. Independent operators are the norm in crop production, even in places like the United States. Apparently, no one has figured out a profitable way to use large sums raised in equity markets to purchase enormous tracts of rural real estate and hire a workforce to grow corn, soybeans, wheat, and other commodities. Perhaps the difficulty of supervising a small number of laborers scattered across a large area is the main impediment to assembly-line production of field crops. Maybe business risks (related to variable commodity prices) and weather and biological risks are too daunting. Regardless, public corporations largely steer clear of farming, thereby leaving this business to individual operators.

A couple of observations about these operators are in order. For one thing, most farmers do not rely heavily on debt, probably because of agriculture's environmental and business risks. Thus, most of the value of their land and capital comprises equity. Moreover, their great numbers—tens of thousands in the United States alone—mean that no single farmer is in a position to influence prices of inputs or output unilaterally, by varying factor employment or production on his or her own. If ever a sector resembled the economic ideal of a perfectly competitive industry, in which no single actor exercises market power, production agriculture is it. Accordingly, the conceptual framework that economists have developed to describe competitive supply serves splendidly for the analysis of trends in food production.

3.2 Increases in Agricultural Supply

As explained in this chapter's appendix, in which the fundamental economics of supply are reviewed, supply can go up for various reasons. One of these is investment, which causes more to be produced at any given price. A decline in input prices also lowers costs and therefore increases supply. In addition, more is produced at any given price as better ways are found to transform inputs into output.

Food supplies have been profoundly affected by the discovery and application of new agricultural technologies. This development is also in line with the contributions of technological improvement to overall economic growth. In the United States, for example, Solow (1957) has found that increases in capital per worker created just 12 percent of the rise in per capita gross domestic product (GDP) between the late 1800s and middle 1900s, while the other 88 percent resulted from improved productivity. Findings such as these have led economists to analyze growth in **total factor productivity (TFP)**—that is, output expansion beyond what can be explained by increases in capital, labor, and other inputs—and its impacts on GDP.

TFP growth can be traced largely to research and development. It also has much to do with education—as Theodore W. Schultz, who was both an agricultural economist and a Nobel laureate, stressed (Schultz, 1961). Something else to recognize about these linkages is that it is not unusual for many years to elapse between investments in research, development, and education, on the one hand, and the resulting effects on productivity, on the other. This lag is readily observed in the agricultural sector.

In the United States, for example, governmental support for higher education benefiting crop and livestock production dates back nearly 150 years. To be specific, the Morrill Act, which was enacted in 1862, provided for the establishment of **land-grant colleges** (as the new institutions were called) thanks to the sale of federal lands in each of the states. Today, the results of this legislation can be seen on the campuses of the Universities of California and Wisconsin, Cornell and Purdue Universities, and similar institutions across the country.

The Morrill Act was complemented in 1887 by the Hatch Act, which authorized the founding of **agricultural experiment stations**. Similar to counterpart institutions in Europe and Japan, these state-level stations are responsible for research and development benefiting the rural economy, including improvements in the biology of major crops and livestock breeds for example. In addition, the Smith-Lever Act of 1914 created an extension service partially funded by the federal government to educate farmers about management and marketing as well as new farming practices.

With time, new technology developed in land-grant colleges and experiment stations and disseminated by the extension service profoundly changed the mix of agricultural inputs in the United States. In 1949, for example, wages and salaries amounted to 48 percent of total production costs and spending on commercial inputs such as **hybrid**

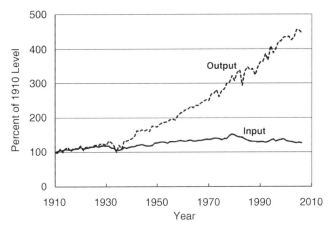

Figure 3.1 U.S. real agricultural output and inputs from 1910 to 2001
Source: ERS-USDA (2002), Table 1; ERS-USDA (2009a), Table 38; and earlier data obtained from the same agency.

seeds[7] and fertilizer, which U.S. farmers began using in major quantities in the middle 1930s, accounted for another 26 percent. During the next 36 years, these shares were nearly reversed, with labor's portion of total costs falling to 27 percent and commercial inputs' portion rising to 40 percent. Capital expenses as a share of total costs went up from 11 to 14 percent. Meanwhile, land's share remained the same, slightly exceeding 18 percent in 1949 and being a little less than 19 percent in 1985 (Craig et al., 1994, cited in Alston and Pardey, 1996, pp. 103–105).

As commercial inputs and machinery have been substituted for labor, overall production costs in U.S. agriculture have changed little. Corrected for inflation, these costs rose by just 7 percent between 1949 and 1985 (Craig et al., 1994, cited in Alston and Pardey, 1996, pp. 103–107). Meanwhile, the value of output has soared (Figure 3.1), mainly due to yield growth. For example, there was no discernible change in corn biology between the end of the Civil War and the eve of World War II, with national yields averaging 25 bushels per acre[8] throughout this period.

---

[7] Hybridization allows for more corn to be produced on each acre, in part because individual plants have more grain-bearing ears. Also, these plants are all of identical dimensions, which allows for higher planting densities. The development of hybrid corn that occurred in the United States during the early 1900s involved the crossing of inbred varieties, many of which had been developed at state experiment stations. However, cross-bred seeds were produced and marketed by private firms, which found this innovation rewarding because hybrids are infertile and therefore cannot be reproduced by farmers. That is, the infertility of cross-bred plants solved the problem of nonpaying beneficiaries of the new technology.
[8] A bushel of corn weighs 56 pounds, or 25.397 kilograms, and, as mentioned in footnote 2, there are 2.471 acres in a hectare. Thus, a yield of 25 bushels per acre is equivalent to 1.569 metric tons per hectare and a yield of 100 bushels per acre is the same as 6.276 metric tons per hectare.

But since farmers started using hybrid seeds and fertilizer, average yields have climbed steadily, rising above 150 bushels per acre in recent years. There have been similar trends in average yields of wheat and other commodities (Gardner, 2002, pp. 19–22).

Since yield growth has exceeded long-term declines in real commodity prices (Chapter 4) and since increased expenditures on machinery and inputs such as seeds and fertilizer have been largely offset by diminished wages and salaries, TFP has increased—by 80 percent between the late 1940s and middle 1980s (Craig et al., 1994, cited in Alston and Pardey, 1996, pp. 103–107). As indicated in Figure 3.1, the inflation-adjusted value of U.S. agricultural output has quadrupled since the 1930s, as real costs have risen moderately.

In light of the gains in agricultural productivity, it comes as no surprise that investments that have made these gains possible have yielded generous payoffs. In a landmark study, Griliches (1958) found that the **internal rate of return** on the research and development that produced hybrid corn was 35 to 40 percent. What this means is that $100 spent on coming up with this innovation subsequently led to a recurring annual benefit (consisting of additional output and cost reductions) of $35 to $40. Another perspective on this internal rate of return is that financing this research and development with a loan would have had a positive payoff even if the interest rate on that loan had been 35 to 40 percent.

More recently, Huffman and Evenson (1989) found that, from 1950 to 1982, the internal rate of return for expenditures on applied research at state agricultural experiment stations and other public institutions exceeded 40 percent. Support for agriculture's scientific base has produced comparable benefits in Australia, Canada, Japan, New Zealand, and Western Europe (Alston and Pardey, 1996, p. 208). Public investment in research, infrastructure, education, and other assets for the rural economy has been similarly beneficial in a number of developing countries (Tweeten and McClelland, 1997).

Even though increases in agricultural TFP have had sizable effects on crop and livestock output, some of the recent growth in food supplies has resulted from increased use of land and other inputs. In the rest of this section, previous trends in farmed area are examined, as are the prospects for future extensification. Also analyzed is the role that yield increases have come to play in the growth of food supplies around the world, with particular emphasis on distinguishing between the impacts of technological improvement and the results of applying more fertilizer and other non-land inputs.

Extensification

As U.S. experience illustrates, augmenting the supply of food by manipulating the biology of agricultural species (plant as well as animal) is a recent phenomenon. The science of **genetics**, which provides a basis for modern crop and livestock breeding, did not exist before Gregor Mendel, an Austrian priest, carried out experiments with garden peas in the middle 1800s. Truth be told, these experiments, which revealed how dominant and recessive genetic traits are passed from generation to generation, had no immediate impact on agricultural yields. Instead, per hectare output started to go up in Germany, Britain, and other parts of Europe as farmers began applying fertilizer. Mendel's genetics had to be rediscovered after his death (Pardey and Beintema, 2001) and one of the first examples of yield growth resulting from scientific breeding occurred in Japan, where the adoption of improved varieties of rice began in the late 1800s (Hayami and Ruttan, 1985, pp. 232–237).

Relative to the impacts of incipient intensification, extensification remained the primary driver of supply growth through the early part of the twentieth century. This was certainly true in the United States, where as already mentioned there were no appreciable increases in crop yields before the middle 1930s. During the 1800s, American farmers, who had started to venture west of the Appalachian Mountains as the eighteenth century drew to a close, pushed the **agricultural frontier** (i.e., agriculture's extensive margin) across the continent. The western frontier vanished in 1890 or so and the area used for crop and livestock production reached an all-time peak around 1920 (Cochrane, 1979, pp. 37–102).

Intensifying agriculture only after opportunities for extensification had been exploited was entirely sensible. In particular, this sequence in the agricultural history of the United States was consistent with the hypothesis of **induced innovation**, which holds that agricultural development is usually accomplished by raising the productivity of land or labor—whichever of these two resources is scarcer (Hayami and Ruttan, 1970). During the nineteenth century, land was abundant and labor was in short supply in North America. This meant that enhancing **labor productivity**, mainly through **mechanization**, was more rewarding than raising land productivity. The invention and adoption of the reaper, the steel plow, and (subsequently) the tractor enabled each farmer to cultivate more land, and hence to raise more crops.⁹ Agricultural mechanization also had an economy-wide impact of great importance. With output-per-farmer rising, labor could move from farms to other parts of the economy, in which employment opportunities were increasing due to the Industrial Revolution (Cochrane, 1979, pp. 189–208).

One hundred years ago, when the human population was less than one-fourth its current size, there were just a few places where intensification was the best response to growing food demands. One of these places was Japan, where innovations aimed at enhancing **land productivity** were induced because land was scarcer than labor. Among these innovations were fertilization, irrigation, and the improvement of crop varieties. After the middle of the last century, land scarcity mounted and intensification became the predominant response to demand growth throughout the world. Consequently, the geographic expansion of agriculture subsided.

Global agricultural land use, including fields planted to crops as well as pastures and other land grazed by livestock, has gone up by little more than 10 percent since the early 1960s (Table 3.1, column 4). This growth was modest compared to demographic expansion in recent decades (Table 2.4, column 4). There was a larger increase, 28 percent, in the area planted to six major commodity groups, which do not include cotton, sugar cane and sugar beets, and commercial tree crops (e.g., coffee and cocoa). Land used for the production of oil crops like soybeans has more than doubled since 1961, largely because demand for soy products has gone up considerably. There has been a comparable increase in the area planted to fruits and vegetables, which comprised a tiny fraction of farmed area 50 years ago, because better living standards have led to dietary improvement and diversification (Chapter 2). On the other hand, there has been modest change in the area used to produce commodities that supply most of the calories that people consume. For roots and tubers (e.g., potatoes), this expansion amounted to 10 percent between 1961 and 2007. Similar increases occurred in plantings of rice, corn, and other cereals, which account directly or indirectly for more than 60 percent of total caloric intake, as well as pulses (e.g., peas and lentils).

---

⁹ By no means was this the first time in history when extensification had been facilitated by technological change. Between 1000 and 1500, for example, farmers in northwestern Europe adopted deep-cutting iron plows in place of wooden implements originally introduced by the Romans. This change made possible the tillage of heavy soils in river valleys and other locations. As a result, agricultural land use increased substantially north of the Loire River between the eleventh and fifteenth centuries (Landes, 1998, p. 41).

Table 3.1 Global agricultural land use (million hectares) in 1961, 1984, and 2007

Crop Group	1961 Area (1)	1984 Area (2)	2007 Area (3)	Percentage Change, 1961 to 2007 (4)
Cereals	648	715	696	7.4
Oil Crops	114	170	249	118.4
Pulses	64	66	72	12.5
Roots and Tubers	48	46	53	10.4
Fruits	25	34	54	116.0
Vegetables	24	28	53	120.8
Subtotal	923	1,059	1,177	27.5
Other Crops	487	821	843	73.1
Total Cropped Area	1,410	1,880	2,020	43.3
Grazing Land	3,087	3,244	3,378	9.4
Agricultural Total*	4,457	4,726	4,932	10.7

*The numbers in the bottom row of this table are not column totals, because some agricultural land is used in two or more ways in the same year. For example, a single cropping cycle may be followed by several months of grazing by livestock, in which case the parcel in question will be counted both as cropland and pasture.
Source: FAO (2009a).

In addition to reflecting changes in human diets, the data reported in Table 3.1 have been influenced by regional trends, which have been far from uniform. This is true of cereal area, for example. Between 1961 and 1984, the global total went up from 648 to 715 million hectares (columns 1 and 2)—a one-tenth increase. Subsequently, a contraction occurred, leaving the area in 2007 (column 3) not much greater than what it had been four decades earlier. Some of this latter contraction took place in Western Europe, due to a partial reform of government policies encouraging excessive production of grain and other commodities (Chapter 10). The decline was even steeper in the Former Soviet Union, where socialized agriculture began to fail before the 1980s and where food demand has been contracting due to population shrinkage and falling standards of living (Chapter 14). Obviously, reductions in cereal area in Europe and Russia between 1984 and 2007 exceeded the expansion that occurred elsewhere in the world.

Global or regional declines in farmed area cannot really be ascribed to urban encroachment on the countryside. Land-use conversion of this sort provokes concern in a number of places, such as the northeastern United States and along China's Pacific coast, and policies have been adopted in various countries to halt the loss of farmland. However, well under 5 percent of the world's land is urbanized, which is small compared to all the world's farms and ranches. Even if urban areas doubled in size, there still would be plenty of real estate for crop and livestock production. Soil erosion and other forms of degradation threaten the natural resources needed for agriculture more. However, land degradation is usually not severe enough to render farm fields entirely useless (Chapter 5).

Could extensification be stepped up for the sake of producing more food? Almost certainly, yes (Box 3.2). However, the benefits of extensification need to be balanced against its costs. On the benefits side, crop yields on parcels not already cultivated might compare poorly with yields on existing farmland, because of arid conditions, soil infertility, and other problems. The costs of agricultural extensification include the expense of providing roads and other infrastructure and also reflect the environmental impacts created whenever forests and other ecosystems contract so as to accommodate the expansion of farming (Chapter 5).

Box 3.2 Where could agricultural extensification take place in the future?

Geographic expansion of agriculture is possible in various parts of the world. Consider, for example, the temperate settings where agricultural land use has declined in recent years. Normile, Effland, and Young (2004) estimate that 18 million hectares have been set aside in the United States and Europe to reduce commodity supplies and prop up prices. This area is nearly equal to the reduction in land planted to cereals around the world between 1984 and 2007 (Table 3.1). There are also extensive tracts of farmland in Russia and the Ukraine that are either underutilized or not planted to any crops at present.

Economist Robert Thompson (cited in World Bank, 2009a, p. 81) estimates that 12 percent of the world's arable land is not being farmed yet is also not forested. Much of this land, which could be sown to crops fairly easily, is in Sub-Saharan Africa. Also, 180 million arable hectares in Brazil are currently grazed by cattle and could be cultivated with little impact on livestock production since **stocking rates** (i.e., ratios of cattle to land) by and large are low. In Africa and South America, additional land currently covered with forests could be harnessed for agriculture (World Bank, 2009a, p. 82). However, commodity prices would have to rise to cover the expenses of extensification, which can be substantial in this sort of setting.

> Recent assessments of options for expanded agricultural land use by the World Bank and others are generally consistent with older assessments, such as a comprehensive study by Crosson and Anderson (1992) that is now nearly 20 years old. This consistency, together with the modest changes in global farmland during the past couple of decades, makes one confident that the potential for agriculture's geographic expansion has been gauged accurately.
>
> But it must be emphasized that the additional output gained because of an increase in agricultural land use would come at the expense of diminished biodiversity and other effects of habitat loss.

A study carried out by Fischer and Heilig (1997) furnishes a perspective on the benefits and environmental costs of increasing agricultural land use in the developing world. Using data from the U.N. Food and Agriculture Organization (FAO) on the agricultural capabilities and current uses of unfarmed land as well as population forecasts from the U.N. Population Division (UNPD), the two investigators forecast agricultural land use in 2050 in Africa, Asia, and Latin America, impacts on agricultural output, as well as losses of forests and wetlands. Most of the arable land in Asia is already being farmed, so projected increases during the next few decades are expected to be modest (Table 3.2, column 1). However, nearly three-quarters of this expansion, which would account for only a tenth of expected growth in agricultural production (column 2), would be at the expense of natural ecosystems (column 3). Much more

Table 3.2 Projected agricultural extensification and its impacts in the developing world from 1990 to 2050

Region	Cropland Increases between 1990 and 2050 (%) (1)	Share of Increase in Production between 1990 and 2050 Resulting from Extensification (%) (2)	Share of New Cropland Obtained from Forest or Wetland Conversion (%) (3)
Africa	96	29	61
Asia	19	10	73
Latin America	49	28	70
All Developing Countries	47	21	66

Source: Fischer and Heilig (1997).

agricultural extensification is anticipated in Africa and Latin America, where rural population densities are much lower than in China, India, and other Asian nations. However, the relationship between benefits and costs is far from encouraging. In Latin America, expanding farmed area by 49 percent, due to widespread encroachment on forests and wetlands, is expected to account for a little less than 30 percent of the growth in that region's harvests. The impact on output would be about the same in Africa, where two-thirds of the expected doubling of agricultural land use is expected to be at the expense of natural ecosystems.

To summarize, increasing the supply of agricultural commodities through extensification, which used to be the rule throughout the world, continues to be an option in many places. But it can be a costly option. To prepare a new field, a farmer must remove trees and other vegetation, improve drainage, and carry out related tasks. Whether or not this preparatory work is warranted depends on the returns coming his or her way. Especially if the additional output won from newly cleared land is modest, as would be the case where an agricultural frontier traverses low-quality land, extensification is rewarding for farmers only if food is becoming more scarce, as reflected in rising prices. But even with rising scarcity, expanding agricultural land use may be undesirable because of the adverse environmental impacts of habitat loss.

Intensification

In light of its costs, environmental and otherwise, it is fortunate that extensification stopped being the main source of increased food supplies decades ago. In developing countries, alone, yield increases accounted for 69 percent of overall growth in agricultural production from 1970 through 1990. Extensification's contribution amounted to just 31 percent (Table 3.3).

To be sure, the relative contributions of intensification and extensification have varied considerably from one part of the developing world to another. During the 1970s and 1980s, yield growth was responsible for nearly three-quarters of increased production in the Middle East and North Africa and more than four-fifths of all output gains in South Asia. In East Asia—outside of China, where yields have gone up markedly (see below)—this share was 59 percent. In contrast, intensification and extensification had comparable effects in Sub-Saharan Africa and Latin America (Table 3.3).

Yield growth and increases in farmed area also have had varied impacts among crop groups (Table 3.4). The land used for fruit and vegetable production has gone up faster than yields for this commodity group. However, the combined area planted to these crops remains small

Table 3.3 Sources of increased crop production in the developing world from 1970 to 1990

Region	Share of Output Growth from Higher Yields (%) (1)	Share of Increased Output from Extensification (%) (2)
Sub-Saharan Africa	53	47
Middle East and North Africa	73	27
East Asia (not including China)	59	41
South Asia	82	18
Latin America	52	48
All Developing Countries (not including China)	69	31

Source: Alexandratos (1995), p. 170.

relative to the farmland planted to, say, cereals (Table 3.1). Pulse yields have not gone up much, but then again the area used to grow peas, lentils, and such has increased even more slowly and is now exceeded by the combined size of all the world's fruit and vegetable farms. For roots and tubers, yield growth has been limited. This has an adverse impact on food security in places like West Africa, where the poor subsist largely on yams, cassava, and other starchy roots. However, even this modest yield growth has exceeded increases in the area planted to these crops.

In the case of cereals, yield growth has been substantial and much greater than extensification. As mentioned already, the area planted to soybeans and other oil crops, which was much smaller than cereals area in

Table 3.4 Yield increases (annual percentages) for six major crop groups in 1961, 1984, and 2007

Crop Group	Yield Growth in 1961 (1)	Yield Growth in 1984 (2)	Yield Growth in 2007 (3)	Increase in Farmed Area, 1961 to 2007 (4)
Cereals	3.17	1.83	1.29	0.15
Oil Crops	4.13	2.12	1.42	1.72
Pulses	0.85	0.71	0.61	0.25
Roots and Tubers	0.72	0.62	0.54	0.22
Fruits	0.60	0.53	0.47	1.74
Vegetables	1.86	1.30	1.00	1.76

Source: FAO (2009a).

the early 1960s, has been increasing rapidly, as production has expanded first in the United States and then in Brazil and other tropical and subtropical settings. Annual yield growth, which is still above 1 percent, has fallen a little below the annual rate of extensification for this crop group since 1961. However, this represents a modest exception to the general trend of supply growth being driven mainly by intensification.

As the figures in Table 3.4 show, annual percentage increases in crop yields have been trailing off for a number of years. What this reflects is linear growth in per hectare output, which has been sustained year in and year out since the early 1960s. Shown in Figure 3.2 is the linear trend for cereals, which continue to be planted on more land than the other five crop groups combined. Of course, dividing a constant annual increment by average yields produces an ever smaller ratio as yields rise over time. This is why percentage yield growth has been declining.

Just as the development of new crop varieties started to drive up per hectare output and add to agricultural TFP in Japan during the late 1800s and in the United States shortly before World War II, the agricultural intensification experienced in various parts of the developing world during and since the **Green Revolution** has been a consequence largely of scientific improvement. Preceded by many years of effort and experimentation, similar to what happened in the United States before hybrids were made available to farmers, this scientific improvement has permitted output to climb faster than expenditures on inputs.

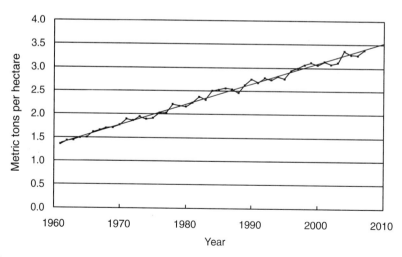

Figure 3.2 Average global cereal yield from 1961 to 2006
Source: FAO (2009a).

During the Green Revolution, new varieties of rice and wheat were developed and disseminated (Box 3.3). One advantage of these new varieties was that they are not as tall as their traditional relatives, which is why they are often described as semidwarf. As can be appreciated, shorter plants with larger (and heavier) grain-bearing **panicles** are less apt to lodge—that is, to be bent to the ground by wind and rain. Another merit of Green Revolution varieties was that their panicles are recessed, which increases exposure of the leaves (where photosynthesis occurs) to sunlight and thereby enhances cereal production. The beneficial trait of greatest importance, however, was that new rice and wheat varieties yield much more grain than traditional strains when supplied with ample nutrients and water. Accordingly, the use of Green Revolution varieties was accompanied by sizable increases in fertilizer use and irrigation (Dalrymple, 1985). Also, since fertilizer, improved seeds, and other commercial inputs must be paid for months before crops are harvested, demand for credit grew.

Box 3.3 The origins of the Green Revolution

Research that was to lead eventually to the Green Revolution was under way before the middle of the twentieth century. In 1944, the Rockefeller Foundation recruited a young plant scientist from the American Midwest, Dr. Norman Borlaug, to join a group of crop breeders in Mexico intent on developing a variety of wheat that could withstand the rust virus. After a few years of concentrating its efforts on disease resistance, the research team became primarily concerned with yield improvement. There were achievements in this latter area during the 1950s.

Supported originally by the Rockefeller Foundation, the **International Maize and Wheat Improvement Center (CIMMYT)** was established outside Mexico City in 1967, seven years after the Ford Foundation founded the **International Rice Research Institute (IRRI)** on the outskirts of Manila. These two institutions subsequently became leading elements of the **Consultative Group on International Agricultural Research (CGIAR)**, which still works with national agricultural research institutes and extension agencies to develop and disseminate improved crops (Dalrymple, 1985).

The success of this venture was commemorated by the Nobel Peace Prize awarded to Dr. Borlaug in 1970.

Especially in Asia, farmers benefited quickly from the Green Revolution. Introduced in the middle 1960s, semidwarf varieties accounted for 30 percent of all the wheat planted in Bangladesh, India, Nepal, and Pakistan by 1970. Little more than a decade later, that share had risen above 70 percent. During the same period, the share of total rice area in South and Southeast Asia planted to improved varieties rose from 10 percent to nearly 40 percent (Figure 3.3). It is also noteworthy that the Green Revolution did not greatly exacerbate income inequality. Larger, wealthier farmers comprised a disproportionate share of those making an early shift from traditional varieties. Within a few years, though, small and medium farmers were about as inclined to use semidwarf crops. Benefits even accrued to landless people, who are among the poorest of the rural poor, because the Green Revolution reduced food prices while strengthening demand for their labor (Anderson et al., 1985).

The linkage between farmers' adoption of improved varieties and increased fertilization and irrigation underlies trends in yields and input use during the last 25 years in the 15 most populous developing countries (Table 3.5). Each of these countries has at least 60 million inhabitants and, as a group, they have 60 percent of the world's 6.83 billion people and nearly four-fifths of the combined population of Asia (outside of the Former Soviet Union), Africa, and Latin America. Needless to say, agricultural trends in China, India, and other large developing nations matter a great deal in terms of how well the entire human race eats.

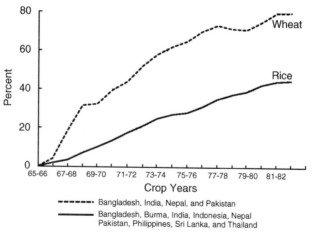

Figure 3.3 Area planted to high-yielding varieties of rice and wheat in South and Southeast Asia from 1965–66 to 1982–83
Source: Dalrymple (1985), p. 1071.

Table 3.5 Agricultural intensification and its underlying factors for the fifteen most populous developing nations

Country	Cereal Yield (kg/hectare) (1)		Per Capita Cropland (hectares) (2)		Fertilizer Use (kg/hectare) (3)		Irrigated % of Arable Land (4)		Value Added per Ag. Worker ($) (5)	
	79-81	04-06	79-81	04-05	79-81	04-05	79-81	04-05	79-81	04-05
Bangladesh	1,938	3,639	0.10	0.05	46	176	17	56	225	343
Brazil	1,496	3,077	0.37	0.32	78	154	3	4	1,020	3,178
China	3,027	5,237	0.10	0.11	149	331	45	36	172	419
Egypt	4,053	7,588	0.05	0.04	286	692	100	100	1,033	2,100
Ethiopia	1,192	1,374	n.d.	0.17	n.d.	15	n.d.	2	n.d.	168
India	1,324	2,405	0.24	0.15	35	121	23	33	252	393
Indonesia	2,837	4,317	0.12	0.11	65	166	16	12	420	590
Iran	1,108	2,426	0.35	0.24	43	99	36	47	1,313	2,596
Mexico	2,164	3,141	0.34	0.25	57	73	20	23	2,109	2,823
Nigeria	1,265	1,434	0.39	0.23	6	6	1	1	n.d.	n.d.
Pakistan	1,608	2,532	0.24	0.14	53	166	73	84	432	702
Philippines	1,611	3,074	0.11	0.07	64	142	13	13	926	1,092
Thailand	1,911	2,963	0.35	0.23	18	131	16	28	380	619
Turkey	1,869	2,582	0.57	0.34	53	111	10	20	1,776	1,883
Vietnam	2,049	4,717	0.11	0.08	30	325	26	34	n.d.	309

Source: World Bank (2009b).

South Asia is one of the most crowded parts of the planet (Table 3.5, column 2), and therefore an ideal setting for the sort of innovations introduced during the Green Revolution. Thanks to new varieties of rice and wheat, irrigation (column 4), and increased fertilization (column 3), average cereal yield (column 1) has gone up by 57 percent in Pakistan, more than four-fifths in India, and nearly 90 percent in Bangladesh. Similar changes have taken place farther to the east in Asia. Due to agricultural extensification in rain-fed areas, the irrigated portion of Chinese farmland has declined, although fertilizer applications have gone up by 122 percent and cereal yields in the world's most populous nation have risen above five tons per hectare. Crop production has intensified as well in Indonesia, the Philippines, and Thailand, which is the world's leading exporter of rice. Particularly impressive has been Vietnam's emergence as a major agricultural producer and exporter after decades of warfare. Since the late 1970s, output per hectare has more than doubled as fertilizer use has gone up eleven-fold and the irrigated portion of cropland has climbed from 26 percent to 34 percent.

Intensification has had similar effects in some parts of the Middle East and North Africa, most spectacularly in Egypt but also in Iran. Yields have increased as well in Turkey, even though incentives to raise land productivity are not especially strong due to rural population densities that are lower than in the other 14 nations listed in Table 3.5. In the global economy, Turkey's comparative advantage lies largely outside of agriculture. The same is true of Mexico, which along with the Philippines was the cradle of the Green Revolution (Box 3.3). Crop yields have increased in the northernmost country in Latin America, though not more than in Bangladesh, China, and a number of other emerging economies. In contrast, yields have more than doubled in Brazil, which in recent years has become a major agricultural exporter.

Finally, Sub-Saharan Africa has benefited little from agricultural advances during and since the Green Revolution. In Nigeria, fertilizer use and irrigation are rare, so average yields were modest three decades ago and have not increased significantly since then. Thirty years ago, Ethiopia was ruled by a Marxist dictatorship, which did not bother to keep agricultural records as it made war against multiple opponents. By the standards of the other 13 countries listed in Table 3.5, agriculture is poorly developed in the two most populous nations south of the Sahara, as well as in the region as a whole. The lack of progress relates in part to opportunities for extensification. At the same time, intensification has been held back by geographical impediments to irrigation development—in some places, a lack of good reservoir sites. Moreover, numerous African governments have neglected agriculture or applied policies injurious to the sector, thereby delaying the spread of improved technology.

Since fertilizer application rates have risen and irrigation has expanded at least as rapidly as yields in a number of countries, should one conclude that intensification during and since the Green Revolution has not constituted technological improvement of the sort that began to occur in the United States before the Second World War? The answer to this question is an unequivocal no. Strong evidence that agricultural TFP has increased in countries embracing the Green Revolution is provided by an analysis of **value added** per agricultural worker. The numerator of this indicator equals the difference between the value of agricultural output and the costs of fertilizer, capital goods, and all other inputs aside from labor. The denominator comprises the entire agricultural workforce, including farmers as well as the people they hire.

As indicated in column 5 of Table 3.5, value added per member of this workforce has gone up since the late 1970s, including in those places where the Green Revolution has had the greatest impact. But the most impressive gains have been registered in Brazil, where the abundance of natural resources has induced mechanization and other measures for boosting labor (not land) productivity. As this has occurred, value added per agricultural worker has tripled.

3.3 Has Intensification Run Its Course?

The Green Revolution that transformed Asian agriculture between the middle 1960s and early 1980s undoubtedly rescued untold millions in the world's most populous continent from death by starvation. Gains in the productivity of agriculture have been similarly impressive more recently. During the 1980s and 1990s, agricultural TFP grew by 1.5 to 2.5 percent a year, with the fastest gains registered in Asia and North America (Coelli and Rao, 2003). Thanks to improved productivity, growth in food supplies has been a consequence mainly of yield increases, not agricultural extensification.

Can this trend be sustained? There is no reason to believe that the limits of scientific improvement of agricultural species have been reached (Box 3.4). As is discussed in Chapter 5, climate change may interfere with the continued intensification of crop production. But even if global temperatures remain the same, yield growth may be limited by the mounting scarcity of water and fossil fuels, which farmers depend on heavily. We now examine the problem of resource scarcity and also assess the support provided for agricultural research and development, which has been the ultimate source of the technological improvement chronicled in this chapter.

> **Box 3.4 Has the biological improvement of crops come to an end?**
>
> There is speculation from time to time that biological limits on yield growth are being approached. If this were true, the prospects for tropical forests and other habitats would be gloomy and alleviating food insecurity would be even more of a challenge.
>
> Particularly worrying are predictions made from time to time by American environmentalist Lester Brown that output per hectare has "hit the wall." However, Danish statistician Bjørn Lomborg criticizes Brown's methodology, which involves "cherry-picking" a recent year when yields were unusually high in places such as the United States or Japan and then raising alarm for as long as these yields are not exceeded.
>
> Lomborg (2001) catalogs a number of Brown's mistaken claims (pp. 96–97). Moreover, his survey of the scientific literature yields no evidence that grain yields have reached peaks that are impossible to surpass (Lomborg, 2001, pp. 97–98).

Irrigation and Water Scarcity

Thanks to the investment of tens of billions of dollars, irrigated area has nearly doubled since the early 1960s. Currently exceeding 260 million hectares, which is a little more than one-eighth of the world's cropland, irrigated farmland is the source of approximately one-third of all agricultural output (FAO, 2009a).

More hydrologic resources could be developed for agriculture. In a 1990 study, the World Bank and the **United Nations Development Program (UNDP)** estimated that 110 million rain-fed hectares could be irrigated, thereby causing agricultural water use around the world to go up by approximately 30 percent. The resulting production would be sufficient to feed up to 2 billion people (World Bank and UNDP, 1990). To put this potential impact into perspective, keep in mind that human numbers are expected to go up by 1 to 3 billion during the next 40 years (Figure 2.4).

However, irrigation development carries a price. Nearly 20 years ago, the per hectare cost of canals and other infrastructure needed to deliver water to farm fields was reckoned at $8,000. If roads, power grids, and other complementary infrastructure were also required, then the estimated cost rose above $18,000 per hectare (FAO, 1992). Due to inflation, these expenses are higher today. Moreover, infrastructure, once it is in place, cannot be operated and maintained free of charge. Expenditures on technology transfer are also required to acquaint farmers accustomed to rain-fed production with best irrigation practices.

Aside from capital, operating, maintenance, and training expenses, competition between agriculture and other water uses can impede the expansion of irrigated farming. This competition is less acute in temperate settings, including most of the world's high-income nations as well as Eastern Europe and the Former Soviet Union. In these settings, annual withdrawals of freshwater from rivers, lakes, and streams and from underground aquifers (Table 3.6, column 2) are greatly exceeded by available resources (column 1). Furthermore, agriculture's share of total withdrawals (column 3) is relatively modest.

Total hydrologic resources are also much greater than freshwater withdrawals for agriculture and all other uses in East and Southeast Asia as well as Latin America and the Caribbean. However, this comparison of aggregates masks acute scarcities in specific places and seasons. For example, much of the hydrologic endowment of the Western Hemisphere is in the drainage basin of the Amazon River, which discharges ten times more water into the Atlantic Ocean as passes from the Mississippi River into the Gulf of Mexico. In contrast, water is in short supply much of the year in Chile, Mexico, and other countries (Chapter 12).

Table 3.6 Water availability and agricultural water use in 2007

	Renewable Freshwater Resources (billion m^3) (1)	Annual Freshwater Withdrawals (billion m^3) (2)	Agriculture's Share of Freshwater Withdrawals (%) (3)
High-Income Nations	9,516	921	43
East and Southeast Asia	9,454	959	74
South Asia	1,819	941	89
Latin America and Caribbean	13,425	265	71
Middle East and North Africa	225	276	86
Eastern Europe and Former Soviet Union	5,167	368	60
Sub-Saharan Africa	3,858	120	87
World	43,464	3,850	70

Source: World Bank (2009b).

As indicated in Table 3.6, water is especially scarce in the Middle East and North Africa, as is the competition over resources between farmers and other users (Chapter 13). Likewise, water withdrawals are sizable relative to available resources, agriculture's share of total water use is elevated, or both in South Asia and Sub-Saharan Africa—a pair of tropical and subtropical settings where food insecurity and poverty are widespread.

Underscoring the mounting competition over increasingly scarce resources, Rosegrant, Cai, and Cline (2002) point out that domestic and industrial demand for water quadrupled between 1950 and 1995, while demand from crop and livestock producers went up by a little more than 100 percent. Given the importance of satisfying unmet needs for clean drinking water in poor countries, they anticipate that, between 1995 and 2025, uses of water other than irrigation would rise by 62 percent and diversions to farmland would go up by just 4 percent. This is obviously a small fraction of the feasible growth of 30 percent estimated by the World Bank and the UNDP (1990). Furthermore, irrigation subsidies, which governments around the world offer farmers to boost agricultural production but which create economic inefficiency and harm the environment (Chapter 5), are bound to come under closer scrutiny as competition over water grows more intense with each passing year.

There are other impediments to irrigation development aside from greater competition over hydrologic resources. Large dams, which create substantial environmental and social impacts, tend to arouse fierce opposition—in affluent places and even in poor settings. When these impacts are evaluated and combined with capital and other costs, many projects begin to look prohibitively expensive. This is why Rosegrant, Cai, and Cline (2002) forecasted that irrigation would expand modestly. The only way to justify the costs, including damages to the environment and local communities, would be if agricultural commodities were becoming much scarcer, thereby causing a major run-up in food prices.

Energy and Agriculture

As pointed out earlier in this chapter, agricultural development in places where human resources are scarcer than land has been accomplished mainly by mechanization, which has enhanced labor productivity in the countryside. But in addition to allowing each farmer to cultivate more ground and grow more crops, the increased use of tractors and other machinery has tied production costs to the prices of diesel, natural gas, and other fossil fuels.

In the United States, for example, the combined expense of energy and agricultural chemicals manufactured from fossil fuels represented 34 percent of the cost of growing corn and 27 percent of the cost of raising wheat in 2007 (World Bank, 2009a, p. 61). Likewise, one-third of the cost of mechanized soybean production in Brazil comprised expenditures on fuel and chemical inputs in 2002 (World Bank, 2007, p. 66), when energy was still relatively cheap.

Energy costs matter even where rural population densities are higher and, as a consequence, agricultural development has involved little mechanization and instead has happened primarily thanks to gains in land productivity. In the Indian Punjab, expenditures on fertilizer, which must be applied in greater quantities in order to boost yields, accounted for nearly one-fifth of the variable cost of growing wheat in 2002 (World Bank, 2007, p. 66).

Linkages between energy and agriculture were obvious in late 2007 and the first half of 2008, when energy prices spiked. As diesel fuel, natural gas, and so forth grew more expensive, the costs of producing edible goods increased, which in turn drove up food prices (World Bank, 2009a, p. 52).

Recent events in energy markets have had another effect on the food economy, namely, the conversion of agricultural commodities into biofuels. Manufacturing ethanol in Brazil involves relatively modest trade-offs, since much of the sugarcane used by the industry can be grown on grazing land with low stocking rates (World Bank, 2009a, p. 62). But in other parts of the world, biofuel development impinges directly on food supplies. This is true in the European Union (EU), which a few years ago set a goal that ethanol and biodiesel (manufactured from oil crops such as rapeseed and flax) would comprise 5.75 percent of all transport fuel by 2010 (Commission of the European Communities, 2006). Food-versus-energy trade-offs are even more pronounced in the United States, where subsidies and trade barriers have encouraged the development of an ethanol industry that uses corn as a feedstock.

Two-thirds of the growth in global corn production since 2004 has been needed to satisfy expansion of the U.S. ethanol industry. Since the United States is the world's leading exporter of corn, there have been direct impacts on the commodity's price (Chapter 4). Other prices have been affected as well. As the area planted to corn in the United States rose by 22 percent in 2007, there was a 16 percent decline in soybean plantings, which drove up the latter's price. At the same time, higher corn prices strengthened the demand for rice and other substitutes, exactly as the microeconomic theory of demand (reviewed in the appendix to Chapter 2) suggests should have happened. This in turn led to a general rise in cereal values (World Bank, 2009a, p. 61). Moreover, higher prices for feed grain, including corn, drove up costs for livestock producers, which caused products such as milk to become more expensive.

As reported in the next chapter, food prices retreated after June 2008, as energy prices declined and as the global recession weakened the demand for edible goods. Moreover, prospects for the U.S. ethanol industry have deteriorated, because corn is not as cheap as it used to be and because fuel values have softened. A revival of the industry cannot be ruled out, however. According to the World Bank (2009a, p. 73), incentives for converting corn into ethanol will remain strong as long as the price of oil stays above $50/barrel. This threshold value is probably too low, if the financial difficulties experienced during the past couple of years by U.S. ethanol plants is any guide. But at some level, the prediction that corn prices will rise by 0.9 percent for every 1 percent increase in the price of oil (World Bank, 2009a, p. 73) cannot be discounted.

Technological innovation could alleviate the trade-offs arising when liquid fuels are produced from agricultural commodities. This would be the case, for example, if cellulosic material such as switch grass could be converted economically into energy products. On the other hand, trade-offs between fuel and food would be aggravated by technological change that enables agricultural commodities to be converted more efficiently into ethanol.

Support for Technological Improvement

In light of the growing competition for water resources as well as the consequences of expensive energy, the approaches that have raised food output in the past cannot be relied on in the future. In labor-scarce settings, additional mechanization will heighten agriculture's vulnerability to spikes in energy prices. Where land is the limiting factor of production, higher fertilizer prices and impediments to irrigation development will interfere with the application of conventional measures for raising yields.

These measures also create risks for the natural environment and public health. The mismanagement of irrigation systems often results in polluted aquifers and surface waters (Chapter 5). The misapplication of agricultural chemicals has the same impact. Also, a number of studies raise concerns about the well-being of farmers who make heavy use of this input. For example, research undertaken in the Philippines suggests that the value of the additional rice harvested due to the application of pesticides may be exceeded by the costs of illness suffered by farmers and other people exposed to harmful substances (Pingali et al., 1995).

The solution is not to halt agricultural intensification. Doing so would inevitably accelerate the food economy's geographic expansion at the expense of forests, wetlands, and other habitats since food demand is still growing. Instead, new avenues of intensification must be pursued that will raise yields while lessening agriculture's water and energy requirements. Accomplishment of this task, like hybrid development in the United States, will require a sustained commitment to research and development.

As reported above, the Green Revolution owed its existence to substantial investments in technological improvement, made initially by the Rockefeller and Ford Foundations and later by the World Bank and other development agencies. At an aggregate level, monetary support for agricultural research and development has continued to grow; corrected for inflation, public-sector budgets in 2000 were 35 percent higher than in 1981. Also, budgetary increases were greater in the developing world than in rich nations (Pardey et al., 2008).

But relative to human numbers, governmental spending on agricultural research and development remains much smaller in the developing world ($8 per agricultural worker) than in rich nations ($697 per agricultural worker). Moreover, expenditures in the former setting are concentrated, with Brazil, China, India, and three other countries (out of 129 for which data are available) accounting for 50 percent of the total and four-fifths of combined spending occurring in the top 24 or 25 nations. No Sub-Saharan nation aside from South Africa is in this group, even though no part of the world has a greater need to boost agricultural output in environmentally sound ways. Another problem is that agricultural research and development by private firms is minimal in developing countries (Pardey et al., 2008), in part because IPR protections are weak.[10]

Major strides in agricultural biotechnology are occurring in the private sector. Genetically modified varieties of corn, cotton, and other crops can defend themselves against insects and other harmful organisms. Agricultural biotechnology also has yielded crops that can withstand the application of herbicides. The environmental advantages of pest resistance are obvious, having to do with diminished use of chemicals with toxic properties. The

---

[10] As mentioned above, hybrid corn, including varieties that are bioengineered to resist rootworms and corn borers, cannot be reproduced. But this is not true of genetically modified soybeans. In the United States, firms that produce and sell the latter item enforce their patent rights vigorously, by prosecuting farmers who purchase and sow bioengineered seeds one year and then use a portion of their harvests as seeds the next year. In countries that do not recognize or enforce patents, the private sector does not supply bioengineered though reproducible seeds. Instead, this input is furnished by government. Examples of this approach include Brazil, China, and India.

planting of herbicide-resistant varieties is also advantageous since doing so allows farmers to cut back on deep plowing and other tillage practices that accelerate soil erosion and other forms of land degradation (Chapter 5).

Highlighting these benefits, Rauch (2003) criticizes those environmentalists who oppose any and all applications of biotechnology to agriculture. Opposition to GMOs has impeded agricultural biotechnology in Europe, as have exaggerated perceptions of risks to human health. A ban on pest-resistant corn adopted by the German government in April 2009 is a recent case in point (BBC, 2009).

Grain production in Europe could rise by as much as 15 percent if restrictions on GMOs were lifted (Collier, 2008). Africa stands to benefit much more from agricultural biotechnology. But aside from South Africa, GMOs are not being used in the world's poorest and hungriest continent, largely because of the efforts of international activists (Collier, 2008; Paarlberg, 2008).

Reflecting on recent spikes in food prices, U.N. Secretary General Ban Ki-Moon has spoken in favor of a substantial increase in agricultural research and development, in particular citing the need for a new "Green Revolution in Africa" (Ban, 2008). As in other parts of the world, such an advance will be difficult to achieve without biotechnology.

3.4 The Food Economy Beyond the Farm Gate

Technological advances such as crop hybridization and the Green Revolution have had a substantial effect on agricultural TFP. Productivity gains have occurred as well in the other parts of the food economy.

In places like the United States, technological progress beyond the "front gate"—at which primary farm output is sold to marketing agribusinesses, which add place (i.e., transportation), form (or processing), and time (i.e., storage, wholesaling, and retailing) utility to edible goods—has been substantial enough to affect the ways people eat. Illustrative of the growing impacts of post-harvest services in the food economy are the changes over time in how potatoes are consumed in the United States.

> Before World War II, Americans ate massive amounts of potatoes, largely baked, boiled, or mashed. They were generally consumed at home. French fries were rare, both at home and in restaurants, because the preparation of French fries requires significant peeling, cutting, and cooking. Without expensive machinery, these activities take a lot of time. In the postwar period, a number of innovations allowed the centralization of French fry production. French

fries are now typically peeled, cut, and cooked in a few central locations using sophisticated new technologies. They are then frozen at −40 degrees and shipped to the point of consumption, where they are quickly reheated either in a deep fryer (in a fast food restaurant), in an oven, or even a microwave (at home). Today, the French fry is the dominant form of potato. . . . From 1977 to 1995, total potato consumption increased by about 30 percent, accounted for almost exclusively by increased consumption of potato chips and French fries (Cutler, Glaeser, and Shapiro, 2006, p. 94).

Advances in the centralized processing and preparation of food and in its transportation and storage have been welcomed by consumers, sometimes at the expense of expanding waistlines (Chapters 10 and 16). In addition, the demand for products that can be prepared quickly has gone up as living standards have improved and the opportunity cost of time has risen. Antle (1999) highlights another consequence of income growth for the food economy, generally, and for marketing, specifically. Rather than simply involving competition over the prices of a limited number of uniform commodities, success in the food economy also requires discerning and catering to demands for myriad specialty products. Among other things, this requires the effective application of information technology—for example, a computerized system that uses bar code data to keep close track of purchasing patterns.

Marketing is characterized by economies of scale and size. The same holds for the business of supplying agricultural inputs, delivered to the "back gate" of farms. As agribusinesses expand to capture scale and size economies, industries become more concentrated, with fewer firms providing a larger combined share of total output. As has happened elsewhere, **concentration ratios** have risen in many parts of the U.S. food economy. The four largest beef-packing firms, for example, produced 83 percent of industry output in 2005, up from 72 percent in 1990. The concentration ratio in the pork-packing industry rose from 40 percent to 64 percent during the same period. There have been similar trends in input markets. For example, 56 percent of the seed corn sold in the United States is supplied by two firms. Throughout the world, one firm, Monsanto, supplies the seeds for nine out of every ten hectares planted to genetically modified corn, cotton, soybeans, and canola (Hendrickson and Heffernan, 2007).

The capture of scale or size economies by large agribusinesses drives down costs, to the benefit of consumers. However, industries dominated by a few firms can behave like **monopolies**, raising prices, reducing output or quality, or a combination of these measures. A monopolized industry that produces something that has close substitutes generally avoids harming consumers. For example, beef-packers know that their customers switch to other livestock products (or even become vegetarians) as they raise prices. Even if an industry dominated by a few firms produces something unique, an economic carte blanche does not necessarily exist. All that is required to avoid monopolistic behavior is for the markets served by the industry to be **contestable**, in the sense that new firms can enter the business at low cost. As long as this is the case, excess profits, deficient service, and the like are kept in check due to the threat of additional competition.

More than any other firm, Monsanto is being scrutinized, with an eye toward ascertaining the contestability of markets for its genetically modified seeds (Gillam, 2009). But on the whole, there is little evidence that agribusinesses behave monopolistically, at least in the United States (Antle, 1999). If they did so, the gap between the retail value of food and the value of unprocessed farm products would be widening over time. This is because, in contrast to aspects of marketing and input supply, the production (or agricultural) portion of the food economy is highly competitive, as noted in the first section of this chapter. Significantly, the information provided in Table 3.7 does not suggest a broad trend toward monopolization of the U.S. food economy. Between 1998 and late 2008 (when retail and farm-level prices were lower than prices earlier in the year), farmers' shares of consumers' expenditures declined for various goods, while holding steady or going up for others. Overall, farmers

Table 3.7 Farmers' shares of U.S. consumers' expenditures on food in 1998 and 2008

Food Group	Average 1998 Share (%) (1)	September–December 2008 Share (%) (2)
Meat	30.3	30.2
Dairy Products	36.0	30.8
Poultry	42.9	39.9
Eggs	42.0	45.0
Cereal and Bakery Products	6.4	8.1
Fresh Fruit	17.3	14.4
Fresh Vegetables	19.6	19.0
Processed Fruit and Vegetables	18.2	16.4
Fats and Oils	21.8	20.6
Overall Market Basket	22.2	21.6

Source: ERS-USDA (2009a), Table 8.

received 21.6 percent of consumers' expenditures on food in late 2008, down slightly from 22.2 percent ten years earlier.

Data of the sort presented in Table 3.7 do not always allay populist suspicions that agribusinesses exploit farmers. Some Americans have these suspicions solely because agribusinesses, not farmers, receive most of what is spent on food, and an even larger share of value added in the U.S. food economy.[11]

But the mere existence or presence of agribusinesses is a bad reason to embrace populism. As is emphasized in the first part of this chapter, agribusinesses make money by providing marketing services and inputs. Where no such services are provided—in the most deprived parts of the African, Asian, and Latin American countryside, to be specific—subsistence farming predominates, with households struggling to survive on what they can grow in an unforgiving environment. Hunger and poverty prevail in these rural settings, where the gains from specialization and trade in the specific form of agribusiness activity are unavailable.

3.5 Trends in Per Capita Production

Modern agriculture, of the kind found not just in wealthy nations but in many parts of the developing world as well, is a far cry from the kind of farming that our ancestors engaged in and that barely kept them fed. Muscle power, from farmers and draft animals, has largely been replaced by machinery driven by fossil fuels. Thus, the cost of this machinery now accounts for an appreciable share of total input expenditures, as does spending on energy. Growth of another category of inputs—land—has not ceased entirely. However, humankind's long experience with agricultural extensification is coming to an end. The alternative, which is intensification and which is made possible by increased specialization and trade as well as investment in public goods, has proved to be an efficient way to increase output. Farmers' embrace of improved technology is indicated by their adoption of improved varieties and increased use of fertilizer and other inputs, provided almost entirely by agribusinesses. As they have done this, agricultural TFP has risen noticeably.

---

[11] Value added in the U.S. food economy amounts to $1,111.2 billion, of which $743.7 billion (66.9 percent) is contributed by the marketing sector (broadly defined to include restaurants and fast-food eateries), $203.5 billion (18.3 billion) by input providers, and $164.0 billion (14.8 percent) by crop and livestock producers (ERS-USDA, 2009a, Table 36).

Table 3.8 Change in per capita food production in various years relative to the level in 1961 through 1965 (%)

Continent	1977 (1)	1987 (2)	1997 (3)	2007 (4)
Africa	−7	−12	−6	−2
Asia	10	32	67	98
Europe	18	27	13	9
North America	19	19	30	31
South America	13	20	41	73
World	6	15	19	33

Source: FAO (2009a).

Mainly because of intensification, food supplies have gone up faster than human numbers in most of the world since the early 1960s. Indeed, per capita production for the entire planet has never ceased growing—either during the 1960s and 1970s, when rates of population growth were at a peak, or since then. As indicated by entries in the last row of Table 3.8, per capita food output in 1987 was 8.5 percent greater than what it had been ten years earlier.[12] Growth slackened during the next ten years to 3.5 percent, but then picked up again, registering a gain of 11.8 percent between 1997 and 2007.

Just as growth in global output per capita has surged periodically, regional trends have varied. Thanks mainly to yield growth during and since the Green Revolution, per capita output in Asia today is double the level in the early 1960s. South America experienced an increase of nearly 75 percent during the same period. There has been less growth in North America and Europe, although demographic expansion and increases in food demand in these regions are much less than what the developing world has experienced. Furthermore, agricultural output used to be excessive, because of subsidies and trade barriers, so recent declines in places like Europe are actually desirable.

The same cannot be said of declines in Africa, where food production failed to keep pace with human numbers between the early 1960s and late 1980s. Per capita output has grown since then, but is still no greater than what it was a little less than a half century ago (Table 3.8).

Africa's problems aside, recent decades' improvements in the availability of food represent quite an accomplishment, especially in light of the predictions of impending worldwide famine made by Paul Ehrlich

---

[12] [1987 output per capita − 1977 output per capita] ÷ 1977 output per capita ≈ 8.5 percent.

and others in the 1960s and 1970s (Chapter 2). In light of this experience, no serious observer has any doubts about humankind's ability to feed itself during the twenty-first century—much more because of yield growth than because agriculture is expanding geographically at the expense of forests and other natural habitats.

If anything, we may now be too complacent that supply increases will continue to exceed our growing demand for the products that nourish us. Ruttan (2002), for example, emphasizes the environmental hurdles to continued intensification, which scientists must help to overcome. In addition, biotechnology has yet to fulfill its promise to create "better, more nutritious crops, which would be drought-resistant, cost-resistant, and salt-resistant" (Carr, 2003). As emphasized in this chapter, this promise will be fulfilled and environmental challenges will be overcome only if technological innovation is allowed to continue.

Key Words and Terms

agricultural experiment station
agricultural frontier
agricultural revolution
antitrust
aquifer
biofuel
capital
change in supply (in appendix)
cobweb cycle (in appendix)
comparative advantage
concentrated animal feeding
   operation (CAFO)
concentration ratio
Consultative Group on
   International Agricultural
   Research (CGIAR)
contestable market
economy of scale or size
extensification
extension
factory farming
fertilization

genetics
Green Revolution
hybrid
induced innovation
intellectual property right (IPR)
intensification
internal rate of return
International Maize and Wheat
   Improvement Center
   (CIMMYT)
irrigation
International Rice Research
   Institute (IRRI)
labor productivity
land-grant college
land productivity
marginal cost (MC) (in appendix)
marginal product (MP)
   (in appendix)
marketing
mechanization
monopoly

panicle
public good
research and development
shifting cultivation
stocking rate
supply elasticity (in appendix)
temperate zone
total factor productivity (TFP)
tropics and subtropics
U.N. Development Program (UNDP)
value added
weed control

Study Questions

1. Why did human numbers grow and urban life arise after the agricultural revolution?
2. Compare and contrast temperate zone agriculture and farming in the tropics and subtropics.
3. Compare and contrast the supply side of the food economy in impoverished settings in the developing world with the supply side of the food economy in affluent places.
4. In what ways are U.S. farms different from the agribusinesses that supply inputs and services to these farms?
5. What are two ways for a government to encourage the production of public goods?
6. Describe the origins of productivity growth in U.S. agriculture and compare this growth to productivity trends in the rest of the economy.
7. Explain why U.S. agriculture was mechanizing with little change in crop yields during the late 1800s and early 1900s, while yield enhancement was taking place with little mechanization in Japan.
8. Describe the relative contributions of increases in planted area and higher yields to growth in global production of various crops in recent decades.
9. Describe the relative contributions of extensification and intensification to increased agricultural output in Asia, Latin America, and Africa.
10. Why were expanded irrigation and increased fertilization major elements of the Green Revolution?
11. What are the prospects for a major increase in irrigated area during the twenty-first century?
12. Identify the consequences for agriculture of higher energy prices.
13. What are the current trends for agricultural research and development in rich countries as well as the developing world?
14. How has technological change affected the marketing sector of the U.S. food economy?

15. Is there conclusive evidence that agribusinesses are exploiting farmers in the United States?
16. Compare and contrast trends in per capita food output in different parts of the world since the early 1960s.

Appendix: The Fundamental Economics of Supply

Agriculture is influenced more by environmental conditions than other sectors are. Also, a host of nonenvironmental factors contribute to crop and livestock production. These include labor, machinery and other sorts of capital, fuel, fertilizers and pesticides, as well as management inputs. The technology for transforming all these resources and inputs into agricultural output also matters. Furthermore, how much farmers produce depends on market values, both the prices they pay for factors of production (i.e., inputs) and those they receive for output.

Similar to the approach used to analyze changes in food demand, trends in food availability can be examined by applying the assumption of *ceteris paribus*. That is, a general production function, in which everything affecting output is represented and freely variable, is converted into a supply curve and then shifts in that curve are analyzed.

Ceteris paribus in production applies in what economists call the short run, which is not long enough either to make an investment in productive capacity (comprising land and capital) or to apply new technology. Under these circumstances, output can be changed solely by increasing or decreasing other inputs (e.g., labor), which by definition are variable. A fundamental feature of the relationship between output and variable inputs is diminishing **marginal product (MP)**. What this term means is that the additional production that results from a slight increase in employment of these inputs grows progressively smaller as the employment level rises. By the same token, the quantity of variable inputs needed to raise output a small amount grows ever higher as production increases.

Diminishing MP has implications for the **marginal cost (MC)** of production as well as supply in the short run. If the prices of variable inputs do not vary (another application of *ceteris paribus*), then the fact that more and more variable inputs are needed to raise output marginally means that MC in the short run goes up as production rises. As is explained in any introductory economics textbook, profit-maximizing firms in a competitive industry, such as farming, seek out the production level (Q_S) at which output price (P), over which no single firm has any control, equals MC. Thus, the

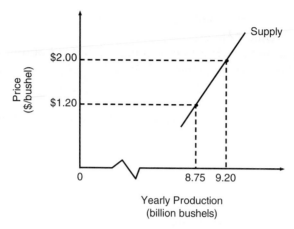

Figure 3.4 Short-run corn supply in the United States

short-run MC function for either a single firm that maximizes profits or an entire industry made up of such firms is equivalent to the short-run supply curve for that same firm or industry.[13] Just as a demand curve's negative slope reflects diminishing marginal utility (Chapter 2), the upward slope of the supply curve corresponds exactly to the positive linkage between the production level and MC, which in turn reflects diminishing MP as more inputs are converted into output.

Unless stated otherwise, all supply curves presented in this chapter and the rest of this book are of the short-run variety, meaning that these are derived by holding productive capacity, technology, and input prices fixed. One such curve, the short-run relationship between price and output of corn in the United States, is depicted in Figure 3.4. In the lower reaches of the curve, near the vertical axis, variable inputs have a high MP, which implies that MC is low. As output rises, MP goes down, which drives up MC. Progressively higher prices are required for producers to justify using ever greater amounts of variable inputs to produce a little more output.

Supply Elasticity

Just as own-price elasticity of demand describes how sensitive consumption is to price changes, **supply elasticity**, which is reflected in the supply

---

[13] To be more precise, the short-run supply curve comprises the upward-sloping portion of the MC curve above the minimum price a firm or industry needs to receive for its output to cover the costs of variable inputs in full. If price falls below this level, then the firm or industry produces nothing at all in the short run.

curve's slope, expresses the responsiveness of output to rising or falling prices—that is, relative change in production divided by relative change in price. Since a higher price induces more output, the elasticity of supply is always positive. Recall from Chapter 2 that own-price elasticity of demand is always negative because consumption goes down as price increases.

If relative change in output exceeds relative change in price, as would be indicated by a gently sloped supply curve, then supply elasticity is high (i.e., supply is elastic). As with own-price elasticity of demand, the threshold for categorizing supply as elastic is 1.00 (unitary elasticity). An example of high elasticity would be an 8 percent increase in output brought about because of a 5 percent rise in price—in other words, a supply elasticity of 1.60. In contrast, an elasticity of 0.80—observed, for example, because a 5 percent decline in price causes production to go down by 4 percent—means that supply is inelastic (i.e., the elasticity of supply is low).

As long as two points have been identified on a single supply curve, one can come up with a measure of supply elasticity. Let P and Q_S represent one such point and P' and Q'_S be another. Supply elasticity, then, is:

relative quantity change ÷ relative price change

$$= \left[\Delta Q \div \text{average of } Q'_S \text{ and } Q_S\right] \div \left[\Delta P \div \text{average of } P' \text{ and } P\right] \quad (3.1)$$

$$= \left[(Q'_S - Q_S) \div \tfrac{1}{2}(Q'_S + Q_S)\right] \div \left[(P' - P) \div \tfrac{1}{2}(P' + P)\right].$$

This formula can be applied to the short-run supply of corn in the United States. As shown in Figure 3.4, cutting the price from $2.00 to $1.20 per bushel causes annual production to go down from 9.20 billion to 8.75 billion bushels. Supply elasticity is:

$[-0.45 \text{ billion} \div \tfrac{1}{2}(8.75 \text{ billion} + 9.20 \text{ billion})]$
$\div [-\$0.80 \div \tfrac{1}{2}(\$1.20 + \$2.00)]$
$= [-0.45 \text{ billion} \div 8.98 \text{ billion}] \div [-\$0.80 \div \$1.60] \approx 0.10.$

This small number is consistent with actual estimates of supply elasticity in agriculture during the short run, which lasts one or two years. Beyond the short run, farmers can respond to price changes by varying more of their inputs, so the elasticity of supply is greater. When farmers have three to five years to react to price changes, supply elasticity is in the range of 0.30 to 0.50. In the long run, which is a decade or longer in agriculture, unitary elasticity is typical. For example, permanently increasing the price of corn by 10 percent will eventually induce a 10 percent increase in output, as the new productive capacity created by investment comes on line (Askari and Cummings, 1976, pp. 87, 117, and 150; Henneberry and Tweeten, 1991).

Changes in Supply

By definition, any adjustment in productive capacity or technology shifts the short-run supply curve, which means that there is a change in quantity produced at any given price. A rise or fall in the prices of variable inputs, which is another way to relax the assumption of *ceteris paribus*, has the same effect.

Since supply reflects marginal costs, a rise in the latter causes a downward **change in supply**. This is sometimes the outcome of disinvestment in productive capacity: switching land to another sector or industry, transferring mobile capital (e.g., vehicles), or gradually depreciating immobile capital (e.g., buildings and heavy equipment with no alternative use). MC also goes up if prices of variable inputs increase: for example, higher costs of crop production, and therefore diminished supplies, resulting from a rise in energy prices. Regardless of the reason, the supply curve moves toward the vertical axis, with less being produced at any given price (Figure 3.5).

Marginal costs can decline for various reasons. One of these is investment, either in land or capital goods, which raises the productivity of labor. MC can also be driven down by technological progress, which allows more to be produced with a given quantity of inputs (or, if one prefers, the same output to be produced with fewer inputs). In addition, lower costs can be the result of inputs being bought and sold at lower market values. If any of

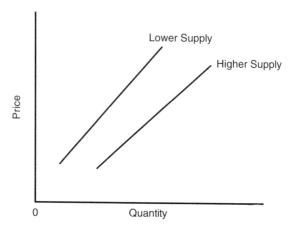

Figure 3.5 A shift in supply

these events happen, the supply curve will move away from the vertical axis, thereby causing more to be produced at any given price (Figure 3.5).

Just as a clear distinction needs to be made between movement along a stationary demand curve and a shift of that curve, care must be taken not to confuse movement along a stationary supply curve with a change in supply. As long as the *ceteris paribus* assumption holds, the short-run supply curve remains in place and output can go up or down only if the price is increasing or decreasing. If there are changes in land, capital, technology, or the prices of variable inputs, then supply itself shifts, with output different at any given price. Clearly, treating supply (a relationship between price and output) as a synonym for production creates as much confusion as treating demand as a synonym for consumption.

Finally, it needs to be acknowledged that analyzing changes in the supply of food can be a little tricky. Whereas a change in population or income obviously causes demand to go up or down, an increase or decrease in supply is sometimes difficult to distinguish from movement along a stationary curve. Consider an increase in crop yields resulting from increased use of fertilizer, which is a variable input. If the change happens only because this input is cheaper, then the *ceteris paribus* assumption has been relaxed and supply has increased. However, the change may be a consequence of a higher price for output, in which case movement along a stationary curve is happening. Unlike the different meanings of increased fertilization, however, there is no ambiguity at all about the consequences of technological improvement, which always lowers production costs and hence always amounts to an increase in supply.

Agricultural Supply: Real and Imagined Characteristics

Like the discussion of food demand in the appendix to Chapter 2, this appendix's treatment of agricultural supply is orthodox. The supply curves derived when the assumption of *ceteris paribus* is applied have an upward slope. Also, these curves shift in and out precisely as expected in response to investment, technological change, and other changes.

Heterodox views of the supply of farm commodities also exist. One thoroughly discredited notion is that agricultural supply curves bend backward, instead of sloping upward. Producers, it is alleged, respond to lower prices by increasing output, not reducing it, because they have mortgage payments and other fixed financial obligations. However, one cannot reconcile this view with the tendency of MC to rise with output, which implies that defying the guideline for profit maximization (i.e., setting output where price equals MC) reduces the financial resources available for paying fixed costs. In any event, empirical investigation consistently yields positive estimates of supply elasticities.

More sophisticated thinking has gone into the explanation of low farm returns, as these are alleged to be. In his technological treadmill theory, Cochrane (1965) contends that improved inputs from science and industry are adopted quickly by farmers. As a result, agricultural labor requirements fall and commodity output rises, thereby driving down prices. Since redundant labor cannot move out of the agricultural sector fast enough, downward pressure is exerted on profits. To maintain economic viability in the face of low returns, individual operators continue to draw on new technology in the hope of lowering production costs. As everyone does this, technological improvement accelerates, output expands, and prices and profits fall. In short, the treadmill continues.

Related to the treadmill view is the fixed asset theory, which explains the alleged propensity on farmers' part to adjust sluggishly or not at all to chronically low returns (Johnson and Quance, 1972). Factors of production are imagined to be immobile—that is, not transferable out of agriculture—and are therefore used even if profits are extremely low. For example, milking machines and combines, which have little value outside the farm economy, are purchased when commodity prices are favorable but continue to be used to generate excessive amounts of output in the face of falling prices.

The treadmill and fixed asset theories do not explain chronically low returns in agriculture for a simple reason, which is that these returns are not chronically or systematically low on commercial farms—as is documented in numerous studies (Hopkins and Morehart, 2002). Even people with small, less efficient farms find ways to take advantage of the tax code or even derive recreational or psychic value from their rural enterprises.

Although it cannot explain a nonexistent problem of chronic low returns on farm resources in developed countries, fixed asset theory yields insights into the annual and cyclical instability that characterizes markets for farm products, in rich and poor countries alike. Months and sometimes years separate the application of inputs from the realization of commodity output. Under these circumstances, a **cobweb cycle** can emerge after a drought or some other shock lowers output and raises prices. Responding to high prices, farmers during the next season increase input use so as to produce more. As a bumper crop is harvested, prices fall, which induces a cutback on inputs and production. This variability, which needs to be distinguished from chronically low returns, can last quite a while.

4

Aligning the Consumption and Production of Food over Time

Broad challenges facing the world food economy have been described in the two preceding chapters. On the demand side, human numbers are not growing as rapidly today as they did during the twentieth century, although demographic expansion will still be going on for at least a few more decades. In addition, living standards, which have been improving in various parts of the world, will continue to do so in the years to come, thereby driving up per capita consumption.

Demand growth is dealt with in some places by expanding the geographic domain of farming. Elsewhere, farmers have occupied practically all the land that lends itself well to crop production. Agricultural intensification, which has replaced extensification as the primary response to rising consumption, is obviously still needed. Otherwise, edible commodities will become scarcer, more people will experience hunger, and more natural habitats will be destroyed.

Significantly, no supreme agency decides on agricultural production in various places or the allocation of available supplies of food among consumers worldwide. Even at the national level, this sort of **central planning** is very much the exception to the rule. Instead, decision making is decentralized. On the demand side, countless households around the world make their own choices about how much of this or that product to buy and eat. These choices reflect individual earnings and tastes as well as prices. Likewise, innumerable farmers decide what and how much to grow and sell entirely on their own, taking into account the resources they possess, the value of inputs they must purchase, as well as output prices. Governments exercise influence, sometimes a lot of influence, by manipulating market forces. However, neither consumers nor producers receive direct orders from above.

At first glance, the lack of central planning seems to be a formula for chaos. One might think that any match struck between the production decisions of hundreds of millions of farmers and the consumption choices made by billions of consumers would be serendipitous. Actually, precisely the opposite is true. Instead of preventing chaos, central planning creates it. This was demonstrated in brutal experiments with bureaucratic coordination conducted during the twentieth century in the Soviet Union, the People's Republic of China, and other communist tyrannies. Secret police forces were deployed in each of these nations to try to enforce planners' edicts and to suppress criticism when planning failed, as it did routinely. The Soviet Union, a nuclear power capable of killing everyone on the planet, collapsed mainly because it could not accomplish the mundane task of matching the goods and services that people desire with production of these same goods and services. The Marxist mandarins of China phased out central planning before that system brought them down; this junking of communist economics created the improved living standards that so far have allowed these rulers to forestall political reform.

What totalitarian authorities sought to accomplish—at the expense of tens of millions of Chinese, Russian, Ukrainian, and other lives—was the elimination of markets. This turned out to be disastrous because markets are not just a setting—geographic, electronic, or otherwise—for exchange between businesses (including farms) with something to sell and their customers. They also comprise a mechanism for bringing production in line with consumption. As explained in this chapter's appendix, this coordination is accomplished with the price signals that markets generate. Prices are a remarkably effective way to guide the decisions of independent economic agents, not least when these decisions have to do with something as basic as food. Indeed, price signals are clearly superior to any alternative coordinating mechanism.

Along with bringing production in line with consumption, markets accommodate change with remarkable ease. This is important since the global food economy is greatly affected by population growth, technological improvement, and other trends examined in this book. How markets adjust to change is a major concern of this chapter.

Some of the discussion that follows, including an analysis of how government policies affect food markets, is conceptual, like the subject matter of a basic class in economics. We also analyze the food price increases that commanded headlines around the world during the past couple of years. In addition, long-term trends in prices are examined, both during the second half of the twentieth century and between now and 2050. As we show, recent years' prices have been much lower in inflation-adjusted terms than 50 to 60 years ago, although there is no guarantee that edible goods will remain cheap.

4.1 The Desirability of Competitive Equilibrium

As explained in the appendices of the two preceding chapters, economists analyze consumption and production in a competitive market, in which there are many buyers and many sellers, first by deriving demand and supply curves—which express simple linkages between consumption and production of a good, respectively, and the good's price—and then by examining shifts in these curves. In the case of demand, quantity consumed at any given price goes up or down in response to variations in income, consumers' tastes, and other factors that must be held constant in order to trace out a demand curve. Similarly, an increase or decrease in supply results from a change in productive capacity, technology for converting inputs into output, or input prices, all of which must remain fixed for a supply curve to be derived.

For any given combination of demand and supply, there is a price at which a competitive market is in equilibrium, in the sense that the quantity that consumers demand at that price equals the quantity that firms supply at the same price. The equilibrating tendencies of competitive markets are examined in this chapter's appendix, as are changes in **market equilibrium** caused by shifts in demand, supply, or both. But there is something else to appreciate about competition. As is explained in the appendix, the difference between the value that consumers place on output—a value comprising nothing other than their **willingness-to-pay (WTP)** for it—and the cost of labor, raw materials, and other variable inputs used to produce that output is maximized.

This difference between WTP and cost, which we can call **net economic value (NEV)**, is distributed between consumers and producers. The former group's portion, called **consumers' surplus (CS)**, equals WTP less the **market value** of output, or price multiplied by quantity. **Producers' surplus (PS)**, which is the latter group's share of NEV, equals the same market value less variable production costs. Both CS and PS can be identified by referring to a standard demand-and-supply diagram, such as Figure 4.1's representation of a hypothetical market for caviar in Astoria-

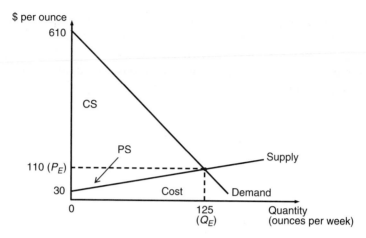

Figure 4.1 Consumers' surplus and producers' surplus

by-the-Sea. Interpretation of the same diagram indicates that the sum of CS and PS, which by definition equals NEV, is maximized if a market reaches competitive equilibrium.

Consider CS first. Astoria-by-the-Sea happens to be inhabited by a wealthy individual with a strong craving for caviar, and the maximum "bid" he would make for the first ounce coming his way every week is high: $610. Since no one else offers more money for this first ounce, the rich individual's bid represents its **marginal value (MV)**, which is where the demand curve intersects the vertical axis (Figure 4.1). By the same token, the MV of the second ounce supplied to the market comprises someone else's highest bid: $606 (i.e., the second point on the demand curve). Though still high, this is under the MV of the first ounce. MV, which is always found by inspecting the demand curve and in this example is a linear function of output,[1] falls as more caviar is made available. This decline reflects diminishing marginal utility (MU), which is a fundamental characteristic of consumers' preferences (Chapter 2).

As indicated in Figure 4.1, MV falls to $110, which is the competitive-equilibrium price (P_E), as weekly caviar purchases reach 125 ounces, which is the competitive-equilibrium level of consumption and production (Q_E). This makes perfect sense. What someone offers for the last, or marginal, unit consumed is precisely equal to its price, neither more nor less. Significantly, however, every buyer faces the same price, $110. There is obviously

---

[1] $MV_Q = 610 - 4Q$, where Q represents weekly output.

a $500 gap between the $610 that the rich caviar addict would pay for his first ounce and P_E, a $496 difference between the MV of the second ounce and P_E, and so on up to no gap at all between MV and price for the 125th ounce. Adding up all these differences, we obtain overall CS:

$$\begin{aligned} CS &= (\$610 - \$110) + (\$606 - \$110) + \ldots + (\$110 - \$110) \\ &= \{\tfrac{1}{2} \times (\$610 + \$110) \times 125\} - \{\$110 \times 125\} \\ &= \tfrac{1}{2} \times (\$610 - \$110) \times 125 = \$250 \times 125 = \$31,250. \end{aligned}$$

As this calculation illustrates, CS is the difference between the sum of the 125 MVs and market value. The former sum is equivalent to WTP, which is represented in Figure 4.1 as the trapezoidal area under the demand curve (which is the same as the MV curve) between 0 and Q_E. Also, the market value of competitive output is represented in the same figure by the rectangle with a height of P_E and a length of Q_E. Accordingly, CS consists of the triangle bounded by the demand (or MV) curve, the vertical axis, and a horizontal line extending out of that axis from P_E.

The description of PS is analogous to that of CS. Just as there is a consumer in Astoria-by-the-Sea who greatly values the first ounce of caviar he eats, there is an unusually efficient firm that can supply an ounce to the same market at a cost that no other firm can match. Let us say that the marginal cost (MC) of this first ounce is $30, which is the supply curve's intercept on the vertical axis. While still low, the second ounce's MC is slightly higher: $30.64. MC, which like demand in this example is a linear function of output,[2] continues to be pulled up as output rises.

In competitive equilibrium, MC equals $110, which of course is P_E. Since every unit of output changes hands for this same price, regardless of its MC, PS accrues:

$$\begin{aligned} PS &= (\$110 - \$30) + (\$110 - \$30.64) + \ldots + (\$110 - \$110) \\ &= \{\$110 \times 125\} - \{\tfrac{1}{2} \times (\$30 + \$110) \times 125\} \\ &= \tfrac{1}{2} \times (\$110 - \$30) \times 125 = \$40 \times 125 = \$5,000. \end{aligned}$$

Variable costs comprise the sum of the first MC through the 125th MC and, since the supply and MC curves are equivalent (Chapter 3), this sum is represented in Figure 4.1 as the trapezoidal area under the supply (or MC) curve between 0 and Q_E. Subtracting this area from the rectangle that stands for the market value of competitive output, one obtains PS: the triangle bounded by the supply curve, the vertical axis, and the horizontal line extending out from P_E.

---

[2] $MC_Q = 30 + 0.64\, Q$, where Q represents weekly output.

Having defined and described CS and PS, we point out that the sum of the two, which to repeat is NEV, is maximized if equilibrium is reached in the competitive market depicted in Figure 4.1. A clear indicator of this efficient outcome is that P_E (equal to the MV of Q_E) is in line with MC. In other words, the value to the consumer who purchases the last unit supplied to the market (in this case, the 125th ounce of caviar) equals the cost to some firm of producing that marginal unit. A lower level of output, at which MV exceeds MC, is inefficient in the sense that consumers' WTP for additional output exceeds the cost of raising production. This outcome, which is typical of monopoly, is examined in this chapter's appendix. In contrast, producing and consuming more than the efficient level means that the value of excess output (i.e., everything over and above Q_E) is less than the cost of same; in other words, NEV would increase if output diminished. This outcome can be observed in markets for agricultural commodities in which governments subsidize production. In addition, output is often excessive because firms do not pay for environmental impacts (Chapter 5).

4.2 Public Policy and Markets for Farm Products

Food prices routinely go up and down, due to shifts in demand, supply, or both. However, the upswings and dips of recent years have been breathtaking. How, one could be excused for asking, has NEV or anything else desirable been maximized in the food economy?

From late 2007 through the first several months of 2008, prices skyrocketed. Wheat, which was bought and sold for $200/ton in March 2007, was changing hands for $440/ton a year later. In April 2008, the price of rice peaked at $900/ton, nearly triple its value six months earlier. But then market values slid back nearly as quickly and as much as they had risen. By late 2008, for instance, the price of wheat was below $250/ton and rice was being bought and sold for not much more than $500/ton.

To understand market gyrations, one's first impulse is to examine Mother Nature's influence, which is often substantial. For example, recent price increases were immediately preceded by droughts in Australia, China, and other places, which cut into harvests. Also, charges of speculation are usually made as prices go up, even though these charges have little basis in fact (Box 4.1). Closer analysis of price trends, including those of 2007 and 2008, usually reveals that the actions of government have a decisive impact.

> **Box 4.1 Speculation and food prices**
>
> Once an allowance has been made for the effects of good or bad weather, high food prices usually lead to attacks on speculators. A number of politicians, including several U.S. senators, condemned these market actors in 2008, as prices were skyrocketing. Likewise, Italy's finance minister vilified them as the "plague of the twenty-first century" (Reuters, 2008).
>
> To be sure, **real-side speculation**, including additional purchases of food by households and others who anticipate further price increases, can add to the upward momentum of commodity markets (World Bank, 2009a, pp. 63–64). However, no evidence has been presented that some person, firm, or group tried recently to buy up all available wheat, soybeans, and so forth in the hope of driving up prices so that inventories could be sold later at a profit. Given the sheer size of the global food economy, it is next to impossible to "corner the market" in this way, as experienced traders know full well.

Why Did Food Prices Spike in 2007 and 2008?

The influence of government on food prices certainly has been apparent recently in the world's largest economy and leading source of agricultural exports. Due to macroeconomic mismanagement, exemplified by a chronic mismatch between bloated government budgets and tax revenues, the U.S. dollar lost much of its value relative to the euro during the first decade of the twenty-first century, falling from a peak of €1.15 in early 2002 to €0.65 six years later. In light of this devaluation, is it any wonder that a lot of dollars had to be offered in 2008 for any given amount of food?

Monetary devaluation in the United States also coincided with higher energy prices. This is because the market value of petroleum is expressed in dollars and, whenever the U.S. currency weakens, exporting countries demand more dollars for every barrel they supply. Crop production requires substantial amounts of diesel fuel, natural gas, and so forth (Chapter 3), so expensive energy translates into high costs. Furthermore, expensive energy encourages the conversion of crops into biofuels, thereby driving up food prices even more.

This linkage is obvious in the United States, which is by far the world's leading exporter of corn and which now converts one-third of its crop into ethanol. The U.S. government supports biofuel development in two ways.

First, a tariff of approximately $0.50/gallon (equivalent to $0.13/liter) is levied on ethanol imports from Brazil and other tropical countries. This policy keeps the U.S. price above the world level, which is generally in line with the relatively low cost of manufacturing biofuel from sugarcane. Second, fuel containing ethanol is exempted from federal excise taxes, which amount to $0.05/gallon. This creates an effective subsidy for ethanol of $0.50/gallon if, as is typical, retail outlets sell a blend consisting of 90 percent petroleum derivatives and 10 percent biofuel. The annual cost to the U.S. Treasury is $7 billion (Doornbosch and Steenblik, 2007, p. 6).

Estimates vary of the market impacts of policy-induced biofuel production, with some experts claiming that this development was responsible for 70 percent of recent increases in the price of corn (World Bank, 2009a, p. 61). A balanced judgment has been offered by the Director General of the **International Food Policy Research Institute (IFPRI)**. Observing that "a moratorium on grain-based biofuels would quickly unlock these commodities for use as food," Joachim von Braun contended that "this measure might bring corn prices down globally by about 20 percent" (2008). Since corn production draws on the same resources used to raise other crops, the value of commodities that are not converted into energy products would be affected as well. For example, von Braun (2008) suggests that a moratorium on making ethanol from corn would diminish international wheat prices by 10 percent. Livestock products would become cheaper as well, as feed grains became less expensive.

Trade barriers and subsidies benefiting biofuel producers in the United States were by no means the only public policy contributing to high food prices in 2007 and 2008. At one time or another, more than three dozen nations around the world—including agricultural exporters such as Argentina, the Ukraine, and Vietnam as well as a number of countries capable of being net suppliers in international markets—failed to exploit the commercial opportunities created by high commodity prices. Instead, the governments of these nations chose to insulate their consumers from international economic conditions, in particular by taxing, regulating, or otherwise discouraging agricultural exports.

This intervention was consistent with attempts made in various parts of the world during the twentieth century to develop at agriculture's expense (Chapter 7). But recent export restrictions had the additional consequence of driving global prices even higher, as the Secretary General of the United Nations underscored in a speech to an international summit in June 2008 (Ban, 2008). Coincidentally or not, some of these restrictions were loosened after this speech.

Relaxation of export restrictions helped cause prices to fall during the next few months. So did the global recession since the demand for food—which had been increasing rapidly in China, India, and other emerging economies—stopped growing as fast. Furthermore, biofuel development, which was robust as long as corn and other commodities were cheap, decelerated sharply, which helped dissipate the momentum for price rises.

Governments and Markets for Agricultural Commodities

By no means were 2007 and 2008 the first time when output levels and prices in the food economy were influenced by governments.

An early attempt to manipulate markets for agricultural commodities occurred in the United States during the 1930s. As part of the New Deal, the federal government engaged in **supply management**, with farmers encouraged to take land out of production (Cochrane, 1979, pp. 286–289). Although the same approach continued to be applied from time to time after the Second World War, supply management became less effective with the passage of time. The reason was that improved seeds (for hybrid corn, for example), agricultural chemicals, and other inputs were being substituted for land, so farmers could withdraw fields from production and still maintain or even increase output.[3]

As supply management lost its effectiveness, the U.S. government interfered with markets in other ways to the benefit of farmers. One option, which was employed not only in the United States but in Europe and other parts of the world as well, was to dictate a minimum market value, or **support price**, above the equilibrium level. As illustrated in Figure 4.2, a support price of P_S that exceeds P_E causes production to go up, to Q_S.

The benefits of this approach for farmers consist of an increase in PS: as defined above, the triangular area bounded by the supply (or MC) curve, the vertical axis, and the horizontal line extending out from the axis from the price level (P_S if government is intervening in the market). Balanced against this, however, is a reduction in CS (again, the triangular area under the demand, or MC, curve and above the horizontal line intersecting the vertical axis at the price level). There is also the expense of public commodity purchases, which are needed to support the price above the equilibrium level. This expense equals P_S multiplied by surplus production, which comprises the difference between output (Q_S) and quantity demanded (Q_D) at P_S.

---

[3] Supply management has been no more effective in other parts of the world and European authorities, for example, have given up trying to do so.

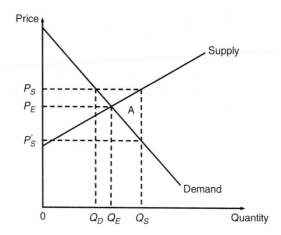

Figure 4.2 Price supports and targets

Moreover, there is a reduction in NEV because output is greater than the efficient level, Q_E. This loss, which is represented by area A in Figure 4.2, consists of the difference between the cost of the excess production (i.e., the trapezoidal area under the supply curve between Q_E and Q_S) and WTP for this same extra output (i.e., the smaller trapezoidal area under the demand curve between Q_E and Q_S).

Aside from public expenditures on surplus production, the government must incur the expense of storing its purchases before disposing of same, often at a loss. To avoid this expense, a government bent on propping up farmers' earnings can resort to the alternative of **deficiency payments**. Under this arrangement, the price set for farmers, P_S in Figure 4.2, is still treated as a guarantee, yet there are no public commodity purchases. Instead, farmers sell all their output for as much as they can, which is P_S' in Figure 4.2, and the government makes up the difference for every bushel or metric ton supplied. Total public expenditures equal Q_S, which is the amount farmers produce at the guaranteed price (P_S), multiplied by the difference between P_S and P_S'.

Note that output is the same, Q_S, regardless of whether a support price is decreed and government buys up the resulting surplus or a target price is guaranteed and deficiency payments are disbursed. Likewise, the two approaches have identical impacts on PS and CS. In addition, the loss in NEV is the same: area A in Figure 4.2.

There is another similarity, which has to do with real estate values. Support prices or deficiency payments enhance the returns to farming, as represented by the enlargement of PS in Figure 4.2. So as to benefit from this enlargement, farmers and others seek to augment their holdings of agricultural land. More to the point, they bid up the price of this resource. As Ricardo (1965) was the first to point out, nearly 200 years ago, the ultimate impact of distortions in food prices engineered by meddling governments is inflated farmland values.

By raising the financial bar for getting started in crop or livestock production, higher land prices burden aspiring new farmers. In addition, public expenditures on the surplus created by price supports, deficiency payments, and the like have adverse economic repercussions, since gross domestic product (GDP) contracts by at least $0.16 for every dollar the government collects in taxes from the general public (Ballard, Shoven, and Whalley, 1985). The sacrifice might be worthwhile, if, as is sometimes claimed, these expenditures accelerate farmers' adoption of new technology. But, in fact, there is no evidence of any such impact (Gardner, 2002, pp. 257–260).

So there is little economic justification for governmental interference with market forces. As public budgets have tightened, expenditures on surpluses and so forth have received greater scrutiny. These expenditures are attracting greater criticism in Europe, for example (Chapter 10). Also, international trade negotiations (Chapter 6) have added to the pressure for reform (Peterson, 2009, p. 153).

But rather than being reformed out of existence, governmental meddling in the food economy has changed its form over the years. In the United States, for example, relief provided to farmers for floods, droughts, and other natural disasters has increased as older categories of support have declined (Westcott and Young, 2000). Biofuel subsidies are another recent development. Truly, the survival and continuing evolution of public support for agriculture in places like the United States owes much to the political resourcefulness of farmers and the groups that represent them. It is also a constant source of frustration for the advocates of reform, who would like markets to be free throughout the food economy.

Proper Roles for the Public Sector

As is illustrated in Figure 4.2, inefficiency is created when government guarantees minimum prices, purchases surplus output, or otherwise intervenes in markets. However, governments also can help markets perform efficiently. Sometimes this help is decisive.

One positive contribution has to do with the suppression of monopoly—through the application of antitrust legislation as well as the promotion of international trade, which enables foreign competitors to keep local monopolists in check. Also, inefficiency can be lessened by curbing the excessive pollution and output that results if people and firms neglect the adverse environmental impacts of production and consumption. This can be accomplished in various ways, such as taxing environmental damage (Chapter 5).

In addition, it is not uncommon for governments to try to dampen fluctuations in food markets. One alternative is to administer public stockpiles, which can be built up when prices are low and drawn down if prices threaten to climb too high. Another is to administer a public fund for buying and selling commodities.

It is unclear that interventions such as these really stabilize prices more than profit-driven releases from and replenishment of private commodity reserves (Williams and Wright, 1991, pp. 410–451). Moreover, governments are often the source of price instability. Export restrictions implemented recently in Argentina, the Ukraine, and Vietnam are a case in point (see above). Likewise, surplus output resulting from support prices is often dumped on international markets, thereby depressing global values artificially and weakening incentives for food production in places where governments are not meddling with market forces.

Compared to the debatable merits of trying to dampen fluctuations in food prices, there is something that government is uniquely suited to provide that is of critical importance. Commonly referred to as the rule of law, this is the institutional framework that markets require for their existence and that involves reliable and even-handed enforcement of contracts and property rights. Without this framework, there are no markets in the true sense of the word, so the sort of efficient outcome depicted in Figure 4.1 fails to materialize. Worse than that, economic chaos of the sort imagined by critics of the market system actually happens.

4.3 Historical Trends in the Scarcity of Agricultural Products

A large number of national markets for agricultural commodities are distorted by government policies, some severely so. But at a global level, market values are mainly determined by demand and supply. To be specific, long-term trends in **real prices** (which by definition have been corrected for inflation) reflect changes over time in the scarcity of food (Box 4.2).

Box 4.2 Trends in food scarcity during the late nineteenth and early twentieth centuries

One of the first empirical studies of long-term trends in the scarcity-value of products from agriculture and other resource-based sectors made use of a time series of prices and production costs in the United States from 1870 through 1957. The two economists who carried out this study noted that prices of farm products had fluctuated markedly. Going down when harvests were good and increasing when output fell short, market values also tended to fall during recessions and rise in wartime. Having declined after the Second World War, market values in 1957 were little changed from levels 87 years earlier (Barnett and Morse, 1963, pp. 211–212).

But while price trends indicated no major change in food scarcity, other evidence suggested that scarcity had diminished. For example, the labor and capital required to produce a unit of farm output went down, not up, from 1870 to 1957. Evidently, mechanization—the substitution of capital for labor—had coincided with a decline in production costs. The study's two investigators also acknowledged the impacts of rising TFP—the difference between the value of crop and livestock output, on the one hand, and the costs of inputs other than labor, on the other (Chapter 3)—caused by the introduction of hybrid seeds and other innovations (Barnett and Morse, 1963, pp. 166–168 and 197–198).

Food supplies have gone up faster than demand in recent decades (Chapter 3). Consistent with this trend toward diminished scarcity, real prices have declined. In the United States, for example, the inflation-adjusted price of corn fell by three-fourths between 1950 and the middle 1980s and then remained low through the turn of the twenty-first century (Figure 4.3). Real prices for two other staples, rice and wheat, followed the same path.

As can be seen in Figure 4.3, there have been exceptions at times to the general trend toward cheaper food. The most important of these was a spike in cereal prices during the 1970s, which related mainly to events in the Soviet Union. Reluctant to cut food supplies to consumers in the face of poor grain harvests and the expanded feeding of livestock, communist authorities abandoned their long-standing policy against imports from places like the United States. Increased international purchases by the

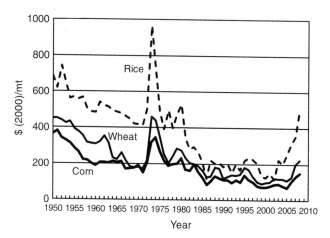

Figure 4.3 Real prices (2000 $) of U.S. rice, wheat, and corn from 1950 to 2008
Source: U.S. Department of Agriculture, prices deflated by implicit GDP.

Soviet Union added substantially to demand in international markets. At the same time, increased earnings from oil-exporting nations, which had benefited from a quadrupling of oil prices in 1973–74 as well as an additional tripling of prices at the end of the decade, were being deposited in international banks, which in turn made loans in Latin America and other parts of the developing world. Some of these loans were used to buy grain on international markets, thereby augmenting demand.

The food crisis of the 1970s brought home the lesson that food availability cannot be taken for granted, even in international markets in which supply interruptions in one part of the world are usually cancelled out by abundance somewhere else. Taken individually, none of the positive shifts in demand or negative changes in supply would have affected prices. But together, the combined demand-and-supply shocks were enough to affect market equilibrium dramatically, driving up grain prices by more than 100 percent in a little over a year.

Another lesson to draw from periods of unusually high prices, both during the 1970s and more recently, is that markets soon return to normal conditions. Just as prices can climb rapidly due to a "perfect storm" of unusually strong demand coupled with supply shortfalls, market values fall back quickly once unusual conditions end. After the 1970s, for example, grain values resumed their long-term downward slide (Figure 4.3).

The speed with which markets recover from shocks is not fully appreciated, in part because spiking prices receive a lot of coverage in the press and from public figures while the subsequent return to normalcy is all but ignored. Furthermore, warnings from the Club of Rome about imminent overpopulation and resource exhaustion (Chapter 2) amplified the distress aroused by the 1970s food crisis. In particular, the doubling of grain values seemed to confirm that mass starvation was just around the corner, as Ehrlich (1968) and others were contending.

In a sense, temporary abatement of the long-term trend toward cheaper food was fortuitous. As prices held steady, or even went up, incentives to adopt Green Revolution varieties of rice, wheat, and other crops were reinforced during the 1970s. This was true at the farm level, since individual producers were more receptive to new technologies when output values were high. It was also true for entire countries, which were more inclined to underwrite the research and extension needed to accelerate technological change if rising prices could be regarded as a signal of mounting food scarcity.

In contrast, technological improvement seems less urgent if farm products are cheap, as was the case for approximately two decades beginning in the middle 1980s (Figure 4.3). Support for agricultural research and development softened during this period, not coincidentally while real food prices were low (Chapter 3).

Spurred for a while by high output prices and having the fundamental advantage of enhancing total factor productivity (TFP) in agriculture, the Green Revolution ultimately saved much of humankind from the cataclysm that many were forecasting in the late 1960s and early 1970s. As agriculture has intensified, increases in the supply of food consistently have exceeded those in demand, particularly in Asia. Per capita food supplies rose (Chapter 3) and, though little or no mention of it was made in the press, the long-term decline in commodity values that had stalled in the early 1970s resumed a decade later. As the century drew to a close, inflation-adjusted prices for the goods that nourish us had reached historical lows.

The easing of food scarcity that has taken place since 1950 is a remarkable achievement. Equally remarkable are the improvements in productivity that are primarily responsible for diminished scarcity, and which also have prevented farm profitability from declining (Box 4.3). Thanks to advances such as the adoption of hybrid corn in the United States and semidwarf crops in Asia and other parts of the developing world, it has been possible for agricultural TFP to multiply precisely as real commodity values declined by 50 percent or more.

Box 4.3 Long-run equilibrium in agriculture

Productivity growth has eased the burden that falling prices otherwise would have created for farmers. In fact, the internal rate of return (Chapter 3) on the assets of commercial farmers has held remarkably steady, averaging 10 percent or so in the United States for many years (Hopkins and Morehart, 2002). This means that, in times past as well as today, $100 in productive capacity regularly has generated annual earnings of $10, year in and year out.

This level of profitability is about what one expects in a competitive market that, in addition to consistently aligning consumption with production, reaches what economists call **long-run equilibrium**. In this equilibrium, the internal rate of return is barely sufficient to maintain existing productive capacity without inducing any new investment (i.e., additions to capacity). A market is out of long-run equilibrium if returns are higher. However, this triggers investment, which augments supply and causes a price decline that in turn diminishes profitability. Conversely, below-normal profits result in disinvestment, which reduces supply and raises prices and profitability. Agriculture's internal rate of return is compatible with the profitabilities of other sectors in which investments are comparably risky.

Staying in long-run equilibrium is more of a challenge than simply aligning consumption and output through price adjustments, as described in this chapter's appendix. Since productive capacity is durable, economic decisions about replacing it, allowing it to depreciate, adding to it, and so on are based on expectations of market conditions in the future, which will affect the value of output that results from harnessing that capacity. If these expectations turn out to be erroneous, under- or overinvestment results. This causes profitability to vary, with the internal rate of return sometimes exceeding the level consistent with long-run equilibrium and other times falling below that level. The finding by Hopkins and Morehart (2002) that **normal profitability** has been sustained in U.S. agriculture is conclusive evidence of the capacity of competitive markets to adjust efficiently to major improvements in productivity that alleviate food scarcity.

4.4 Outlook for the Twenty-First Century

In light of agriculture's track record over many decades, occasional spikes in commodity values, either those of the 1970s or those of the past few years, are not an accurate guide to broad price trends. Instead, real prices in the future will depend on changes to come in fundamental market forces. If demand outpaces supply, then prices will rise. But if the reverse is true, supply increases that exceed growth in demand will allow humankind to continue along the path of diminished food scarcity.

Our analysis of future trends focuses on cereal crops. Food grains are the source of many of the calories that people consume directly. When indirect consumption of feed grains through livestock products is factored in, cereals account for at least 60 percent of total caloric intake. Examining trends in the demand, supply, and prices of grain has another merit, which is that information about yields and other production variables is more reliable than data for other agricultural commodities.

The projections of real price changes provided in Table 4.1 correspond to different assumptions about population growth and trends in consumption per capita between 2009 and 2050. We use low, medium, and high forecasts of human numbers provided by the U.N. Population Division (UNPD) (Figure 2.4). In addition to investigating the scenario that per capita consumption goes up at an annual rate of 0.3 percent, as it has done for a number of years, we examine the consequences of slower growth, 0.2 percent per annum, since Engel's Law (Chapter 2) informs us that continued improvements in earnings should cause the income

Table 4.1 Growth in food demand, supply, and prices from 2009 to 2050

Population in 2050 (billions) (1)	Annual Growth in Per Capita Consumption (%) (2)	Demand Growth, 2009 to 2050 (%) (3)	Supply Growth, 2009 to 2050 (%) (4)	Price Change, 2009 to 2050 (%) (5)
7.96	0.2	27	52	−50
7.96	0.3	31	52	−42
9.15	0.2	45	52	−14
9.15	0.3	52	52	0
9.15	0.4	58	52	12
10.46	0.2	66	52	28
10.46	0.3	73	52	42

elasticity of food demand to diminish over time. Also on the demand side, we look at a combination of medium demographic expansion (from 6.85 billion in 2009 to 9.15 billion in 2050) and annual growth of 0.4 percent in per capita consumption. This scenario incorporates biofuel development, which would have about the same effect as a 0.1 percent annual increase in food consumption per capita.

We make a single projection of supply growth, one that is quite simple. We suppose that there will be no overall increase in agricultural land use, which is what happened for a decade or so beginning in the middle 1990s. We also assume that linear increases in per hectare output, which have been observed since 1961 (Figure 3.2), will continue. Yields may stagnate because irrigation does not increase. Alternatively, they may shoot up in the future, if there is a bonanza of biotechnological innovation, for instance. In any event, we assume linear growth in yields and, because there is no extensification, in output as well.[4]

Finally, our projections of changes in real prices between now and the middle of this century rest on the assumption that, because elasticities of demand and supply are both low, real prices will go up or down by 2 percentage points for every difference of 1 point between demand growth and supply growth.

The long-term forecasts provided in the right-hand column of Table 4.1 should allay concerns that spiking prices in 2007 and 2008 signaled the beginning of a sustained rise in food scarcity. To the contrary, diminished scarcity is possible during the next four decades. For example, real prices should be 50 percent lower in 2050 if human numbers go up by a little more than 1 billion between now and then and if annual increases in per capita consumption average 0.2 percent. Inflation-adjusted values would also decline, though not by as much, with more demographic expansion or with faster growth in per capita consumption.

If real prices are not lower 40 years from now, they will probably be about the same, although fluctuations will occur because of higher or lower energy values and other developments. As indicated in Table 4.1, increases in supply will be in line with demand growth if the population in 2050 is 9.15 billion, which is the medium UNPD forecast, and if per capita consumption keeps on growing by 0.3 percent a year. With faster demographic expansion, food will become scarcer, with inflation-

---

[4] There is a superficial resemblance between our supply forecast and Malthus's view of agricultural production. Both are obviously linear. But while we assume equal additive growth in yields and production, Malthus did not consider changes in per hectare output. Instead, he supposed that agricultural output would go up arithmetically due exclusively to extensification. Of course, Malthus also contended that human numbers increase exponentially, which is different from recent trends (Chapter 2).

adjusted values rising by 28 to 42 percent, depending on growth in average food intake. Finally, real prices ought to rise by 12 percent or so during the next four decades if more grain is converted into ethanol.

Notwithstanding its simplicity, our approach to forecasting yields estimates of price changes that are generally consistent with other projections. For example, the World Bank (2009a) expects demand growth to decelerate (p. 64), as we do, and also points out that there are ample opportunities to boost production (pp. 79–83). Accordingly, real prices go down under its "baseline" scenario (p. 85). However, a 30 percent increase is possible if an ambitious program of biofuel development is pursued (p. 9). Food could also become scarcer if global warming impairs agricultural productivity later this century (pp. 84–85).

The general lesson to be drawn from the price projections offered by the World Bank as well as those reported in Table 4.1 is that neither alarm nor complacency about food scarcity is in order. If recent trends in population, living standards, and agricultural production continue, then inflation-adjusted prices ought to hold steady. Dramatic declines in real market values, of the sort that occurred between 1950 and the middle 1980s, are not out of the question, although counting on such an outcome would not be prudent. Neither is a run-up in prices impossible. Increased scarcity could be self-containing, in the sense that sharply higher prices would spur capital formation and technological innovation. However, mounting scarcity would also create a heavy burden for hundreds of millions of people in Sub-Saharan Africa, South Asia, and other parts of the world—people who were food-insecure even at the low real prices sustained for two decades beginning in the middle 1980s (Chapter 8).

Key Words and Terms

central planning
consumers' surplus (CS)
deficiency payment
International Food Policy Research
  Institute (IFPRI)
long-run equilibrium
marginal value (MV)
market equilibrium
market value
net economic value (NEV)

price discrimination (in appendix)
producers' surplus (PS)
normal profitability
real price
shortage (in appendix)
speculation (real-side)
supply management
support price
surplus (in appendix)
willingness-to-pay (WTP)

Study Questions

1. Did central planning in the Soviet Union, China, and other communist states contain economic chaos or create it?
2. How does the achievement of competitive equilibrium affect the difference between what consumers are willing to pay for the output they purchase and use and the variable costs of producing that output?
3. Who captures net economic value and in what form?
4. What were the reasons for spiking food prices in 2007 and 2008?
5. Compare and contrast supply management, support prices, and deficiency payments.
6. Describe trends in inflation-adjusted grain prices since 1950.
7. Explain the causes of high grain prices during the 1970s and describe the impacts of high prices on the Green Revolution, which was under way that same decade.
8. What combination of demand and supply trends would cause real prices to decline during the next few decades? What combination of circumstances would result in price increases?

Appendix: The Coordination of Decentralized Decision Making

Consumption of any good or service is influenced by various factors. The amount of food people eat depends on their numbers and incomes. Choices among food items also reflect individual tastes. In addition, purchases of any single item depend on its price as well as the prices of a host of other goods (Chapter 2).

Food output is affected by a number of variables as well (Chapter 3). Productive capacity (i.e., land and capital) obviously has an effect. So do the prices of labor, materials, and other inputs. Another determining factor is the existing state of technology for transforming factors of production into agricultural output. Furthermore, output depends on the price of the good being produced, just as consumption is.

Underlying the alignment of consumption and production, then, are various adjustments by households and firms. To understand how all these adjustments end up being coordinated, some abstraction is required. As explained in the preceding two chapters' appendices, it is convenient to employ the assumption of *ceteris paribus* to create a demand curve, which relates consumption of a good to its price alone, as well as a supply curve, which relates output of the same good only to its price.

After doing this, one can analyze equilibrium in the market where consumers and producers do business, with primary focus on the role of prices in the achievement of that equilibrium.

The Alignment of Production and Consumption

There are two special sorts of market equilibrium. One of these occurs if production is economically infeasible, in which case nothing at all is exchanged between producers and consumers. Making caviar in Sierra Leone would be a good example. Infeasible production is signaled by the supply curve's intersection with the vertical axis (on which monetary values are measured) above the demand curve's intercept on the same axis, which means that the maximum amount that any consumer would pay for just one unit of output is less than the cost of producing the same unit. Nonscarcity is signaled if the supply curve rises out of the horizontal axis (on which quantity of output is measured) to the right of where the demand curve descends to that axis, in which case the good is free. Though vitally important, the air we breathe is an example of something that is not economically scarce.

Depicted in Figure 4.4 is the more typical case: a market for a good, like corn in the United States, that is both scarce and feasible to produce. With the demand curve sloping downward and the supply curve sloping upward, there is one and only one point where the two intersect. This point, at which production and consumption both amount to Q_E, is the market equilibrium. Let us say that Q_E in the U.S. market for corn is about equal to 9.20 billion bushels per annum.

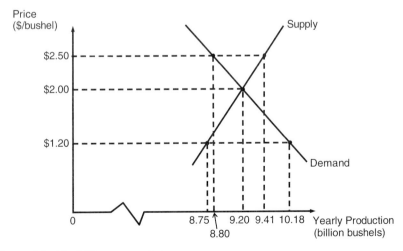

Figure 4.4 The U.S. corn market

Price adjustments drive the market inexorably toward equilibrium. As shown in Figure 4.4, the price, P_E, that causes annual consumption as well as yearly output of corn to equal 9.20 billion bushels is $2.00 per bushel. Consider what takes place if the price is higher, $2.50 per bushel for instance. At this higher value, production (9.41 billion bushels) exceeds consumption (8.80 billion bushels). In other words, there is a **surplus** of 0.61 billion bushels. With unsold inventories accumulating, producers cut prices. As the market value of the good being exchanged falls toward P_E, production goes down and consumption goes up, thereby causing the surplus to dwindle. Price adjustments cease entirely as the difference between output and consumption disappears—in other words, as equilibrium is achieved.

The tendency toward market equilibrium is similarly robust if price starts out below P_E. If a bushel of corn changes hands for $1.20, what people want to consume (10.18 billion bushels) exceeds what farmers want to produce (8.75 billion bushels). Responding to this **shortage** of 1.43 billion bushels, consumers bid the price up, which causes the gap between consumption and production to close. Once again, equilibrium is achieved once bushels of corn are changing hands for $2.00 apiece.

Shifts in Demand and Supply

The equilibrating tendencies of markets are powerful, indeed—so powerful that markets adjust quickly to changes in demand or supply caused by things like population growth and technological innovation. That is, there is an expeditious adjustment from one equilibrium to another in response to a shift in demand, supply, or both. Almost always, the adjustment has an impact on output as well as price.

The easiest changes in equilibrium to analyze are those that result when either the demand curve or the supply curve shifts and the other curve remains in place. Consider a market in which demand has increased due to a rise in consumers' earnings. The price that formerly aligned consumption and production no longer does so. With no change on the supply side, production at the old equilibrium price is exactly the same. However, there is more consumption at that price because of the demand shift. The shortage induces a price rise, which ceases once consumption has fallen sufficiently and production has gone up enough for the two to equal one another. The end result of this adjustment is a higher price and more output.

Another example of a simple change in market equilibrium is a supply increase not accompanied by any alteration in demand. Suppose farmers have access to more land. This increase in productive capacity leads them to produce more crops at any given price. At the old equilibrium price, there is a surplus. To be specific, consumption is unchanged while output is greater. As always, a surplus induces price-cutting, which causes consumption to grow and production to fall. Equilibrium is reestablished at a lower price and higher level of output.

Changes in price and output are easy to determine when just one of the two curves, either demand or supply, shifts. With a change in demand, price and output always move in the same direction. If demand goes up, both are higher at the new equilibrium. If demand goes down, then these two variables decline. With supply changes, the two always move in opposite directions. An increase in supply causes output to rise and price to fall. The opposite happens if supply declines.

At least as interesting as the general direction of price and output adjustments in these simple cases are the relative magnitudes of these adjustments. As with so much else, the determining factor here is elasticity. Whether the relative change in price resulting from a shift in demand exceeds the relative change in output or vice versa depends on the elasticity of supply. Similarly, the sensitivity of consumption to price variations determines which changes more because of a supply shift, price or output.

The importance of elasticity is illustrated in Figure 4.5. Shown in Part A is the shift in the spinach market's equilibrium caused by reduced prices for lettuce. Since the substitute good is cheaper, demand for spinach falls, which obviously causes price and output to decline. Supply being elastic, relative change in the latter variable exceeds relative change in the former. The interpretation of all this is that farmers can move in or out of spinach production quickly and easily, which makes their output sensitive to price changes. It is only to be expected, then, that the decline in demand results mainly in diminished production, with modest price effects.

Represented in Part B of Figure 4.5 are the consequences of an improvement in agricultural technology, which increases food output at any given price. For a typical food item, own-price elasticity in the domestic market is fairly low, as indicated by the demand curve's steep slope. Given the limited responsiveness of consumption to price changes, technological change does not have much impact on equilibrium quantity. Instead, the main consequence is a price reduction, which implies that consumers are the main beneficiaries of supply growth.

Of course, things are more complicated if there are simultaneous shifts in demand and supply, as occurs all the time in real markets. If these both decline, one cannot automatically say that equilibrium price, which is

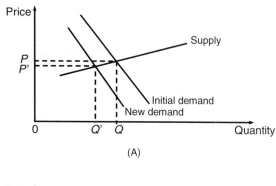

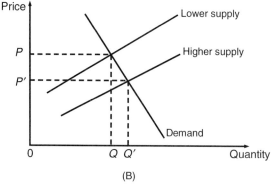

Figure 4.5 Elasticity and changes in market equilibrium
A. A shift in spinach demand with elastic supply
B. A change in food supply with inelastic demand

pulled down by decreased demand though pulled up by decreased supply, will be higher or lower. If demand rises while supply falls, equilibrium output can increase, decline, or stay exactly the same. Empirical investigation is needed to assess the net impacts of the changes taking place.

It is opportune at this point to repeat admonitions from Chapters 2 and 3 to be precise in the use of economic vocabulary. In particular, demand (a functional relationship between price and consumption) should not be treated as a synonym for consumption, just as supply (a functional relationship between price and output) should not be used as a synonym for production. Failure to avoid this error creates confusion, especially when simultaneous changes in demand and supply can result in variations in production and consumption that are either positive or negative.

In places like the United States, where countless markets reach competitive equilibrium routinely, bringing production and consumption into line with each other is taken for granted. Indeed, students in economics classes, who are tested on their understanding of demand-and-supply diagrams like Figure 4.4, sometimes seem to be the only people thinking about market equilibrium. But even if it is taken for granted, a system that coordinates the choices of producers and consumers and readily accommodates changes in demand and supply has enormous merit. As emphasized at the beginning of this chapter, various nations with totalitarian regimes have learned the hard way that suppressing markets creates economic chaos.[5] People have every reason to find this intolerable.

Net Economic Value and Its Maximization

Compared to the chaos created by central planning, competitive equilibrium is eminently satisfying for just about everyone concerned. In particular, the fact that NEV is maximized when consumption is in line with production in a market in which there are many buyers and many sellers can be demonstrated with reference to the hypothetical market for caviar in Astoria-by-the-Sea.

As indicated earlier in this chapter, the value that consumers place on the competitive level of output (Q_E) comprises the sum of MVs, which is represented as the area under the demand (or MV) curve between 0 and Q_E:

$$\text{WTP} = \text{MV}_1 + \text{MV}_2 + \cdots + \text{MV}_{125} = \$610 + \$606 + \ldots + \$110 = \$45{,}000.$$

Also, the variable cost of producing Q_E is represented by the area under the supply (or MC) curve between 0 and Q_E and comprises the sum of MCs:

$$\text{Cost} = \text{MC}_1 + \text{MC}_2 + \cdots + \text{MC}_{125} = \$30.00 + \$30.64 + \cdots + \$110 = \$8{,}750.$$

Subtracting variable cost from WTP yields NEV:

$$\text{WTP} - \text{cost} = \$45{,}000 - \$8{,}750 = \$36{,}250,$$

which is the triangular area between the demand-and-supply curves. As demonstrated above, CS and PS equal $31,250 and $5,000, respectively, under competitive equilibrium, so the two indeed add up to NEV.

---

[5] Totalitarian nations are not the only places where economic chaos is created as governments override markets, with public authorities substituting their "wisdom" for the outcomes of unimpeded commerce. For example, U.S. politicians decided during and since the 1990s to broaden home ownership by tacitly guaranteeing mortgages taken out by households that otherwise would never have qualified for home loans. The long-term result of this has been severe financial turmoil (Chapter 10).

To appreciate the efficiency of producing 125 ounces of caviar, consider the NEV of a lower production level, 120 ounces for instance. Now the gap between WTP,

$$MV_1 + MV_2 + \cdots + MV_{120} = \$610 + \$606 + \cdots + \$130 = \$44,400,$$

and variable cost,

$$MC_1 + MC_2 + \cdots + MC_{120} = \$30.00 + \$30.64 + \cdots + \$106.80 = \$8,208,$$

comes to $36,192. Also, overproduction is every bit as inefficient as underproduction, in the sense that NEV is not maximized. If, say, 130 ounces are produced, the difference between WTP,

$$MV_1 + MV_2 + \cdots + MV_{130} = \$610 + \$606 + \cdots + \$90 = \$45,500,$$

and variable cost,

$$MC_1 + MC_2 + \cdots + MC_{130} = \$30.00 + \$30.64 + \cdots + \$113.20 = \$9,308,$$

is $36,192, which is lower than the net value ($36,250) of 125 ounces of sturgeon eggs.

Inefficient underproduction always occurs if a market is monopolized. Consider, for example, how much caviar is supplied—Q_M—in Astoria-by-the-Sea if only one firm produces sturgeon eggs and that firm can raise or lower the price, P_M, as it sees fit. If the monopolist cannot engage in **price discrimination**, which means charging different customers alternative prices, then the difference between the monopolist's revenues (price times output, or $610Q - 4Q^2$) and its variable costs is maximized by producing 67 ounces and selling caviar for $342 per ounce (i.e., the MV of the 67th ounce).[6]

The foregone NEV resulting from this underproduction is represented in Figure 4.6. At Q_M, a positive margin exists between MV and MC: $342 versus $72.88. In other words, what someone bids for the 67th unit of output (i.e., the last unit supplied by the monopolist) exceeds the cost of producing it. Though somewhat smaller, the margin between MV and MC for the next unit (i.e., the 68th) is likewise positive. The same goes for every other unit up to Q_E. Adding up these margins yields the loss in NEV resulting from monopoly. As shown in Figure 4.6, this loss shows up as the shaded triangular area between the MV and MC curves and

---

[6] To find Q_M, one starts by differentiating revenues, which equal $610Q - 4Q^2$, with respect to Q in order to find the marginal revenue (MR) of output. Profit maximization requires that MR (i.e., $610 - 8Q$) equal MC (i.e., $30 + 0.64Q$). Solving for output, one finds that Q_M is approximately 67. Plugging this output level into the MV equation, one obtains P_M: $610 - 4 \times 67 = 342$.

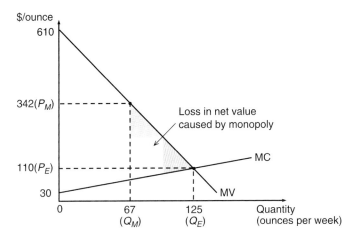

Figure 4.6 Monopoly in Astoria-by-the-Sea's caviar market

between Q_M and Q_E:

$$\tfrac{1}{2} \times [\$342 - \$72.88] \times [125 - 67] = \$7{,}804.48.$$

Finally, we observe that, although NEV is always distributed between consumers and producers, the relative sizes of CS and PS depend entirely on elasticities of demand and supply. In the hypothetical market for caviar depicted in Figure 4.1, consumption is much less sensitive to price changes than production. To be specific, the price of sturgeon eggs must rise by $4.00 to cause weekly consumption to decline by one ounce while a price change of $0.64 is enough to cause a one-ounce variation in output. With supply much more elastic than demand, consumers capture the lion's share of NEV, in the form of CS. If consumption responded more than production to price changes, then the portion of NEV going to producers as PS would exceed the portion distributed as CS.

5
Agriculture and the Environment

Given probable trends in consumption and production, food supplies ought to keep up with demand during the next few decades, when human numbers will still be increasing and when per capita consumption will continue rising because of income growth. Food might be scarcer in 2050, although a decline in prices is about as likely.

If food grows scarcer, the cause will probably be environmental. The World Bank (2009a, pp. 84–85), for example, qualifies its prediction of lower prices in the future with the caveat that climate change could combine with increased water scarcity to reduce agricultural yields significantly, which would make food more expensive.

Why are environmental threats to the food economy so serious? One reason is governmental manipulation of market forces, irrigation policy being an obvious case in point. Water use by agriculture dwarfs all other uses of the resource, especially in the tropics and subtropics (Table 3.6). This happens largely because the prices paid by farmers benefiting from public irrigation projects have little to do with water values. Indeed prices in many places amount to a tiny fraction of the cost of delivering water to their fields.

In Ecuador, for example, irrigation fees cover less than half the expense of operating and maintaining public systems. This means that farmers contribute nothing to capital costs, which are sizable. Paying low prices, they have little incentive to use water wisely, which is a major reason why the returns on irrigation investment have been disappointing. As a rule, no more than one dollar's worth of benefits, in the form of additional agricultural output, has been generated for every two dollars spent on public projects. In addition, farmers lobby assiduously for new dams, canals, and related infrastructure, knowing they will pay only a fraction of the cost. Most of this investment benefits people who are relatively well off, as opposed to the rural poor, so income disparities in the countryside are aggravated. Furthermore, farmers care little about conserving water if they purchase it at low prices. Also, there is a chronic lack of support for technical assistance on the proper use of irrigation water and the management of irrigated soils. Accordingly, hydrologic resources are wasted and a lot of farmland is damaged (Southgate and Whitaker, 1994, pp. 62–67).

The misallocation caused by regulations, subsidies, and other sorts of governmental interference with market forces is referred to as **intervention failure**, or policy failure or even government failure. This is curious language in a way. After all, the farmers lobbying for government favors like subsidized irrigation are perfectly aware of what they stand to gain. So are the public officials who bestow these favors, often in exchange for construction kickbacks or after having bought unirrigated land at low prices before an irrigation project is announced publicly so the same land can be sold afterward at a premium. The actual failure associated with policy-induced distortions in markets relates to losses suffered by society as a whole, including the higher taxes paid by the general population because of subsidies and graft. As illustrated by Ecuador's experience with subsidized irrigation, which is entirely representative of the same policy's effects throughout the world, losses also take the form of environmental damage.

Intervention failure is pervasive and its consequences are felt far and wide. However, it is not the primary focus of the literature on environmental economics, most of which addresses the marketplace's tendency to treat many resources as worthless. This **market failure** and the waste and misallocation it creates have been analyzed in detail. Economists also have helped design policies to correct market failure, for efficiency's sake.

The importance to agriculture of land, water, and energy resources having been examined in Chapter 3, we turn in the pages that follow to an analysis of three issues: global climate change, farmland degradation, and the loss of tropical forests and other natural habitats due to agriculture's geographic expansion. Aside from outlining the dimensions of these problems, we point out how intervention failure and market failure contribute to the misuse of natural resources, and could diminish food availability.

First, though, a few economic concepts need discussing. Most of what one has to know about intervention failure is summarized in Chapter 4. However, analysis of environmental issues cannot proceed without an understanding of market failure and its remedies. We now turn to this topic.

5.1 Diagnosing and Correcting Environmental Market Failure

In any introductory economics class or textbook, the main focus is on exchanges in markets between households and firms. Some of these exchanges have to do with labor and other inputs, or **factors of production**, which households provide and which firms convert into goods and services. Other market interactions have to do with finished products, which people purchase from businesses using the income they earn by supplying inputs.

As long as the natural resources used by firms in the manufacture of finished products are owned by individuals or well-defined groups (e.g., the shareholders of a corporation), then markets for these factors have much in common with other input markets. One difference has to do with issues of timing, which are unimportant for many nonenvironmental factors but are central to the allocation of most natural resources.

Inter-temporal trade-offs relate in part to differences in the accessibility and quality of resources. For example, pumping oil from beneath the sands of the Middle East happens to be cheap. In contrast, production expenses are elevated in recently developed off-shore sites, such as those near Newfoundland. An important part of the opportunity cost of current extraction in places like Saudi Arabia, then, comprises increased dependence in the future on more expensive sources of oil. However, resource heterogeneity is not required for the existence of inter-temporal trade-offs. All that is needed is simple economic scarcity, in the sense that untapped stocks would not fully satisfy current and future demands if the resource in question were free.

A market economy provides guidance for balancing **conservation** (i.e., delaying extraction until a future date) and **depletion** (i.e., bringing forward the schedule of extraction). This guidance takes the specific form of the **real interest rate**, which can be defined simply as the difference between **nominal interest** (i.e., the rate observed directly in financial markets) and inflation. As explained in Box 5.1, resource owners and others who must reckon with inter-temporal trade-offs use the real interest rate, which reflects the compensation savers demand in return for postponing consumption as well as the returns on investment, to compare the near-term value of a little more depletion with the future value of a little more conservation.

Box 5.1 The economics of conservation and depletion

Analyzing inter-temporal trade-offs, McInerney (1976) and other economists show that the owners of scarce resources decide on a time schedule of extraction by considering current and future prices of unextracted deposits as well as the inflation-adjusted return on financial holdings.

For example, an owner has no incentive to leave a deposit untapped if the rate of increase in its real in situ value (i.e., its delivered price less the cost of extracting it) is less than the real interest rate. Under these circumstances, he or she is better off accelerating current extraction and buying bonds and stocks with the proceeds. But if the unextracted value is rising rapidly, there are incentives for conservation. Owners find that they make more money by holding on to resources and letting their worth appreciate rather than extracting them and investing the resulting income. Schedules of development are stable, not to mention efficient, if the pace at which inflation-adjusted values of resources are rising exactly matches the real interest rate.

Market signals, including real interest, guide the development of privately owned resources. But what of resources in which no **property rights** have been established? Actually, the latter are not at all rare. Sometimes the absence of ownership is a consequence of public policy. At an extreme, private property was forbidden in the Soviet Union, which meant that the use and management of natural resources were not influenced at all by market forces. The result was gargantuan waste and misuse (Yergin, 1991, pp. 779–780).

Other resources are unowned, even though no public authority is actively suppressing private property. The reasons for **open access**, which is a term used to describe the absence of ownership, are economic, having to do with resource values that fall short of the costs of delineating, recording, and enforcing property rights. This relationship between values and costs exists for international fisheries. It also exists for much of the world's rangeland, which because of dry conditions cannot support livestock in high densities. As a rule, the value of unowned rangeland, which depends directly on the meat and other products that could be produced if forage and other resources were well managed, is exceeded by the costs of demarcating properties, registering deeds, and interdicting trespassers, all of which are required for the existence of property rights. In contrast, open access gives way to property rights once these costs fall below resource values (Box 5.2).

> **Box 5.2 The benefits and costs of private property**
>
> History provides interesting lessons about the economics of ownership, including changes over time in the benefits and costs of converting open-access resources into private properties.
>
> For example, private ranches did not come into being immediately after Americans began driving herds of cattle and sheep onto the western Great Plains, in the late 1800s. Rather, the region's grassland remained available for any and all to use as long as fencing, which was needed to demarcate individual holdings, was expensive, as it was in dry areas with little timber. However, the invention of barbed wire reduced fencing costs, which in turn caused much of Montana, Wyoming, and neighboring states to be claimed by individual ranchers (Anderson and Hill, 1975).
>
> This specific sequence of events illustrates a broad, long-term trend of converting open-access resources into private holdings. Economic historian and Nobel laureate Douglass North has shown that this conversion is partly a consequence of increasing resource scarcity, which is an outcome of population growth and economic expansion. It is also the result of technological advances, exemplified by the invention of barbed wire, which lower the cost of asserting, recording, or protecting ownership rights (North, 1990). These advances continue to the present day, thereby making possible the expansion of private property in various parts of the developing world where this legal arrangement previously was rare (de Soto, 2000).

The consequences of open access are easy to appreciate. Referring still to the case of unowned rangeland, we can think about a single herder who might conserve forage, perhaps by removing his or her cattle before they graze excessively. Significantly, that individual bears the entire cost of conservation, in terms of less weight gained or milk produced by his or her herd, while sharing the benefits of forage improvement with all herders in the vicinity. On the other hand, the benefits of additional beef and milk production resulting from depletion are captured by a single herder who has let his cattle overgraze, while the costs of rangeland degradation are an **externality** as far as he or she is concerned.

To summarize, property rights are a prerequisite for market allocation. Where the environment is concerned, open access makes the pricing of resources impossible and causes resource users to disregard the benefits of conservation as well as the costs of depletion. Since these economic consequences are externalities, there is chronic overexploitation and, as was demonstrated nearly 60 years ago in a seminal economic analysis of overexploitation of open-access fisheries, destruction of whatever economic worth unowned resources might have (Gordon, 1954).

The waste and misallocation created by open access are not limited to unowned resources, themselves. Distortions also arise in markets in which production or consumption has an impact on resources that anyone can use for free. One case in point would be dairy farmers who pay nothing if manure from their fields and barnyards ends up in rivers, lakes, and streams, thereby exacerbating water pollution. As is explained in this chapter's appendix, unpriced use of the environment to dispose of wastes amounts to a subsidy for dairy production, with more milk produced at any given price. This subsidy causes competitive-equilibrium output to be excessive and the market price to be too low—precisely the opposite of the inefficiency created by monopoly (Figure 4.6).

As is also explained in the appendix, governments can alleviate the inefficiencies resulting from market failure in various ways. **Command-and-control** is one option, with a regulatory authority deciding on emissions from each source. Alternatively, a **pollution tax** can be applied. Yet another way to correct market failure is to institute **cap and trade**, in which government creates a limited quantity of tradable permits to pollute, distributes these permits through an auction or in some other way, and then allows a market to emerge in which polluters bid among themselves for these permits. The United States has utilized cap and trade to reduce emissions of sulfur dioxide, thereby limiting acid rain (Kerr, 1998). The same approach is also being considered for the containment of **global warming**.

5.2 Agriculture and Climate Change

The earth's atmosphere is an unambiguous example of an open-access resource. Nobody pays anything for emitting **greenhouse gases (GHGs)**, including carbon dioxide and methane. Likewise, the atmospheric flux that results as GHGs accumulate is, from every individual's perspective, an externality. Activities that create emissions, such as **deforestation** (see below) or the combustion of fossil fuels, consequently tend to be excessive. So are the climatic impacts.

Climate change can also be an outcome of governmental actions. Examples include the underpricing of fossil fuels as well as the support that some authorities have provided for deforestation. As is the case with many other environmental problems, then, climate change results from a combination of market failure and intervention failure.

Organized by the United Nations, the **Intergovernmental Panel on Climate Change (IPCC)** has issued forecasts of the global warming likely to result later this century as GHG concentrations in the atmosphere rise.

> For the next two decades, a warming of about 0.2°C per decade is projected for a range of emissions scenarios. Even if the concentrations of all GHGs and aerosols had been kept constant at year 2000 levels, a further warming of about 0.1°C per decade would be expected. Afterwards, temperature projections increasingly depend on specific emissions scenarios (IPCC, 2008, p. 45).

Aside from attempting to predict global warming, the IPCC has examined possible effects, including in the food economy. The U.N. panel expects temperature increases in the range of 1 to 3°C to have a modest yet beneficial impact on agricultural productivity and crop yields in temperate settings as well as the high latitudes, such as Canada and Siberia. But closer to the equator, especially in dry areas, productivity and yields are expected to fall even if temperatures increase by just 1 to 2°C (IPCC, 2008, p. 48).

Other investigators agree that global warming will probably harm agriculture more in the low latitudes than in temperate settings. In one recent study, African, Asian, and Latin American yields in 2050 have been modeled (Nelson et al., 2009). Reported in Table 5.1 are possible results

Table 5.1 Potential percentage changes in crop yields in 2050 due to global warming

	Rice (1)	Wheat (2)	Corn (3)
East Asia	−11 to −8	+2	−13 to +9
Latin America and Caribbean	−22 to −19	+11 to +17	−4 to 0
Middle East and North Africa	−40 to −33	−9 to −5	−10 to −7
South Asia	−15 to −14	−49 to −44	−19 to −9
Sub-Saharan Africa	−15	−36 to −34	−10 to −7

Source: Nelson et al. (2009), p. 9.

for the case in which temperatures rise yet farmers do not respond by changing technology or inputs.

Identified in the same study are the measures required for adaptation to climate change. In Asia, for example, annual investment in irrigation systems and other public works would have to increase by as much as $2.5 billion to avoid a 15 percent decline in average calorie availability and a 24 percent reduction in per capita grain production. Likewise, annual investment in roads and other rural infrastructure might have to go up by $3 billion to help Sub-Saharan Africa cope with the potential agricultural impacts of climate change. Yields in the Middle East and North Africa could be hit hard by drier conditions during the next 40 years. In contrast, a little more agricultural research, irrigation efficiency, and transportation infrastructure should suffice to counteract the milder agricultural impacts of climate change in Latin America and the Caribbean (Nelson et al., 2009, p. 16).

Taking into account diminished crop yields, the rise in sea levels that would happen if polar icecaps melted, and other consequences of climate change, economist William Nordhaus has plotted an optimal time-path for carbon values.[1] Its main feature is that the cost of sequestering a ton of carbon (in living vegetation, for example) always equals the **present value**[2] of subsequent economic damages resulting from the emission of an additional ton into the atmosphere. Equal to approximately $30/ton in 2005, the optimal value reaches $100/ton (in today's money) around 2055 and rises above $200/ton at the end of this century (Nordhaus, 2008, p. 94).[3]

With few exceptions around the world, values of the sort modeled by Nordhaus (2008) are treated as externalities. Furthermore, incentives are weak for a single government or nation to deal on its own with the market failure of global warming. The problem is indistinguishable from what a single herder with access to unowned rangeland faces. If a country curtails

---

[1] Like the results of other economic studies, Nordhaus's estimates are based on climatic projections provided by the IPCC, which are not necessarily the final word on the subject.

[2] The present value of X (measured in inflation-adjusted dollars) t years from now is $X \div (1 + r)^t$, where r is the real interest rate. Suppose, for example, that nominal interest is 8 percent and inflation is 3 percent, in which case the present value of $100 two years from now is: $100 \div (1.05)^2 = \$100 \div 1.1025 \approx \$90.70$. Of course, the combined present value of benefits and costs in two or more periods can be obtained by aggregating the corresponding series of present-value calculations.

[3] Note that these prices are expressed in terms of dollars per emitted (or sequestered) ton of carbon—not the value of carbon-dioxide emissions, which is the measure used most often by the IPCC and other international bodies. Dividing a carbon price by 3.67 yields the equivalent carbon-dioxide value. For example, Nordhaus's (2008) estimate that a ton of carbon was worth $30 in 2005 is the same as a payment the same year of $8.17 per ton of carbon dioxide.

its GHG emissions unilaterally, the benefits are shared with the entire world while the costs are internalized. By the same token, emitting compounds such as carbon dioxide and methane creates national benefits in the form of gross domestic product (GDP), while all humankind suffers the costs of climate change. Acting solely in its own interest, every country tries to be a **free rider**, counting on other nations to control pollution (at an expense) while doing nothing to curb its own emissions.

To try to overcome disincentives for unilateral national action on climate change, international negotiations have been pursued since the 1990s. But even in Kyoto, Copenhagen, and other places where negotiators have gathered, agreement has been hard to reach. The Europeans, for instance, demand that the corrective measures they plan to adopt be matched by the United States. The latter nation, in turn, is reluctant to bear significant costs as long as emerging economies such as China and India appear to be free-riding. The U.S. position is not hard to understand. If the country implements cap and trade for the sake of curbing GHG emissions, the financial burden for a typical family of four will be comparable to an income-tax hike of 50 percent (Feldstein, 2009). If emissions are not curbed as well in China and India, then the actual impact on global climate will be negligible.

As is happening elsewhere in the world, much of what the United States is doing about climate change has modest effects, serves other purposes, or is a combination of the two. For example, the country's farmers are expected to reduce annual emissions of carbon dioxide by as much as 430 million tons, mainly by planting trees, if they are offered a price of $15/ton (USEPA, 2005, pp. 4–24). While this reduction might represent an important part of a national strategy for curtailing GHGs, it barely exceeds 1 percent of current emissions for the country as a whole. Furthermore, there is a chance that farmers will receive payments for measures such as **reduced tillage** that profit them and that they consequently adopt on their own.[4] When and where this happens, there is no real impact on GHGs, so the payments are nothing more than an agricultural subsidy.

Given the unlikelihood of unilateral action to arrest global warming as well as the obstacles to an international agreement that is truly effective, individual countries and their citizens should anticipate climate change and adapt to it. Mendelsohn (2006) makes a compelling argument that

---

[4] An alternative to conventional tillage, which involves overturning soil with a moldboard plow, reduced tillage involves minimum disturbance of the soil. Switching from the former system to the latter reduces expenditures on fuel and other inputs, diminishes erosion, and also allows for additional carbon sequestration. Non-tillage, which means the complete absence of plowing and cultivation, sequesters the most carbon and minimizes soil erosion. With this system, crop residues are left on the surface of fields and weeds are controlled by applying herbicides.

adaptation is best accomplished if international trade is unrestricted. For the purpose of illustration, he analyzes a pair of hypothetical trading partners: one that is farther from the equator and where global warming strengthens its comparative advantage (Chapter 6) in wheat farming and the other closer to the equator and where warmer weather enhances its comparative advantage in corn production. Drawing on standard economic insights of the sort examined in the next chapter, he demonstrates that free trade promotes the best possible adaptation to climate change. To be specific, private incentives to specialize according to comparative advantage are reinforced, particularly since price changes are dampened.[5] In contrast, prices fluctuate more and efficient adaptation to new environmental conditions does not happen if trade is restricted (Mendelsohn, 2006).

Trade barriers are not the only impediment to adaptation to climate change in the food economy. Irrigation subsidies are another obstacle, as can be seen in India. Farmers in the world's second most populous nation pay nothing for water diverted from rivers and reservoirs or extracted from underground aquifers. Public authorities also provide them the electricity needed to run pumps and other equipment at little or no charge. With irrigation agencies collecting no revenues from farmers, maintenance of dams and other infrastructure is deficient. Moreover, price supports (Chapter 4) combine with irrigation subsidies to encourage the production of rice, which needs a lot of water, and accelerate groundwater depletion in some of the driest parts of the country (*The Economist*, 2009c).

All of this would be bad enough if the global climate were stable and benign. But since climate change is likely to make India hotter and drier, the penalty suffered because of intervention failure could be existential. Also, unpriced water and other policy-induced distortions complicate the task of convincing wealthier nations to provide technical and financial assistance to India, which some recommend to deal with climate change.

5.3 Farmland Degradation

Global warming is not the only threat to agricultural output. According to the **Global Land Assessment of Degradation (GLASOD)** sponsored by the United Nations, 22 percent of the Earth's agricultural land, including

---

[5] With free trade, the higher-latitude country exports some of its wheat, which holds down prices of the commodity in the importing nation. Likewise, some of the corn raised in the second trading partner is exported to the first country, which diminishes price increases there. With price changes dampened and climate change making corn farming more expensive in the first country and wheat production costlier in the second, incentives to specialize according to comparative advantage strengthen.

38 percent of the entire area planted to crops, shows signs of one sort of damage or another. Of the 1.9 billion hectares affected, more than four-fifths suffer primarily from soil **erosion** (i.e., the displacement of soil by wind and rain). Another eighth has been affected by **salinization** (i.e., the buildup of minerals harmful to plant growth in the rooting zone), **acidification**, and other adverse changes in soil chemistry. The other 5 percent has physical problems, such as soil **compaction** and **water-logging**.

Each year, 5 to 6 million hectares (equal to 0.3 to 0.5 percent of the world's arable area) are lost to agriculture due to erosion and other sorts of damage (Oldeman, Hakkeling, and Sombroek, 1991). This estimate is consistent with Scherr and Yadav's (1997) finding that annual additions to the category of severely degraded farmland range from 5 to 10 million hectares.

The toll land degradation takes on agricultural output varies from place to place. In temperate settings, where good soils comprise a large portion of agriculture's resource base (Wood, Sebastian, and Scherr, 2001, pp. 45–48), the negative impacts of resource deterioration on crop output have been small relative to the yield gains resulting from technological improvement. Having surveyed available research, Mitchell, Ingco, and Duncan (1997) conclude that 100 years of erosion in the United States have lowered yields by 3 or 4 percent (p. 54). This impact is tiny relative to the yield increases brought about by hybridization and other advances (Chapter 3).

Land degradation is more worrying in the tropics and subtropics, where the erosive impact of rainfall concentrated in heavy storms is high and where many soils have one or more deficiencies. Applying the GLASOD methodology for resource assessment, Van Lynden and Oldeman (1997) find "moderate" production impacts on 24 percent of the agricultural land categorized as degraded in South and Southeast Asia and "strong" impacts on an additional 13 percent in the same part of the world. Scherr and Yadav (1997) conclude that erosion and chemical and physical deterioration have lowered global agricultural output by 5 to 15 percent.

Sometimes, land degradation is a consequence of intensified production. For example, soil compaction can occur if heavy farm equipment lacking flotation tires is driven frequently across a field. Also, excessive fertilization changes soil chemistry for the worse. For example, applying ammonium nitrogen can diminish soil pH, perhaps enough to hinder crop growth. Irrigation can create problems as well. If a field is not properly drained, soils are apt to become water-logged. Also, excessive irrigation combined with poor drainage can cause the subterranean water table to rise, thereby bringing soluble minerals up into the rooting zone or even to the surface. The resulting soil salinization may grow so extreme that crop production is preempted.

Where agricultural intensification coincides with resource deterioration, intervention failure is often the culprit. Fertilizer subsidies in India are an example of this. The prices for nitrogen paid by the country's farmers cover just half the average cost of the input. In contrast, other nutrients, such as phosphorus and potassium, are not subsidized. As a result, fertilization is unbalanced, which damages farmland and significantly reduces crop production (Anand, 2010). It has been estimated that elimination of the nitrogen subsidy, which would encourage farmers to apply nutrients in better proportions, would cause annual production of rice and wheat in India to increase by 160 million metric tons and 25 million metric tons, respectively (Roy, 2003).

Throughout Asia, governments spurred the Green Revolution (Chapter 3) by subsidizing fertilizer and irrigation water. The resulting misuse of these inputs damaged natural resources in many parts of the continent. Land degradation has happened for entirely different reasons in places bypassed by the Green Revolution. Irrigation development has not proceeded far in Sub-Saharan Africa, for example, so problems like water-logging are uncommon. Furthermore, the region's farmers, rather than applying fertilizer at high rates, use that input sparingly.

Where fertilization falls short of the uptake of nitrogen, phosphorus, and potassium by crops and other vegetation and where nutrients are being lost for other reasons, **nutrient mining** is occurring. If soil fertility falls enough, plant growth suffers, which in turn leads to erosion because soil is left exposed to the elements.

Differences in fertilization in various parts of the world are striking, as are the changes that have occurred in recent years. Three decades ago, application rates were excessive in the old Soviet Union and its satellites (Table 5.2, column 1). More than anything else, this reflected startling inefficiencies in the use of agricultural resources (not to mention other things) under communism. Between the collapse of communism and a fall in human numbers, which has lowered food demand, fertilization plummeted during the next quarter century, reaching 43 kilograms per hectare of cropland in 2003–2005 (column 2). During the same period, application rates increased modestly in affluent nations.

In most of the developing world, fertilization has gone up substantially (Table 5.2). Already high 30 years ago, application rates in East Asia have risen to levels not observed in any other part of the world, rich or poor.

Table 5.2 Average fertilizer application rates in 1978–1980 and 2003–2005

Region	Kilograms per Hectare of Cropland in 1978–1980 (1)	Kilograms per Hectare of Cropland in 2003–2005 (2)
East Asia	99	263
Eastern Europe and Former Soviet Union	145	43
Latin America and Caribbean	55	111
Middle East and North Africa	38	106
South Asia	34	117
Sub-Saharan Africa	13	12
Western Europe, USA and Canada, Australia and New Zealand, Japan and South Korea, etc.	136	146

Source: World Bank (2009b).

Formerly low, fertilization levels in South Asia more than tripled between the late 1970s and the early twenty-first century and are now approaching levels in wealthy nations, although it bears repeating that intervention failure results in inefficiently unbalanced fertilization in India. Significant increases also have occurred in Latin America and the Caribbean and in the Middle East and North Africa. Sub-Saharan Africa has been the lone exception in the developing world to the upward trend. The region had the lowest application rates 30 years ago and those rates are the same today.

Henao and Baanante (2006) conclude that nutrient mining is taking place on about 85 percent of all farmland in the world's second-largest continent. The problem is acute in the plains of West Africa as well as the East African highlands, where annual losses of nitrogen, phosphorus, and potassium equal or exceed 60 kilograms per hectare. The economic dimensions of soil erosion and related environmental damages remain largely unstudied (Box 5.3). But according to Henao and Baanante (2006), agricultural losses associated with farmland degradation could cause imports of grain by Sub-Saharan Africa, which amounted to 19 million tons and cost $3.8 billion in 2003, to reach 34 million tons (worth $8.4 billion if prices do not change) in 2020.

> **Box 5.3 The costs of soil erosion in Mali**
>
> Mainly because of data limitations, attempts to place a price tag on farmland degradation have been few and far between, especially in Sub-Saharan Africa. Of the small number of available studies, one undertaken in Mali has yielded solid evidence that the costs of soil depletion are, indeed, significant.
>
> In this study, the **Universal Soil Loss Equation (USLE)**, which was developed originally to estimate erosion on fields in the United States (Wischmeier, 1976), was adapted to African conditions. In addition, data from the **International Institute for Tropical Agriculture (IITA)**, which is located in Nigeria and is part of the Consultative Group on International Agricultural Research (CGIAR), were used to analyze the linkage between erosion and crop yields. It was found that losses in agricultural output resulting from erosion were equivalent to 1.5 percent of Mali's GDP and well above the cost of conserving soil (Bishop and Allen, 1989).

Findings like this have given impetus to initiatives to reverse land degradation in Sub-Saharan Africa and other parts of the world. Some of the people and agencies involved in these initiatives have specific ideas about nutrient replenishment. Advocating manure use and the planting of trees and other vegetation that fix atmospheric nitrogen in the soil, they have undertaken pilot projects to make the case that alternatives to commercial fertilizer are viable. Although some of these projects yield interesting results, the general case for alternative fertilization, which is a key feature of organic farming, is far from overpowering. In many areas affected by nutrient mining, manure is in short supply. Livestock herding is impeded by tsetse flies and other pests in many parts of Africa. In India, manure from livestock is available, although a lot of it is gathered for cooking fuel. In either setting, it is difficult to envision manure being spread in sufficient quantities to arrest land degradation.

Regardless of the benefits of nutrient replenishment, commercial fertilizer is not free. Also, fertilizer prices are higher now than they were in the early 1980s in a number of African countries because national governments, obliged to reduce public spending and fiscal deficits, have cut subsidies. However, there is a more fundamental reason for the lack of fertilization, namely insufficient investment in the public goods required for economic progress in the countryside. Inadequate transportation infrastructure, for example, makes fertilizer and other purchased inputs expensive for farmers. Input demand is also depressed because poor roads diminish the prices growers receive for their crops. Where public goods are lacking entirely, the bulk of the rural population has no choice other than to engage in subsistence farming—using no purchased inputs and producing barely enough to survive, if that. A downward spiral of mounting poverty and land degradation often takes hold.

5.4 Agriculture and Deforestation

If a rural family cannot escape this downward spiral either by raising agricultural productivity or by abandoning subsistence farming in favor of some other line of work, it may forestall a deterioration in living standards by relocating to a place where land degradation has not proceeded far. This response to poverty and farmland degradation underlies much of the agricultural encroachment on tropical forests and other natural habitats that is happening in Africa, Asia, and Latin America.

To be sure, deforestation sometimes has nothing to do with agriculture. Hydroelectric development, for example, results in the permanent flooding of some tree-covered land. Also, logging often harms the vegetation remaining after timber has been harvested. Moreover, rural people desperate to escape poverty are not the only agricultural agents of deforestation. In particular, commercial farmers and ranchers are responsible for some of the geographic expansion of crop and livestock production. Rather than being concerned about mere survival, they are driven by profits.

Brazil is one place where encroachment on natural habitats has at least as much to do with commercial gain as with the escape from rural poverty. No tropical or subtropical nation has more tree-covered land. In addition, no country, either near or far from the equator, loses more hectares of forest annually (FAO, 2009c, p. 114). However, agricultural land use is expanding mainly in the *Cerrado*, a vast savannah of grasslands southeast of the Amazon Basin. In 1990, the **Empresa Brasileira de Pesquisa Agrícola (EMBRAPA)**, which is the state agency responsible for agricultural research and development in Brazil, estimated that 136 million hectares in the region were suitable for large-scale farming, of which 47 million hectares were being cultivated at the time. To expand agriculture, land preparation is needed to overcome soil acidity, aluminum toxicity, and other limitations. However, commercial farmers are dealing with these environmental limitations. Much of the recent expansion in the area planted to soybeans in Brazil—from less than 4 million hectares in the early 1970s to more than 12 million hectares in the late 1990s—has taken place in the *Cerrado* (Schepf, Dohlman, and Bolling, 2001).

In the Brazilian Amazon, intervention failure was singled out during the 1980s as a cause of excessive deforestation. Subsidized loans were criticized, as were tax credits given to companies investing in livestock operations in the region (Mahar, 1989). But by the middle 1990s, it had become clear that the land-use changes resulting from these policies had not been as great as many had thought. Having examined data both on deforestation in different places and on the geographic incidence of cheap loans and tax credits, Schneider (1995) reported that agricultural land-clearing had been especially rapid in the southern reaches of the Amazon Basin, which had not received many subsidies. In contrast, subsidies had been directed mainly to the eastern part of the watershed, which had experienced less deforestation.

The same investigator pinpointed another intervention failure of greater importance for Brazil's forests. Thanks in no small part to generous tax treatment given to agriculture, values of rural real estate soared in southern Brazil after the early 1970s. Farmers with small holdings, whose incomes were not high, were not much interested in competing for land in order to take advantage of provisions such as accelerated depreciation and the deduction of farming losses from nonagricultural earnings. Many of them found it more rewarding to sell their farms in the south and move to agricultural frontiers in the Amazon, where real estate remained fairly cheap. As Schneider (1992) argued, it was no coincidence that, as land prices in southern Brazil rose, immigration and agricultural land-clearing in the Amazon increased.

Similar policy regimes have had the same impacts in other countries. In Colombia, competition for prime farmland is biased by the favorable tax treatment of agriculture as well as the fact that owners of large holdings are normally first in line for loans that carry a low interest rate, are apt to be forgiven, or both. Each of these benefits matters most to the wealthy, which causes them to bid more for rural real estate. In contrast, disadvantaged small farmers are consigned to inferior parcels, along Andean hillsides and in tropical forests for example (Heath and Binswanger, 1996).

Driven up by government policies in various places, the net returns captured by those who migrate to agricultural frontiers to exploit the productive potential of natural habitats are also enhanced because the environmental consequences of this activity are not internalized. Some of these consequences are local. For example, removing trees and other vegetation from the upper reaches of a watershed can lead to increased flooding and sedimentation downstream. Also, deforestation can cause the local climate to grow hotter and drier. Other consequences of agricultural land-clearing are of global importance. One of these is diminished biological diversity. Increased atmospheric concentrations of GHGs are another.

Many environmental impacts are difficult to evaluate. This is true of the costs associated with reduced biodiversity (Box 5.4). In contrast, putting a price tag on the global warming that some contend is accelerated by deforestation is fairly straightforward. A simple measure of these costs is found by multiplying the damages resulting from the emission of one ton of carbon (in the form of carbon dioxide) by the carbon released into the atmosphere due to deforestation. Based on Nordhaus's (2008) estimate of the current price of carbon, $30/ton (see above), as well as Sohngen and Mendelsohn's (2003) estimates of emissions, which average 105 tons per deforested hectare, the global warming resulting from the conversion of forests into cropland and pasture costs around $3,150/hectare.

Box 5.4 The costs of destroying biodiverse tropical forests

There is little doubt that tree-covered ecosystems close to the equator harbor a large share of the world's flora and fauna and that, as these ecosystems are encroached on, some species are driven to extinction. All agree that a cost of this encroachment relates to the fact that pharmaceutical advances are impeded because of a reduction in the supply of biological specimens collected in natural habitats. However, this cost may not be large.

Three economists have estimated the one-time payments that the pharmaceutical industry would make for species-rich forests in the tropics. Assuming a high probability of discovering something unique in the wild that yields a valuable drug, they have found that payments of this sort may exceed $10 per hectare in a handful of places—in particular, where there is rapid destruction of ecosystems with many endemic species (i.e., plants and animals found nowhere else). But as a rule, the value of tropical forests as a site for collecting biological specimens amounts to just a few dollars per hectare (Simpson, Sedjo, and Reid, 1996).

To deal with the market failure that results because farmers and other agents of land-use change do not internalize global warming damages and other environmental costs, these agents could be taxed for every hectare they clear. Another option would be to pay landowners to sequester carbon in forests. If taxes or payments reflected the growth in carbon values that Nordhaus (2008) and other economists are anticipating, landowners would respond by planting trees. The world's forests could expand from 3.5 billion hectares today to 4.0 to 4.5 billion in 100 years (Sohngen and Mendelsohn, 2003).

For the time being, mechanisms to pay for carbon sequestration, biodiversity conservation, and other environmental values provided by tree-covered land remain poorly developed. Nevertheless, as indicated in Table 5.3, deforestation is limited in some parts of the world. Forests are spreading in wealthy places, where agricultural development driven by extensification ended long ago. The annual increase in the United States is nearly 160,000 hectares. Forested area in Western Europe is currently growing each year by a little less than 600,000 hectares, with Spain accounting for half this total (FAO, 2009c, pp. 113 and 115).

More extensive than those of any other nation, Russia's forests are currently expanding by nearly 100,000 hectares per annum (FAO, 2009c, p. 112). Throughout Eastern Europe and the Former Soviet Union, food demand is stable and inefficient modes of production characteristic of socialism are being abandoned, so forests are not being lost due to

Table 5.3 Forested area in 2005 and annual deforestation from 2000 to 2005

Region	Thousands of Forested Hectares in 2005 (1)	Average Annual Hectares Gained or Lost from 2000 to 2005 (2)
East and Southeast Asia	451,883	+713
Eastern Europe and Former Soviet Union	885,647	+86
Latin America and Caribbean	924,163	−4,743
Middle East and North Africa	36,563	+60
South Asia	79,239	−88
Sub-Saharan Africa	626,420	−4,095
Western Europe, USA and Canada, Australia and New Zealand, Japan and South Korea, etc.	948,109	−557

Source: FAO (2009c), pp. 109–115.

agricultural extensification. In the Middle East and North Africa, forests are spreading, albeit from a small base (Table 5.3).

There have been interesting changes in East and Southeast Asia. Annual deforestation exceeds 1 percent in several countries. One of these is Indonesia, where encroachment by farmers and others on tree-covered land approaches 2 million hectares per annum (FAO, 2009c, p. 111). Aside from deforestation in Brazil, this is the highest rate in the world. In contrast, China has encouraged the planting of trees to protect water sources in the upper reaches of drainage basins since the 1980s (Richardson, 1990, p. 25). Since the country's forests are currently increasing each year by more than 4 million hectares (FAO, 2009c, p. 110), the total area for the entire region is stable (Table 5.3).

In India, tree-covered area is neither increasing nor decreasing. However, annual deforestation exceeds 1 percent in three of its neighbors: Nepal, Pakistan, and Sri Lanka (FAO, 2009c, p. 111). The net result for South Asia as a whole is an annual loss of a little under 100,000 hectares (Table 5.3).

More forests are lost each year in Latin America and the Caribbean than in any other part of the world (Table 5.3). This mainly reflects deforestation in Brazil, which exceeds 3 million hectares per annum. In relative terms, forest losses are more dramatic in Ecuador and a number of Central American nations. However, tree-covered area is increasing in a few countries, most notably Chile (FAO, 2009c, pp. 113–115).

Nowhere is tree-covered land more threatened than in Sub-Saharan Africa. Deforestation exceeds 1 percent per annum in Ethiopia, Nigeria, and many other countries. Rwanda, which is densely populated and where agricultural encroachment on natural habitats is at an advanced cumulative stage, is one of the few places where forests are expanding (FAO, 2009c, pp. 109–110). Deforestation for the region as a whole, which is approximately 0.7 percent a year (Table 5.3), is an environmental symptom, along with farmland degradation, of the cycle of nutrient mining, soil exhaustion and abandonment, and conversion of natural habitats into new farmland that has ensnared millions of destitute households in the Sub-Saharan countryside.

5.5 Agricultural Development and the Environment

In impoverished rural settings, machinery, agricultural chemicals, and other commercial inputs are not used. Instead, edible output depends on whatever natural fertility the soil may hold, the amount of rain that happens to fall, seeds saved from last year's crop, and of course the muscle power supplied by farm households. Yields are modest.

Though not a basis for widespread wealth, this sort of farming can go on for a long time, and indeed did so throughout the world for millennia. If soil fertility, which is depleted by a few years of cropping, is replenished during a fallow period of adequate length, then the entire cycle of shifting cultivation (Chapter 3) is sustainable. People can live as well as their grandparents did, and their grandparents before them.

However, shifting cultivation becomes unsustainable as demand for food grows, due to an increase in population or economic activity. Land lies fallow for fewer and fewer years between one cropping period and the next, so soil fertility diminishes. As long as crops are produced in traditional ways, farmers can compensate for fertility decline by relocating to other settings, often encroaching on forests as they do so. The alternative is to make a break with traditional modes of production, changing input mixes so as to raise yields.

At least during the initial stages of this development, the environment may well suffer. The application of commercial fertilizer, for example, is often accompanied by increased runoff of nutrients into rivers, lakes, and streams, which creates water pollution. Likewise, the introduction of irrigation where farmers are accustomed only to rain-fed modes of production often leads to water-logging, salinization, and other problems. Market failure is partly to blame for this damage, since agriculturalists do not internalize all environmental costs of crop and livestock production. However, deterioration of natural resources can also result from governmental interventions, such as fertilizer and irrigation subsidies, intended to accelerate the transformation of farming. But regardless of whether market failure or intervention failure is to blame, environmental quality is not moving in the same direction as GDP per capita, which goes up as agricultural development makes food more abundant.

A point is eventually reached at which increased affluence ceases to carry an environmental cost and beyond which growth in GDP per capita actually coincides with improvements in natural resources. This change is easy to understand in the case of agriculture. Income growth in a prosperous setting allows for expanded investment in research and development, which reduces the environmental impacts of crop and livestock production in various ways. For example, the dissemination of **Bacillus thuringiensis** (Bt) cotton, which resists insect attacks, has allowed U.S. farmers to reduce annual pesticide applications by nearly 1,000 tons since 1996. In 2003, regulatory approval was given to Bt corn that resists rootworm, which used to do more damage to U.S. corn harvests than any other pest. As this genetically modified variety has been adopted, annual pesticide applications have gone down, by as much as 7,000 tons (Rauch, 2003).

Genetic modification can also help to conserve soil, which is the single greatest environmental challenge facing agriculture if the resource assessments cited in this chapter are to be believed. Much of the challenge relates to weed control in fields that are not plowed or otherwise tilled. However, biotechnology has produced crop varieties that are not harmed by herbicides, which break down into harmless residuals soon after application. Using these varieties, farmers can maintain yields while avoiding the damage that tillage does to their fields (Rauch, 2003).

Technological progress is an important reason why economic growth can have beneficial consequences for natural resources. But there is another explanation of the positive linkage in affluent places between living standards and environmental quality, which relates to changing preferences. The income elasticity of demand for environmental quality tends to go up as living standards rise.[6] This implies that in countries like the United States, where there is little reason to worry about the adequacy of food supplies, the value of a little more commodity output is often well below the value of a marginal improvement in environmental quality.

This change in preferences affects public policy. For example, support for things like farm subsidies, which stimulate agricultural production, weakens relative to demands for governmental action to limit environmental market failure. This change seems to be under way in the United States, where agriculture is a major source of water pollution. Of 3,692,830 miles (5,942,871 kilometers) of streams and rivers in the country, environmental assessments have been carried out for 699,946 miles, or 19 percent of the total. Of the assessed portion, 269,258 miles, or two-fifths, have been found to have impaired water quality. Moreover, agriculture is the leading source of pollution for 128,859 (48 percent) of these miles (USEPA, 2000, p. 14). In light of these facts, containing the runoff of fertilizer, manure, and other pollutants has become a central focus not just of environmental policy, but also of agricultural policy—which these days is less concerned than it used to be with raising production.

The tendency of increasing living standards to coincide initially with mounting pressure on natural resources and later with improved environmental quality is expressed by the **Environmental Kuznets Curve**

---

[6] Demand for environmental quality obviously contrasts sharply with food demand. According to Engel's Law, the income elasticity of food demand declines as earnings rise (Chapter 2).

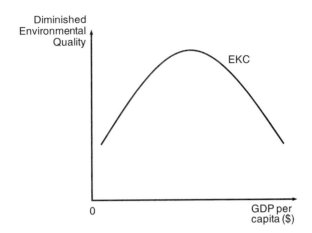

Figure 5.1 The Environmental Kuznets Curve

(EKC), which is illustrated in Figure 5.1. Shaped like an inverted U, this function is inspired by the proposition that income inequality rises in societies emerging from penury but declines after GDP per capita passes a certain threshold (Kuznets, 1955). Of course, an EKC's specific dimensions matter a great deal. For example, a high peak implies that resource degradation grows severe as economic expansion takes place in a setting where living standards are modest. Another possibility is for the peak to be low, though occurring at a high level of average income. Under this circumstance, mild trade-offs between environmental quality and economic growth are a prolonged feature of the development process.

Surveying the available literature, Dasgupta et al. (2002) conclude that EKCs for deforestation, industrial pollution, and other sorts of environmental damage peak when GDP per capita is between $5,000 and $8,000. This is worrying in the sense that living standards are below the level of environmental turnaround in much of the developing world. However, the same investigators identify a number of reasons, all related to globalization, for being optimistic. One of these is the free flow of information, which makes people more aware of pollution and more willing to do something about it. Another is trade liberalization, which has discouraged governments from subsidizing and protecting heavy industries that use a lot of energy and are the source of much pollution. Yet another reason for optimism is the spread of cost-effective approaches to environmental regulation.

There is no basis for supposing that the environmental benefits of globalization apply any less to agriculture than to the rest of the economy. Aside from promoting greater awareness of resource deterioration and its consequences, ready access to information facilitates the development and spread of technology for raising crops and livestock in environmentally friendly ways. In addition, technological change is facilitated by the income growth that results from the expansion of commerce. Ultimately, resource conservation in much of the developing world depends not on maintaining the self-sufficiency of farmers in Africa, Asia, and Latin America, but instead on ending their economic isolation. Insofar as this isolation continues, a large share of the world's rural population will continue to engage in nutrient mining, thereby degrading land resources. Their chances of rising out of poverty will be slim to none.

Key Words and Terms

acidification
Bacillus thuringiensis (Bt)
cap and trade
command-and-control
compaction
conservation
depletion
deforestation
Empresa Brasileira de Pesquisa Agrícola (EMBRAPA)
Environmental Kuznets Curve (EKC)
erosion
externality
factor of production
free rider
Global Land Assessment of Degradation (GLASOD)
global warming
greenhouse gas (GHG)
Intergovernmental Panel on Climate Change (IPCC)
International Institute of Tropical Agriculture (IITA)
inter-temporal trade-off
intervention failure
market failure
nominal interest
nutrient mining
open access
pollution tax
present value
property rights
real interest rate
reduced tillage
salinization
water-logging
Universal Soil Loss Equation (USLE)

Study Questions

1. Discuss the impacts of irrigation subsidies, which are an intervention failure, in terms of economic efficiency, the distribution of income, and the natural environment.
2. Do individual users of an open-access resource have any economic incentive to conserve the resource?

3. What can government do to curb deterioration of open-access resources?
4. Compare and contrast the economic inefficiency caused by non-internalization of environmental costs with the economic inefficiency caused by monopoly.
5. Why is reaching a global agreement to arrest global warming difficult?
6. In terms of the area affected, which sort of land degradation is the most globally prevalent?
7. Analyze the causes of inadequate fertilization in Africa.
8. What kinds of intervention failure and market failure contribute to tropical deforestation?
9. Explain the Environmental Kuznets Curve, both generally and with reference to the food economy.

Appendix: Market Failure and Its Remedies

As observed at the beginning of this chapter, economists' contributions to the understanding of environmental issues focus largely on market failure, which arises if economic agents do not fully internalize the impacts of production or consumption on natural resources.

Market Failure and the Inefficiency It Causes

To put the inefficiencies of market failure into perspective, let us consider first a hypothetical dairy industry that internalizes all its environmental costs. In practical terms, internalization could take the form of farmers' payments to neighbors who enjoy their backyards less on warm days because of odors and flies from nearby herds as well as monetary compensation to fishermen, boaters, and swimmers for the disutility they suffer because of water pollution. Alternatively, internalization could occur as dairy farmers spend money on covered lagoons, manure injection equipment, and other measures. As discussed below, environmental costs might also be internalized because pollution is taxed, the farmer must purchase pollution permits issued by the government, or the public sector takes some other action.

Regardless of how internalization is accomplished, the resulting relationship between the price of milk and the production level, which reflects the marginal cost (MC) of output (Chapter 3), is represented by supply curve S in Figure 5.2. Given the demand for milk, which is curve D in the same figure, market-clearing equilibrium occurs at P_E, with

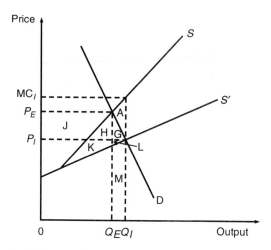

Figure 5.2 Market failure and its impacts on economic well-being

production equal to Q_E. Price represents the marginal value (MV) of output (Chapter 4). Since P_E also equals MC, Q_E is efficient.

Now consider what happens in the market if environmental damages are not internalized. In effect, dairy farms are receiving a subsidy, from those who bear the (uncompensated) costs of foul air, flies, and manure runoff, to be specific. Any subsidy augments supply, and the implicit subsidy of non-internalization is no exception. As shown in Figure 5.2, farmers' neglect of environmental costs shifts the supply curve out to S', with more output at any given price. The short-run equilibrium price, P_I, is lower than the efficient price, P_E. It is also below the MC of Q_I, which is MC_I in Figure 5.2. Thus, output is inefficiently high: Q_I when it should be Q_E.

The inefficiency that results when dairy farms treat—or, better to say, are allowed to treat—the environmental damages of production as an external cost (or a negative externality) is precisely the opposite of the inefficiency associated with the lack of competition. A monopolist produces too little and charges too much (Figure 4.6). In contrast, non-internalization of costs causes output to be too high and prices to be too low, as illustrated in Figure 5.2. The losses in net economic value (NEV) associated with the excessive production, Q_I minus Q_E, comprise the difference between the variable costs of this output (including environmental damages) and consumers' willingness-to-pay (WTP) for the same output. Variable costs amount to the area under the efficient supply (or MC) curve, S, between the two production levels: areas A, G, L, and M in Figure 5.2. WTP consists of the area under the demand curve between Q_E

and Q_I: areas G, L, and M in the same figure. NEV losses, then, are represented by triangle A, which is bounded from above by supply curve S and from below by the demand curve, D.

Actual differences between efficient and inefficient prices and levels of output, which result from negative externalities, depend not just on the magnitude of non-internalized environmental costs but on demand elasticities as well. Suppose consumption varies little as prices go up or down, as indicated by the steeply sloped demand curve in Figure 5.2. Under these circumstances, a negative externality is advantageous for consumers. But if consumption is very sensitive to price changes, relative changes in price and quantity are quite different. Just as an increase in commodity supply has little effect on price if demand is elastic, the difference between P_E and P_I is slight if price changes cause milk consumption to vary a lot. Instead, there is a substantial difference between Q_E and Q_I.

Of course, neither the price gap nor the difference in output levels is sizable if environmental costs are small. This possibility is illustrated by the convergence of the two supply curves, S (which, to repeat, reflects all costs, including environmental damages) and S' (which does not reflect these damages), near the vertical axis in Figure 5.2. This convergence reflects the fact that, at low levels of output, emissions from a polluting industry or sector are correspondingly low and, thus, the environment's capacity to receive and neutralize wastes is not being overburdened.

But as output keeps on rising, waste flows go up as well and the environment deteriorates more. Furthermore, as pollution increases, the marginal utility (Chapter 2) of environmental quality also increases. So at high levels of commercial output and waste flows, the environmental impacts of a marginal increase in production are sizable, as is the disutility that people associate with these impacts. Under these circumstances, inefficiencies may be large enough to induce a response from government.

Correcting Market Failure

As indicated in any textbook on environmental economics (Tietenberg, 2003), governmental responses to non-internalized damage to the environment take various forms. One way to remedy market failure is to deal with its underlying cause, which is the lack of resource ownership (Coase, 1960). However, creating and enforcing property rights are not free. Boundaries need to be surveyed and a registry established where a record of holdings is kept. Furthermore, a reasonably inexpensive way must be found to demarcate property lines, so that owners can ward off interlopers. For some resources, the combined expense of doing all this is prohibitive, since it exceeds the economic losses caused by missing property rights.

For this reason, access remains open to some resources. This is true, for example, of marine fisheries. Likewise, it is difficult to establish effective ownership of the environment's capacity to neutralize waste by-products. If resources such as these are to be conserved, other approaches have to be used. More often than not, regulation has been favored, with limits set on pollution and other forms of resource degradation and fines charged for exceeding these limits. A more direct application of the "polluter pays principle" is to tax pollution, levying a fee on every discharged ton for example. As this principle is applied, the supply curve approaches S in Figure 5.2. If the pollution tax causes MC to be fully internalized, output is efficient.

The market-level impacts of a pollution tax, one on the output of a polluting industry rather than on waste flows themselves, are easily represented for a special case—one in which wastes are directly proportional to marketable output and the marginal environmental damage of discharges is constant. Under these circumstances, the pollution tax is indistinguishable from a charge on the output of the polluting industry.[7] Consider how such a tax would work in a steel industry that, in the absence of governmental intervention, pays nothing for discharging contaminants into the atmosphere.

Seeking to halt the resulting deterioration in air quality, government imposes a tax, T, on every ton of steel. This causes a parallel shift of the supply curve toward the vertical axis, signifying that less is produced at any given market price. As is shown in Figure 5.3, the vertical distance between the supply curve with the tax, S_T, and the curve without the tax (and no other payments for environmental damages), S', equals T. This is because manufacturers, when deciding on production, pay attention to what they will actually receive per unit of output, which is the difference between the market price paid by consumers and T.

Any decrease in supply, caused by taxation or anything else, causes the market-clearing price to go up and equilibrium output (and emissions) to fall. Demand in Figure 5.3 being fairly inelastic, consumers end up paying most of the tax. To be precise, the difference between the equilibrium price, P_T, they pay if manufacturers face a pollution tax and P_I, which they pay if there is no tax or internalization of environmental costs, is just

---

[7] As a rule, taxing a polluting industry's output is a poor way to reduce environmental damage. For example, a tax on each bushel of wheat to reduce soil erosion hurts farmers who conserve their soil every bit as much as those allowing their land to degrade. Incentives to conserve resources would be strengthened by taxing erosion, not wheat.

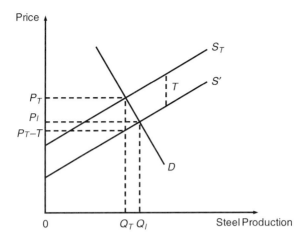

Figure 5.3 Taxing the output of a polluting industry

a little smaller than the tax, itself. Meanwhile, the net price received by producers, $P_T - T$, is only slightly lower than what they receive if nothing is done about environmental market failure, P_I. In other words, producers pay a small share of T.

As can be easily verified, distribution of the tax between consumers and producers is entirely different if demand is elastic. If consumption is highly sensitive to price changes, taxing the output of a polluting industry has little impact on equilibrium price. With consumers paying a small portion of T, most of the burden falls on producers, in the form of a large gap between P_I and $P_T - T$. In contrast, there is a sizable reduction in output, and therefore pollution.

Aside from imposing regulations or levying taxes, governments seeking to reduce overuse of a natural resource that no one owns can create a market mechanism for this purpose. To deal with pollution, for example, an environmental agency can issue transferable (i.e., tradable) emissions permits, the specific number of which corresponds directly to the desired discharge level. These permits partially resemble the licenses that regulatory bodies give to firms to emit specific quantities of pollutants. However, licenses to pollute are no more marketable than licenses to drive. When permits are bought and sold, companies that produce a lot of commercially valuable output with minimal damage to the environment bid pollution rights away from firms that are less adept at doing so. With this approach, commonly referred to as cap and trade, any given level of environmental quality (i.e., the cap) can be achieved with maximum efficiency. In addition, polluting firms treat the market value of permits as a part of production costs. Hence, the goal of internalizing environmental damages is accomplished.

The efficacy of cap and trade is not just a matter of theoretical conjecture. In the 1990 Amendments to the Clean Air Act, the U.S. Congress mandated cuts in emissions of sulfur dioxide (a chemical precursor to acid rain). A regulatory approach, with nonmarketable licenses for specific amounts of pollution issued to individual plants, was not adopted. Instead, polluters as a whole were given an overall emissions target, with a corresponding quantity of permits issued. Firms then proceeded to buy and sell transferable permits among themselves. Sulfur dioxide emissions were subsequently reduced throughout the country by 50 percent and the cost, $1 billion, was below even the most optimistic projections made when the program got under way (Kerr, 1998).

Yet another way to deal with environmental market failure is to invest in the development of technology that reduces the environmental impacts of economic activity. This sort of development narrows the gap between market supply with and without internalization—S and S', respectively, in Figure 5.2—and therefore the differences between efficient and inefficient prices and production. Improvements in agricultural technology often have precisely this impact.

It needs to be stressed that governmental intervention to correct market failure is never free. The administrative resources of the state are scarce, so regulating a polluting industry, taxing emissions, or implementing cap and trade always creates trade-offs. Also, raising taxes to pay for public programs diminishes private economic activity. Clearly, the benefits of intervention need to be compared to its costs.

Furthermore, the benefits of a single corrective intervention may not be as great as one might hope or expect. In the real world, there are multiple inefficiencies, related to market failure, lack of competition, regulations, and subsidies. Some of these inefficiencies are mutually reinforcing, as happens if a polluting industry is subsidized. But it is also possible for one inefficiency to counteract another one. This is the case if monopolization, which results in prices that are too high and output that is too low, coincides with negative externalities, which pull prices down and cause production to be excessive. Under this circumstance, imposing a tax to deal with externalities might actually detract from economic well-being, not enhance it (Davis and Whinston, 1967).

Safe to say, dealing appropriately with environmental market failure can be a tricky business.

6
Globalization and Agriculture

A recurring theme of this book is that specialization and trade create significant benefits, not least the improvements in living standards that drive up the demand for environmental quality and enable its creation. These benefits are detectable within households, as is pointed out in *The Wealth of Nations* (Smith, 1964, p. 401). Likewise, development of the food economy has had much to do with farmers' specialization in crop and livestock production and their trade with agribusinesses that supply inputs and post-harvest services.

In the global context, any single country finds it rewarding to provide a limited set of goods and services in international markets and import a much wider array of items, from places where these are produced more efficiently. Accordingly, trade across boundaries and frontiers has been happening for centuries, even millennia. In modern times, this commerce has flourished, as technological improvement has driven down transportation and communication costs. During the nineteenth century, for example, metal-hulled vessels powered by steam started to replace sailing craft made of wood. This development reduced the expense of transoceanic shipping appreciably, which stimulated trade and accelerated economic growth.

Even more goods and services, not to mention capital and other resources, are moving around the world these days. One reason for **globalization**, to use a recently coined term, has been cheap energy, which kept shipping inexpensive from the middle 1980s until early in the twenty-first century. In addition, a revolution has occurred in information technology, thereby making communications nearly free. Money can now be sent from one place to another with a few keystrokes on a computer. So can knowledge. Likewise, it has become much easier for buyers and sellers, including participants in global commodity markets, to deal with one another.

Many are uncomfortable about proliferating international flows of goods, services, money, and ideas. Globalization's opponents are not limited to unruly protesters, who have tried to disrupt every economic summit since a 1999 meeting of world leaders in Seattle. Foreign trade often imposes a temporary sacrifice on one part of the economy or another. Even though this sacrifice is usually small relative to the benefits of commercial exchange, those bearing costs often clamor for restrictions. Whenever the opponents of trade succeed, and international commerce is suppressed because of tariffs and other barriers, living standards suffer.

As did its predecessor, the **General Agreement on Tariffs and Trade (GATT)**, the **World Trade Organization (WTO)** serves as a bulwark against **protectionism**. The negotiations it sponsors, which are examined in this chapter, now focus largely on agriculture—and for good reason, in light of the economic case for **free trade** that is nearly as old as the "dismal science" itself and to which we now turn.

6.1 The Theory of Comparative Advantage

Stanislaw Ulam, a mathematician, once challenged economist Paul Samuelson to name "one proposition in all of the social sciences which is both true and non-trivial." The recipient of a Nobel Prize, Samuelson (1969, p. 1) confesses that a suitable response did not occur to him immediately. But he eventually thought of one: David Ricardo's **theory of comparative advantage** (1965, pp. 77–93).

Now nearly two centuries old, the theory goes beyond the argument for free trade offered in *The Wealth of Nations*, which stresses the virtues of trade but is vague about how individual economic actors choose to specialize. Part of Ricardo's contribution was to distinguish between **absolute advantage**, which reflects productive capacity, and **comparative advantage**, which depends on opportunity costs. He also showed that competitive markets foster specialization according to comparative (not absolute) advantage. Furthermore, he added, our collective well-being improves as long as economic actors allow themselves to be guided by market forces.

Samuelson is convinced that Ricardo's case for free trade is the most significant idea ever to come out of the social sciences: "That it is logically true need not be argued before a mathematician; that it is not trivial is attested by the thousands of important and intelligent men who have never been able to grasp the doctrine themselves or to believe it after it was explained to them" (Samuelson, 1969, p. 1).

Should Tiger Woods Mow His Own Lawn?

The theory of comparative advantage is captured by a vivid illustration for which we can thank Gregory Mankiw, a leading U.S. economist. The illustration, which might have to be modified after a nocturnal misadventure in November 2009, has to do with Tiger Woods, who has a number of skills and capabilities—including though not limited to his being one of the best golfers of all time. Mankiw (2001, pp. 56–57) supposes that Woods can earn $10,000 in two hours by filming a television commercial, which is (or used to be) a conservative estimate of the compensation he would receive. Another option for him is to spend the same two hours mowing his own lawn. In light of his athleticism and intelligence, no one else can do the job as quickly, so we conclude that Woods has an absolute advantage in lawn-mowing. In contrast, Forrest Gump is less productive, requiring four hours to accomplish the same task. During those four hours, Gump's best option is to work at a fast-food restaurant for a $20 paycheck.

Specializing according to absolute advantage is clearly inefficient. If Woods does the job, he incurs an opportunity cost: to be specific, the $10,000 that would come his way for filming a commercial. In contrast, Gump, who has no absolute advantage in lawn-mowing, holds a comparative advantage, as indicated by his opportunity cost of just $20. As long as Gump charges less than Woods's opportunity cost (to repeat, $10,000), Woods gains by hiring Gump. Gump gains as well, provided Woods offers more than $20 (i.e., Gump's opportunity cost). Hence, the incentive for the two men to strike a bargain is compelling (Mankiw, 2001, pp. 56–57).

Specializing according to comparative advantage and trading amount to what economists call a **positive-sum game**—that is, interchange that rewards multiple participants simultaneously. This is obviously true if two individuals, like Woods and Gump, are dealing with each other. The same holds for entire countries. Some of globalization's opponents deny that small, poor countries can benefit by trading with larger nations that are more productive. This is tantamount to saying that Forrest Gump gains nothing by being hired to take care of Tiger Woods's lawn. Others do not see how a rich country ever profits from commercial exchange with less productive places. This is equivalent to doubting that Woods is better off if he hires Gump. The two groups of opponents cannot be simultaneously correct. More than that, the theory of comparative advantage reveals that both camps usually are simultaneously wrong.

Distribution of the Net Benefits of Trade

As is demonstrated in the chapter's appendix, commerce with foreigners benefits producers that enjoy a comparative advantage. But the gains are even greater for consumers. Together, these two benefits exceed the costs of trade, suffered mainly by producers that lack comparative advantage. Each of these impacts can be understood by thinking about what happens in a country that emerges from **autarky** (i.e., economic isolation) and starts trading.

Without trade, prices in competitive markets reflect domestic costs of production: domestic marginal costs (MC), to be specific (Chapter 4). Since costs are low for producers with a comparative advantage, the prices they receive under autarky are correspondingly low. But once they gain access to other markets, they find themselves selling more at higher prices and there is more producers' surplus (PS), which comprises the difference between the market value of output and production costs (Chapter 4).

Trade benefits consumers because they can purchase imported goods at prices reflecting costs of production in exporting nations: costs that by definition are lower than in importing countries. Increased consumption at lower prices translates into additional consumers' surplus (CS), which consists of the maximum bids that consumers would make for goods coming their way less actual payments, or market value (Chapter 4).

Meanwhile, industries and sectors that lack comparative advantage, owing to elevated costs of production, find that prices for their output fall, as cheaper foreign goods are imported. As this happens and as these industries and sectors reduce output, PS is lost. Modest amounts of CS are sacrificed as well, because prices go up for exported goods.

But as is made clear in the appendix, these last two consequences of trade are outweighed by the aforementioned gains in PS and CS. Thus, overall well-being is enhanced.

6.2 Trade Distortions and the Economic Impacts

Regardless of the case for commercial exchange across national frontiers, costs are still incurred. Transferring factors of production from one part of the economy to another is not free, so there are adjustment expenses in any industry or sector lacking comparative advantage. For example, a worker might have acquired skills that apply only to a particular job. If foreign competition causes demand for these skills to contract, or if the job is eliminated, then his or her wages suffer. For someone going

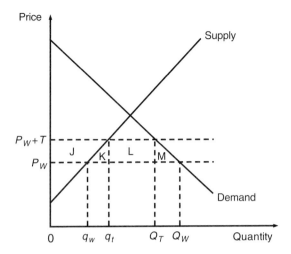

Figure 6.1 Impacts of protectionism

through this, knowing that trade serves broad national interests is cold comfort. He or she eagerly seeks protection from imports.

Protection comes at a price. If imports are reduced due to **tariffs** (i.e., taxes) or **quotas** (i.e., quantitative restrictions), domestic prices rise. Higher prices are a boon to domestic producers, who respond by raising output. Likewise, the difference between domestic and international prices is appropriated either by governments, if tariffs are levied, or by those acquiring import permits, if there are quotas. However, the sum of tariff revenues (or quota rents) and additional PS is less than losses in CS, which is the other consequence of higher prices.

This net result can be understood by using the conceptual framework introduced in Chapter 4 to describe net economic value (NEV) and its components. Depicted in Figure 6.1 is one possibility, which applies to the common case in which purchases of a globally traded commodity by an importing nation comprise a tiny portion of total supplies and therefore have no impact on the international price of that commodity. Imposing a tariff (T) causes the domestic price to rise from the world value (P_W) to $P_W + T$. Due to this increase, domestic production rises, from q_w to q_t, and national consumption falls, from Q_W to Q_T. Government collects revenues (area L) on imports, which are lower ($Q_T - q_t$) with the tariff than without ($Q_W - q_w$). Additional PS (area J) is created as well. However, the sum of tariff revenues and the incremental PS falls short of the reduction in CS (areas J, K, L, and M). The bottom line is that NEV declines, with the net reduction being represented by areas K and M.[1]

Protection of the kind shown in Figure 6.1 is ubiquitous in the global food economy. The United States and **European Union (EU)** keep high-cost sugar producers in business by limiting imports from the tropics and subtropics, where the commodity is produced cheaply. Similarly, Japan grows nearly all its own rice by keeping out imports from Thailand, Vietnam, and other places with a comparative advantage in the crop. The resulting loss of CS is greater than the gain in PS, much of which is expressed as increased values of real estate (Box 6.1).

Box 6.1 Agricultural protectionism and real-estate prices

It is not coincidental that David Ricardo, who gave us the theory of comparative advantage, also sharpened our understanding of land values. In his day, Britain's corn laws—agricultural protectionism put in place to prop up commodity prices after the Napoleonic Wars came to an end—were hotly debated. The policy's advocates contended that high prices were needed because costs of inputs, including real estate expenses, were elevated in the country.

Ricardo countered that this view of cause and effect was erroneous, and that land rents comprise a residual return left over after the costs of labor and other non-land inputs are deducted from the market value of output. His position, and that of other economists, was that prices and specialization are best left to free markets and that, as these are determined, rents adjust upward or downward as appropriate, with no effect on output (de Vivo, 1987). While caveats can be made to this argument, its fundamental logic is irrefutable and remains compelling today.

By no means is distortion of international agricultural trade confined to the protection of producers that lack comparative advantage. In the

---

[1] From the standpoint of domestic producers and consumers, a quantitative restriction on imports has exactly the same impacts as a tariff. For example, limiting purchases of foreign goods to $Q_T - q_t$ in Figure 6.1 causes the domestic price to go up to $P_W + T$. The increment in PS is area J and CS losses comprise areas J, K, L, and M in the figure. However, area L does not end up in government coffers, as happens with a tariff. Instead, it is captured by people allowed to import the restricted good, and perhaps by officials suborned as importers acquire their licenses.

United States and elsewhere, price supports have been used to enhance the incomes of farmers who compete internationally with ease. This policy stimulates output and exports, which in turn depresses the prices that other countries' producers receive in global markets. Evaluating the overall impacts of subsidized exports on NEV can be a little complex. In importing countries, there is a gain, with additions to CS exceeding losses in PS. In nations that subsidize exports, PS grows more than CS declines. However, there is another negative impact, associated with using government funds to distort commodity markets. By and large, this last cost, which includes the losses in economic well-being suffered by households that must pay higher taxes, exceeds the net gains in PS and CS in both exporting and importing nations.

International trade is not distorted solely for the sake of producers, either those facing competition from imports or those desiring a boost in exports. Consumers are sometimes accommodated, as in the case of food aid that is dumped on the commercial market of a recipient country rather than being donated to those who cannot afford an adequate diet (Chapter 8). Also, governments often restrict exports, even if the farmers they supposedly represent have a comparative advantage. This distortion, which keeps food cheap for domestic consumers while driving up global prices, happened in 2007 and 2008 in more than three dozen countries, including Argentina, the Ukraine, and Vietnam (Chapter 4).

Consumers also benefit from **currency overvaluation** (Perkins, Radelet, and Lindauer, 2006, pp. 647–749), which makes imports artificially cheap and which often comes about because a country's inflation is not matched by a devaluation of its currency. Overvaluation and the **import bias** it creates can be illustrated with a hypothetical example, one having to do with an imaginary trading partner of the United States called Macondo. Suppose that U.S. inflation is zero and Macondo's rate is 50 percent. In other words, prices of all goods and services in Macondo rise by one-half during the calendar year, while on average U.S. prices do not vary at all. If the **exchange rate** between Macondan pesos and U.S. dollars does not adjust in response to this difference in inflation, then Macondan output will become relatively more expensive than U.S. goods in both national markets.[2] This change in relative prices, or **terms of trade**, causes

---

[2] Suppose, for example, that Macondo imports wheat from the United States, paying $5/bushel. If the exchange rate is 10 Macondan pesos per U.S. dollar, then the price of U.S. wheat is: 10 pesos/dollar × $5 bushel = 50 pesos/bushel. This price remains the same if the exchange rate does not change and if there is no inflation in the United States. Since Macondan goods are growing more expensive due to inflation, wheat and other U.S. products are becoming relatively cheaper.

Macondo's imports from its trading partner to rise and its exports to the United States to shrink.

Consumers who are affluent enough to indulge their appetites for foreign goods welcome distortions, including in the exchange rate, that keep these goods cheap. Since these consumers tend to be influential, political pressure for currency overvaluation can be intense. However, this distortion makes things difficult for the producers of tradable commodities, including farm products. Competing against imports is more of a challenge. So is exporting from a country with an over-valued currency. Of course, simultaneously subsidizing consumers and implicitly taxing producers is unsustainable. Countries that do so lose foreign exchange as production for foreign markets is discouraged and as people rush to purchase imports that are artificially cheap. Once reserves of foreign exchange are depleted, major changes, not least devaluation of the national currency, are imperative to eliminate the import bias.

6.3 The Debate over Globalization

Advantageous though international trade is, including in the food economy, its expansion obliges selected industries and sectors to retool, relocate, and otherwise adjust, always at a cost. Since greater adjustments are required as globalization accelerates, complaints can multiply.

Multinational Firms

Some of the unease about globalization relates to large companies operating in more than one country. There are vague accusations that these companies control culture and access to knowledge, often at the expense of free expression and local advancement. Multinationals are also accused of exploiting workers and ravaging the environment.

Businesses may be willing to shift employment from high-wage to low-wage settings, provided differences in worker productivity are not too great. However, it is difficult to make the case that, within any single country, multinationals' employees are treated unfairly. On average, wages paid by companies with operations in multiple countries are 50 percent above the compensation offered by enterprises that operate in just one place. The discrepancy is especially pronounced in low-income settings, where multinationals' wages are approximately double average levels of compensation (Crook, 2001).

Environmental damage can sometimes be traced to multinationals. A spectacular example, one with direct relevance to the food economy, was the Bhopal disaster of 1984, when an explosion at a pesticide plant in India owned by Union Carbide killed thousands of people and left many more permanently injured. Thanks to subsequent legal action, the company paid fines and restitution amounting to millions of dollars and agreed to a range of safety measures.

Incidents such as Bhopal, which cannot be excused by pointing out that state-owned firms and public agencies routinely do worse environmental damage, demonstrate the need for policies to contain market failure (Chapter 5). The real question where multinationals are concerned relates to the pressure they put on governments not to apply such policies, usually by threatening to relocate to places where environmental controls are more lax. At worst, governments can engage in a **race to the bottom**, hoping to attract and retain the investment, jobs, and exports that mobile enterprises bring by turning a blind eye to pollution or offering other inducements.

The race to the bottom is not as relentless as one might think. Indeed, it is implausible if governments are reasonably representative and if environmental impacts are mainly confined within national frontiers. Why, one wonders, would a representative government waive environmental controls for a polluting industry if the benefits of industrial expansion are outweighed by environmental damages suffered by the government's constituents? If these damages spill across national frontiers, then a government that truly represents its constituents while neglecting other countries' interests might indeed engage in a race to the bottom, creating substantial **trans-boundary externalities** along the way. Of course, a government that is not representative, because it is corrupt or dictatorial for example, can also be induced to compete by despoiling nature. The point is that the race to the bottom is not ineluctable. Rather, it happens because mechanisms for dealing with trans-boundary spillovers are faulty, governments are inordinately swayed by special interests, or both.

Multinational firms certainly have been known to exploit deficient arrangements for the internalization of environmental impacts (be these local or trans-boundary) and to strike deals with unscrupulous officials. However, these companies are not always enthusiastic participants in a race to the bottom. For one thing, many of them are susceptible to consumer boycotts organized to protest natural resource degradation or coziness with despots. Moreover, with information flowing as freely as it does in a globalized world, the risks and expenses of doing business in places where the rule of law is weak are becoming clearer every day, not least to multinationals. Insofar as these firms stay away from such places, pressure grows to strengthen the rule of law. Multinationals' behavior, then, sometimes promotes a race to the top, not the bottom.

Food Self-Sufficiency and Export Cropping

Aside from looking askance at multinational corporations, many critics of globalization are generally suspicious of international trade. In particular, these critics propose that poor countries foreswear export cropping for the sake of **self-sufficiency** in food production, regardless of any comparative advantage they may have in agriculture.

Self-sufficiency is distinct from **self-reliance**, which a country achieves by growing some of its own food efficiently and importing the rest, using the money earned by exporting goods in which it holds a comparative advantage. Indeed, striving for self-sufficiency can impair a country's self-reliance, if producing food inefficiently diminishes the funds needed to pay for imports that are needed when domestic harvests fail.

The hunger and poverty that are experienced if a poor country's comparative advantage in export cropping is not exploited has been amply demonstrated. Surveying economic research carried out in the Gambia, Guatemala, Kenya, the Philippines, and Rwanda, the director general of the International Food Policy Research Institute (IFPRI) concluded that "agricultural commercialization raises the income of the rural poor, thus improving their food security" (von Braun, 1989, p. 4). Cash cropping is not a cause of hunger, but rather an important part of the cure.

Another shortcoming of agricultural self-sufficiency is the risk it creates of output shortfalls. The chances of weather-related crop failure in specific places or even entire nations are higher than in the world as a whole. In contrast, shortfalls in some parts of the world are almost always offset by above-average output elsewhere, which dampens swings in global prices.

International agricultural trade is opposed categorically on other grounds. Some critics worry about the energy required to move commodities from place to place, although this is not a problem if fuel prices are not too distorted. Other critics worry that women, who often specialize in the production of food crops, are disadvantaged if cash cropping is emphasized, even though there is no evidence to support this claim. At the end of the day, there is no escaping the conclusion that self-sufficiency is a recipe for deprivation. Food security requires trade.

Free Trade versus Fair Trade

Less extreme than categorical opposition to export cropping is the stance that trade, though generally desirable, should be "fair" rather than "free."

In principle, the difference between the two is clear enough. Free trade can be defined as international commerce that is unencumbered by any sort of barrier, with the resolution of social and environmental issues left to affected countries. In contrast, **fair trade** is supposed to be consistent with the protection of workers, women, minorities, children, and the environment. One way to accomplish this protection would be to establish social and environmental rules (or chapters) that would be enforced throughout the world. For example, Runge et al. (2003, pp. 172–176 and 205) call for a global environmental organization that would address trans-boundary pollution and discourage relaxation of national ambient standards in a race to the bottom, much as the WTO deals with multilateral trade issues.

Such an organization would have to be careful about avoiding harm to the interests it purported to serve. Consider the challenge for international bureaucrats charged with setting and enforcing fair wages. If they set wages too low, living standards would be unduly depressed. More likely, minimum compensation would be too high, in which case investors would be discouraged from creating jobs and household earnings would be hurt by unemployment. It must also be remembered that international trade rules that would enforce wage standards by curtailing imports from noncomplying poor countries would hurt consumers elsewhere, by raising prices.

Furthermore, the pursuit of fair trade, regardless of how commendable its objectives might be, can be damaging if trade **liberalization** is impeded. It is notable that efforts to begin a new round of multilateral trade negotiations in 1999 broke up when the EU and the United States failed at the global summit in Seattle to convince low-income nations that environmental and social (labor) chapters should be included in the negotiating agenda. The ultimate result of this failed attempt to implement fair trade was to postpone the beginning of multilateral talks for two years.

6.4 Potential Gains from Agricultural Trade Liberalization

Most of the world's farmers, including many in wealthy countries, have a positive interest in free trade. To be sure, prices of cotton and sugar in the United States would fall if there were no import restrictions. Likewise, rice would be cheaper in Japan and South Korea. On the other hand, global commodity prices would increase by 5.5 percent on average (World Bank, 2007, p. 10). A greater overall impact, of 11.6 percent, has been estimated by Diao, Somwaru, and Roe (2001), who conclude that half the anticipated changes in global prices would result from scrapping tariffs, a third from eliminating price supports in the United

States and Europe, and another sixth from jettisoning export subsidies by the EU.

In the United States, which has a comparative advantage in corn and other cereals, annual farming receipts would go up by $22.4 billion (Diao, Somwaru, and Roe, 2001), after averaging $193.2 billion from 1998 through 2000. This potential gain exceeds the $16 billion per annum in deficiency payments and other governmental support (Chapter 4) that U.S. farmers received during the same period (ERS-USDA, 2001, p. 58). Even where farmers experienced temporary losses due to falling prices, there nevertheless would be a net improvement in NEV, as consumers gained access to cheaper food.

Trade barriers in prosperous nations belonging to the Organization for Economic Cooperation and Development (OECD) have been criticized, and rightly so. However, barriers are worse in other parts of the world. In 1995, low-income countries faced tariffs averaging 15.1 percent on their exports to rich nations, but 18.3 percent on exports to other poor places (Hertel, Hoekman, and Martin, 2002). Duties on farm imports exceeded 60 percent in the Caribbean, Sub-Saharan Africa, and North Africa, and the average agricultural tariff in South Asia exceeded 110 percent (Gibson et al., 2001). In these settings, liberalization of trade would lower food prices, much to the benefit of poor people who spend large shares of their earnings to satisfy basic dietary needs.

In nine recent studies reviewed by Huff, Krivonos, and van der Mensbrugghe (2007), estimates of NEV gains for affluent countries are as small as $2 billion and as great as $80 billion. For the developing world, which produces a wide variety of fruits and vegetables as well as coffee, tea, cocoa, sugar, and rice efficiently, estimates range from annual losses of $3 billion to annual gains of $75 billion. Consistent with these findings is Diao, Somwaru, and Roe's (2001) investigation of the improvements in well-being that would result in the near term from the elimination of tariffs, subsidies, and other distortions in agricultural markets. This short-term improvement, called **static resource allocation gains**, would amount to $31 billion for the world as a whole. Most of this amount would accrue to taxpayers and consumers in rich nations (Table 6.1, column 1). Outside of Japan, Norway, and the EU, free trade would cause commodity values to rise. Higher prices would, of course, benefit farmers in poor countries, especially if the developing world became a net exporter of agricultural commodities. On the other hand, nonagricultural households in these places would lose purchasing power. But since this latter impact would be exceeded by the benefits of specializing according

Table 6.1 Annual welfare gains from agricultural trade liberalization (billions of 1997 dollars per annum)

	Static Resource Allocation Gains (1)	Static Gains plus Dynamic Benefits (2)
Entire World	31.1	56.4
Affluent Nations	28.5	35.2
United States	6.6	13.3
Canada	0.8	1.4
European Union	9.3	10.6
European Free Trade Area	1.7	0.2
Japan and South Korea	8.6	6.2
Australia and New Zealand	1.6	3.5
Developing Countries	2.6	21.3
China	0.4	2.2
Rest of Asia	1.5	5.1
Southern Africa	0.3	0.8
Mexico	−0.2	1.6
Rest of Latin America	3.7	6.1
Other	−3.1	5.4

Source: Diao, Somwaru, and Roe (2001), pp. 37–39.

to comparative advantage, static resource allocation gains would be positive outside the OECD (Diao, Somwaru, and Roe, 2001).

Diao, Somwaru, and Roe (2001) also have analyzed the **dynamic benefits** of liberalized trade in farm products. These benefits arise because savings, investment, and productivity change follow a more efficient course over time if trade barriers and other distortions are removed. Impacts of this sort, which are difficult to estimate,[3] were modeled for a 15-year period. Interestingly, dynamic benefits were found to be nearly as large as static resource allocation gains. In addition, the former would be shared fairly evenly between OECD members and the rest of the world: $35.2 billion per annum for affluent countries and $21.3 billion per annum for developing countries (Table 6.1, column 2).

Anderson and Martin (2006, pp. 12–13) offer two useful observations about potential gains in economic welfare captured by 2015 (relative to

---

[3] Quantitative models underestimate gains from freer trade because increased competition, the capture of economies of size, and greater innovation are all difficult to evaluate. Trade also motivates and enables infrastructure development, although this impact is hard to evaluate as well.

base conditions in 2001) due to free trade. First, 63 percent of global gains come from agriculture, 14 percent from textiles and clothing, and 23 percent from other sectors. Second, 93 percent of the gains in agriculture would come from better market access, as import tariffs and quotas are removed. Five percent would result from ending domestic farm price and income supports, which the United States and others are trying to accomplish by **decoupling** payments to farmers from market prices and levels of production (Box 6.2). Another 2 percent would relate to the removal of export subsidies. These findings are broadly consistent with Huff, Krivonos, and van der Mensbrugghe's (2007) conclusion that 84 percent of the gains of liberalizing agricultural trade would result from the elimination of tariffs and quantitative restrictions and another 15 percent from other reforms.

Box 6.2 Decoupled support for agriculture

In the United States and other affluent nations, there have been attempts in recent years to provide decoupled income supplements to farmers that ostensibly are related to neither current prices nor levels of production (Tweeten and Thompson, 2002, p. 10). Even if this approach is only partially successful, then market forces are reinforced in the food economy and the agricultural subsidies that some governments provide distort global prices less.

One must recognize, however, that complete decoupling is difficult to accomplish. As farmers appreciate full well, payments at any given date are influenced in one way or another by production, areas planted, or both in the recent past. Also, payments that are supposed to be decoupled (e.g., the relief provided after a natural disaster) can loosen credit constraints and therefore expand farmers' purchases of production inputs. As a result, output is affected.

Finally, higher average food prices would undoubtedly burden nonfarm households in the developing world. However, this group would still benefit from agricultural trade liberalization. During the 1980s, economists Rodney Tyers and Kym Anderson estimated that commodity prices would be more stable with free trade, with the coefficient of variation declining by 50 percent for beef and lamb and by 75 percent or more for dairy products, rice, and wheat (World Bank, 1986, p. 131). Diminished variability would be a boon especially for impoverished consumers. Millions of these people can barely afford an adequate diet under normal market conditions but go hungry when the cost of food spikes, as happened in 2007 and 2008 (Chapter 4).

6.5 Multilateral Trade Negotiations and Agriculture

Notwithstanding the resistance that remains in the agricultural sector, international trade has increased markedly since the middle of the twentieth century, just as it did during most of the preceding 150 years. Exactly as predicted by the theory of comparative advantage, living standards have improved as well, especially in the world's most open economies.

Trade has been allowed to grow since the Second World War partly thanks to important lessons learned painfully during the 1930s. At the beginning of that decade, leading industrial nations veered sharply toward autarky as protectionism in one country begot protectionism elsewhere, all in vain hopes in each nation of maintaining output and jobs in industries facing foreign competition. The ensuing collapse of global commerce was largely responsible for the length and severity of the Great Depression, which did not end in the United States until global conflict induced an industrial recovery in the early 1940s.

Acutely aware of the linkage between protectionism and economic collapse, which can in turn threaten world peace, 44 countries, including the noncommunist powers, took steps toward liberalization at a meeting in Bretton Woods, New Hampshire, in July 1944, many months before the final defeat of Germany and Japan. The WTO would not come into being for another 50 years. But the GATT, its precursor, was established in the late 1940s.[4]

Before the Doha Round

The GATT promoted the cause of unencumbered commerce among nations. During the first round of negotiations, which got under way in Geneva in the late 1940s, the 23 participating countries agreed to cut average industrial tariffs by 20 percent. Subsequent talks halved these average duties and then halved them again, to around 4 percent in 2000. In the **Uruguay Round**, which ended in 1994, agreement was reached to establish the WTO, with mechanisms for dispute settlement and

---

[4] Negotiators at Bretton Woods also agreed to create the International Monetary Fund, which provides guidance and financing to countries experiencing chronic outflows of foreign exchange because of persisting trade deficits, as well as the World Bank, which underwrites development projects.

sanctions for the violation of trade accords that were unavailable under the GATT. The Uruguay Round also obtained commitments from participating governments for **tariffication**—that is, the conversion of nontariff barriers into tariffs, which are more transparent and easier as a rule for negotiators to ratchet downward.

As protectionism was rolled back, international commerce expanded. From 1961 through 2001, growth in trade outstripped increases in global output. In keeping with this broad pattern, world agriculture became more trade-intensive, with commodity exports going up at a yearly rate of 4.0 percent during the last four decades of the twentieth century as output growth was averaging 2.3 percent per annum (World Bank, 2003, p. 110). However, agriculture was not at the forefront of trade intensification. Amounting to 27 percent of global merchandise trade in 1961, exports and imports of farm products had fallen below 10 percent of the world total by the turn of the twenty-first century (FAO, 2009a) and agricultural trade consistently grew at a slower rate than imports and exports of manufactured goods after 1980 (Table 6.2).

One impediment to the liberalization of agricultural trade was the memory of food shortages during the 1940s. Rich and poor nations alike sought food self-sufficiency and well-placed farm lobbies wielded their influence to protect domestic producers. In addition, agriculture's trade performance was weak because most developing countries, including several that would have challenged agricultural protectionism, kept their distance

Table 6.2 Growth in agricultural and manufactured exports, 1980 through 2001

	Annual Export Growth, 1980 to 1990 (percent) (1)	Annual Export Growth, 1990 to 2001 (percent) (2)
Entire World		
Agriculture	4.3	3.6
Manufacturing	5.9	4.8
Developing Countries		
Agriculture	3.4	4.8
to Developing Countries	3.6	7.8
to Developed Nations	3.4	3.3
Manufacturing	7.6	8.9
to Developing Countries	7.3	10.0
to Developed Nations	7.8	8.3

Source: World Bank (2003), p. 110.

from the GATT. The reason for this is that Latin America pursued development after the Second World War by favoring manufacturing over agriculture and other primary sectors (Chapters 7 and 12). The same approach was subsequently adopted in Asia, the Middle East, and Sub-Saharan Africa, as European colonialism came to an end. Since industrial protectionism would have conflicted with the obligations of GATT membership, most of the developing world refrained from joining. Moreover, exports of farm products suffered due to the net taxation of agriculture in poor countries: 28 percent as recently as the early 1980s (World Bank, 2007, p. 10).

This net taxation subsequently declined, averaging just 10 percent from 2000 to 2004 (World Bank, 2007, p. 10). This left trade restrictions as the main barrier to globalization of the food economy. The average global tariff on agricultural commodities was 62 percent at the turn of the twenty-first century (Gibson et al., 2001) and growth in agricultural exports from developing countries to affluent nations did not accelerate in the 1990s (Table 6.2, column 2), mainly because the latter nations' farmers were still protected from foreign competition.

The Doha Round

Protectionism in the food economy having been discussed for the first time during the Uruguay Round, the next series of trade talks got under way in the capital of Qatar in November 2001. By this time, more than 150 nations, most of which can be classified as poor or developing, belonged to the WTO. In addition, many parts of Africa, Asia, and Latin America still have a comparative advantage in farming. As a result, an agricultural focus has been inevitable during and since the **Doha Round**. Emphasized as well have been labor-intensive lines of manufacturing (e.g., textiles and clothing), which now provide three-quarters of total exports from the developing world (Hertel, Hoekman, and Martin, 2002).

The most unequivocal advocate of free agricultural trade in the WTO has been the **Cairns Group**, which includes Argentina, Australia, Brazil, New Zealand, and other major commodity exporters. One of its specific concerns has been **dirty tariffication**, which satisfies the letter of the approach embraced during the Uruguay Round but violates its spirit by setting duties that distort trade more than the nontariff barriers being replaced. To avoid this pitfall, the Cairns Group has insisted that tariffs be pared from **applied** (or actual) levels instead of **bound** (or maximum allowable) levels, which in many instances are much higher. It also has proposed across-the-board reductions of at least 25 percent in agricultural duties (Williams, 2009).

Allied frequently with the United States, the Cairns Group has been opposed by the **Group of Ten (G10)**, which comprises ten affluent nations (e.g., Japan and Norway) that produce agricultural commodities at a high cost. Joining forces at times with Canada and the EU, this block generally favors liberalization. However, the group would allow individual countries to designate their own **sensitive products**, which would not be subject to tariff reductions. Examples of these products would include rice in Japan and South Korea and dairy goods in Europe.

The **Group of Twenty (G20)**, which comprises 20 emerging economies led by Brazil and India,[5] generally supports broad liberalization of the sort advocated by the Cairns Group (to which 11 members of this block belong) though with an exception for **special products**. A larger collection of developing nations, the **Group of Thirty-Three (G33)** supports liberalization less. Enactment of its proposals would give national governments wide latitude to insulate industries and sectors from foreign competition (Williams, 2009).

As often happens in trade talks, participants in the Doha Round fixated on sensitive and special products and tended to lose sight of the broader interest the world as a whole has in free trade. Pressure was even exerted on U.S. representatives—by domestic growers who are threatened by foreign competition. The liberalizing vision of the Cairns Group gradually faded and negotiations collapsed in July 2008 without a new trade agreement being reached. The immediate cause was a dispute between India and the United States, with the latter nation viewing a proposal made by the former to shield poor farmers from surges in agricultural imports as unacceptable protectionism.

Why Negotiations Failed

Success in the Doha Round was never guaranteed. To repeat, more than 150 nations now belong to the WTO, which is more than seven times the GATT's charter membership. Rarely does a group this large agree on anything substantial.

Adding to the challenge has been the diversity of the WTO's current membership. As noted above, 23 affluent nations, all broadly supportive of trade liberalization, founded the predecessor organization more than six decades ago. The same nations continued to dominate the GATT for

---

[5] As of 2009, there is an entirely new G20, comprising the old G7 (Canada, France, Germany, Italy, Japan, the United Kingdom, and United States) as well as Argentina, Australia, Brazil, China, India, Indonesia, Mexico, Russia, Saudi Arabia, South Africa, South Korea, Turkey, and the European Union. After four summits, from 2008 through mid-2010, this new group has yet to point to a single great accomplishment.

much of its existence. But this is no longer the case. As scores of African, Asian, and Latin American states have joined the WTO, the **Global South**, as some call it, has gained power and influence.

The Global South is by no means monolithic. Neither are the various stances attributed to it thoroughly coherent. With respect to agriculture, exporters such as Brazil and Thailand favor the removal of distortions that depress global prices, so have aligned themselves with the Cairns Group. But many more developing nations import food. Consumers in these nations benefit from artificially low prices, and also reap rewards if surpluses created by prices supports in places like Europe and North America are donated to them.

Further diminishing the support for worldwide liberalization have been the concessions that many developing countries have received unilaterally from wealthy nations. Since 1974, when the United States put in place the **Generalized System of Tariff Preferences**, duties have been terminated on more than 2,700 items that poor countries export to the world's largest economy. In 1975, the EU enacted the **Lomé Convention**, which created similar preferences for former colonies in Africa, the Caribbean, and the Pacific.

Support for worldwide liberalization of the sort sought by the WTO also has been diminished by the sense that China would capture most of the benefits, in many instances at the expense of the rest of the developing world. Insofar as this fear is well founded, it is because many African, Asian, and Latin American nations have yet to adopt measures required for broad-based economic progress, generally, and success in international markets, specifically (Tweeten, 2007, p. 144).

For these and other reasons, Vanzetti and Sharma (2007) have found that more than half the 161 developing countries they studied would not actually gain from the worldwide liberalization of agricultural trade. As a group, these countries would lose $1.6 billion in annual income if liberalization occurred only in affluent nations. If the same countries allowed farm products to be traded freely among themselves, their combined incomes would go up by $11.1 billion per annum. However, progress along these lines has been slow.

Enthusiasm for liberalized agricultural trade also has waned in wealthy nations. Part of the problem has been the continuing opposition of sugar growers in the United States, rice farmers in Japan, and other producers in affluent nations who have built up political support over the years to overcome their lack of comparative advantage.

In addition, wealthy countries at times have exhibited more interest in bilateral and regional trade pacts, which can put the greater efficiencies of global liberalization farther out of reach. A particular shortcoming of these pacts is that many are reached only because contentious issues, such as agricultural trade, are dodged (Good, 2007). However, it is at least as worrying when key nations appear indifferent not only about WTO negotiations, but about bilateral and regional deals as well. It is feared that the United States is one such nation, as indicated by agreements it has reached with Colombia, Panama, and South Korea but has left unratified.

So the Doha Round has plodded along since 2001. It is now on life support, with a faint pulse and unlikely to revive in the foreseeable future. If the economic research cited in this chapter is to be believed, tens of billions of dollars in annual net gains are being left uncollected.

6.6 The Case for Free Trade Still Stands

Whether individual persons or entire nations are involved, voluntary exchange of goods and services creates prosperity. But as the failure of the Doha Round demonstrates, economic evidence of the benefits of liberalization does not always convince national governments. Indeed, protectionism continues to be chosen all too often in response to the complaints of industries and sectors that lack comparative advantage and are reluctant to bear the costs of adjustment. Barriers to trade are also erected to create profits for the favored few.[6]

In the world food economy, self-sufficiency continues to be an aspiration (Box 6.3), even though it is in no country's interest. The waste resulting from this approach is demonstrated year in and year out in wealthy nations belonging to the G10. The consequences are even worse in less prosperous places. This is because protecting inefficient parts of the agricultural sector reduces the ability to pay for imports, which are needed particularly when domestic harvests fail.

Autarky in the food economy would be especially disastrous if climate change interferes with crop production throughout the tropics and subtropics, as many anticipate will happen (Chapter 5). Always a key feature of broad-based economic progress, trade will be needed to avoid mass hunger in the tropics and subtropics, which many expect will become warmer and drier later this century.

---

[6] Just as George W. Bush began his presidency in 2001 by restricting steel imports, Barack Obama imposed a 35 percent tariff on tires imported from China within a year of being inaugurated.

> **Box 6.3 Transnational purchases of farmland**
>
> Food self-sufficiency is clearly beyond the reach of a number of countries, yet several remain leery of agricultural imports and have been inventive in their search for alternatives to trade.
>
> Consider the recent purchases by China, South Korea, and oil exporters in the Persian Gulf of large tracts of farmland in Sub-Saharan Africa and other settings (*The Economist*, 2009b). Provided that the buyers and sellers involved are all willing, these purchases do not deserve to be criticized. Indeed, real-estate acquisition is often a feature of agro-export development that benefits the host country and the foreign investor alike.
>
> However, many purchases do not meet the willing-seller standard. This is particularly true where the property rights of rural people are not respected by national governments that receive payments for turning over large holdings to foreigners.

Key Words and Terms

absolute advantage
applied tariffs
arbitrage (in appendix)
autarky
bound tariffs
Cairns Group
comparative advantage
currency overvaluation
decoupling
dirty tariffication
Doha Round
dynamic benefits
European Union (EU)
exchange rate
fair trade
free trade
General Agreement on Tariffs and Trade (GATT)
General System of Tariff Preferences
Group of Ten (G10)
Group of Thirty-Three (G33)
Group of Twenty (G20)
globalization
Global South
import bias
liberalization
Lomé Convention
positive-sum game
production possibilities frontier (PPF) (in appendix)
protectionism
quota
race to the bottom
self-reliance
self-sufficiency
sensitive products
special products
static resource allocation gains
tariff

tariffication
terms of trade
theory of comparative advantage

trans-boundary externality
Uruguay Round
World Trade Organization (WTO)

Study Questions

1. Why do Tiger Woods and other people suffer economic losses if he mows his own lawn?
2. How are consumers, producers with a comparative advantage, and producers without a comparative advantage affected by freer trade?
3. How do the benefits of an import tariff, which comprise increased producers' surplus as well as tariff revenues, compare to the lost consumers' surplus resulting from a tariff?
4. Briefly describe the causes and impacts of currency overvaluation.
5. Under what circumstances do governments engage in a race to the bottom, so as to attract multinational firms?
6. How might an emphasis on food self-sufficiency at the expense of export cropping aggravate hunger problems?
7. Why is fair trade, as opposed to free trade, difficult to implement?
8. Summarize the economic consequences of agricultural trade liberalization in different parts of the world.
9. Why were the United States and its western allies wary of protectionism at the end of the Second World War and what steps were taken on behalf of freer trade?
10. Compare and contrast the negotiating stances of different participants in the Doha Round.
11. Why did the Doha Round fail?

Appendix: Two-Country Illustrations of Comparative Advantage

The theory of comparative advantage is often explained with reference to a pair of countries. For example, one might consider Argentina and Mozambique, which we suppose for simplicity's sake produce just two goods: beef and sugar.

Let us say that Argentina has more agricultural resources than Mozambique. The resulting absolute advantage is indicated by the location of a **production possibilities frontier (PPF$_A$)**, which depicts how much beef and sugar can be produced in the country (Figure 6.2). Beef output is 1,000,000 head (i.e., the vertical-axis intercept of PPF$_A$) if all inputs are

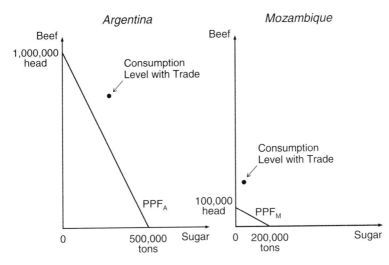

Figure 6.2 Production and trade possibilities

employed in ranching. Likewise, sugar production amounts to 500,000 tons (PPF$_A$'s intersection with the horizontal axis) if Argentina's agricultural capacity is dedicated entirely to raising the crop. Linear combinations of the two commodities are also feasible: 500,000 head of cattle and 250,000 tons of sugar, for example. In contrast, Mozambique lacks any sort of absolute advantage. With limited resources, it can raise 100,000 head of beef and no sugar or 200,000 tons of sugar and no beef or a linear combination of these two extremes, as represented by PPF$_M$.

Even though monetary values do not appear in Figure 6.2, opportunity costs are readily identified. In particular, marginal cost (MC), which consists of the units of one commodity that must be foregone for the sake of a small increase in the other good, is expressed by the PPF's slope. In Argentina, increasing sugar production by one ton requires a two-head reduction in beef output. Conversely, the MC of beef is half a ton of sugar. The PPF$_A$ is steeper than the PPF$_M$, so MCs are different in Mozambique. There, a one-ton rise in sugar production requires that beef output be cut back by half a head, just as increasing beef production by one head involves a two-ton sacrifice of sugar.[7]

---

[7] Most introductory texts in economics feature a PPF or two in the first few pages. Limits on productive capacity are depicted not by straight lines, as in Figure 6.2, but instead by curves bowed away from the origin. Such a curve indicates that, as production of a commodity goes up, its MC rises; for example, the beef given up for a unitary increase in sugar output would grow as sugar production rose. But with linear PPFs, the MC of sugar (i.e., sacrifices in beef production) does not vary at all as sugar output increases or decreases.

In light of these MCs, raising sugar in Argentina would be inefficient, as costly as a decision by Tiger Woods to mow his own lawn. This is proved by demonstrating that both countries can reach superior levels of beef and sugar consumption if each produces the commodity in which it holds a comparative advantage and imports the commodity in which it is not so favored. Argentina should specialize in ranching, because the MC there of beef production (i.e., half a ton of sugar) is lower than the MC in Mozambique (i.e., two tons). Likewise, sugar's MC is lower in the African nation, so it should produce nothing but that crop. Combined production in the two trading partners, then, consists of 1,000,000 head of beef and 200,000 tons of sugar.

One possible allocation of this output would result if Mozambique exchanged 150,000 tons of its sugar for 250,000 head of Argentine beef. Significantly, each country would achieve consumption levels beyond its respective PPF, which represents limits on what people eat (not just what they produce) in the case of autarky. Clearly, 250,000 head of beef and 50,000 tons of sugar cannot be consumed in Mozambique if the country does not trade. By the same token, 750,000 head and 150,000 tons are outside PPF_A, which means that this consumption level is out of reach if Argentina is an autarky. Obviously, trade is mutually beneficial—a positive-sum game.

The mutual benefits of trade are bound to materialize given the prices that will be observed if international commerce is unimpeded. Specific prices for beef and sugar cannot be identified in the hypothetical case of Argentina and Mozambique. However, something can be said about the terms of trade: the price of sugar (P_S) divided by the price of beef (P_B).

Without trade, P_S/P_B in each country would reflect local opportunity costs. This represents equilibrium for a closed economy since no sugar grower can gain by switching to beef production and no cattle rancher has an incentive to sell his herd and start planting sugarcane; both activities are equally profitable. If Argentina were an autarky, for example, P_S/P_B would equal 2/1. Likewise, the terms of trade in Mozambique would be 1/2 if that country were economically isolated. Each of these two price ratios corresponds exactly to the slope of the national PPF, of course.

If trade begins, merchants will buy each good where it is relatively cheap and ship it where the good is relatively expensive. With beef initially worth 1/2 ton of sugar in Argentina, that country, which as we have shown already has a comparative advantage in beef production, will be an international source of that commodity. For identical reasons, Mozambique will export sugar, in which it has a comparative advantage.

As beef moves out of Argentina, P_S/P_B falls in that country. Similarly, the price ratio goes up in Mozambique. In addition, these changes in the terms of trade are reinforced by the Latin American nation's imports of sugar and the African country's imports of beef. Consumers in each trading partner pay less for the good in which the domestic economy lacks comparative advantage. One cannot say what exactly the terms of trade will be with international commerce, but it is certain that P_S/P_B will be lower than 2/1 (relative prices in Argentina if the country is economically isolated) and above 1/2 (i.e., the price ratio in Mozambique if it is an autarky).

Once trade occurs and prices adjust, specializing according to comparative advantage is irresistible, in the sense that this is the only way to cover opportunity costs. Suppose that, with trade, P_S and P_B equal $750 and $450, respectively.[8] Terms of trade, then, would be 5/3, which is obviously between 2/1 and 1/2. Producing one head of beef in Argentina carries a sacrifice of 1/2 ton of sugar (see above), which is worth $375 in the marketplace. Since beef sells for $450/head, the rancher finds the trade-off profitable. By the same token, the market value of a ton of sugar produced in Mozambique exceeds the opportunity cost of this output (1/2 head of beef multiplied by $450). In contrast, growing sugar in the South American country is unprofitable, as is raising cattle in the African state.

So market forces oblige everyone to specialize according to comparative advantage. The only way for producers lacking comparative advantage to survive is to petition government for protection. But whenever petitioning of this sort is successful, the net gains from trade are lost.

NEV Gains from Trade

For another perspective on these net gains, we apply the conceptual framework introduced in Chapter 4 to describe NEV and its components.

Shown in Figure 6.3 are two national markets for the same commodity, wheat. In one country, the exporter, the good would be less scarce in the absence of trade. This is reflected by the relatively low price, P_{EA}, that would be observed there under autarky. In contrast, the price in the importing country would be relatively high, P_{IA}, if it were economically

---

[8] Note that, with these prices, the value of each country's exports equals the value of its imports: $750/ton × 150,000 tons of sugar = $450 × 250,000 head of beef. Thus, the mutually beneficial pattern of specialization and trade depicted in Figure 6.2 is financially feasible for both Argentina and Mozambique.

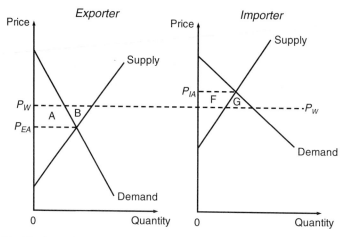

Figure 6.3 Trade and net economic value

isolated. NEV in each country would consist of the area between the domestic demand and supply curves if there were no international commerce.

Suppose now that economic isolation ends. Merchants engage in **arbitrage**, buying wheat where it is cheap (in the exporting country) and selling the commodity where it is dear (in the importing nation). As this happens, prices in the two markets converge to a single value, P_W, which is above P_{EA} and below P_{IA}. At P_W, the excess of production over consumption in the exporting country is the same as the excess of consumption over production in the importing country. The comparative advantage of the exporting country is indicated by the fact that the international price exceeds what the MC of wheat would be in the country if there were no trade. The latter would equal P_{EA} if the market were competitive. Also, the importer's lack of comparative advantage is reflected by the international price's being lower than P_{IA}, and therefore MC without trade.[9]

The increments in NEV created in both countries by commercial exchange show up in Figure 6.3, as do changes in consumers' surplus (CS) and producers' surplus (PS) in each trading partner. With the price

---

[9] As this example makes clear, comparative advantage depends not just on the supply side of the economy, but on the demand side as well. With enough demand growth in an exporting nation, the price observed there without trade rises above the world price, thereby causing that nation to become an importer instead.

of wheat going up in the exporting nation, there is an unmistakable gain in PS, represented by areas A and B. Part of this growth comes at the expense of consumers, the decline in CS being represented by area A. However, the gap between the former gain and latter loss is positive, obviously comprising area B. The changes in NEV in the country with a comparative advantage in wheat production are the mirror image of the impacts on CS and PS in the trading partner. In the latter, a declining price creates additional CS (areas F and G) while simultaneously reducing PS (area F). But since the former impact is greater than the latter, there is a net gain: area G.

To summarize, trade creates two costs, one resulting because consumers must pay more for the goods their country sells internationally and the other comprising losses for producers that are obliged to compete with cheap imports. But these two costs are less than the benefits of trade, which accrue to consumers given access to low-priced imports as well as producers given access to foreign markets with higher prices. There is additional NEV in each trading partner: areas B and G in Figure 6.3.

7

Agriculture and Economic Development

In a market economy, individuals and countries specialize according to comparative advantage and trade with one another, so prosperity spreads. In contrast, economic well-being is diminished where trade and markets are suppressed, as happens too often in the food economy.

But free commercial exchange does not amount to a comprehensive program for broad-based development. Impoverished households may trade frequently with one another. Indeed, a lot of sensible exchange happens in poverty-stricken communities, which should surprise no one. Survival itself can hang in the balance, so poor households do their level best to strike sensible deals. If their living standards do not improve much, their inability to offer much of value is usually to blame.

In remedying this deficiency of limited productive capacity, development inevitably must deal with the foundation on which economic performance ultimately rests, a foundation comprising three pillars. One of these consists of human, natural, and other resources: factors of production, in other words. A second pillar is institutional: the rules, informal and formal alike, that govern one individual's dealings with others. Without a framework of rules that define contracts, property rights, and so on, the marketplace simply cannot exist and mutually beneficial exchange is preempted. An economy's institutions both reflect and influence its third foundational pillar, which is the prevailing culture. For example, consistent enforcement of contracts and property rights, which is a hallmark of the rule of law, creates a culture of trust, which facilitates trade as well as productive investment. In contrast, deficiencies in the rule of law create distrustful attitudes. As a result, commercial exchange and the formation of productive assets are hindered and the prospects are dim for an institutional framework conducive to markets (Tweeten and Brinkman, 1976, p. 60).

If an economy's institutional and cultural underpinnings are sound, then the stage is set for capital formation, or investment, which enables people to produce more and therefore achieve better standards of living. This whole process requires savings, which are left over after consumption expenditures are subtracted from income. An economy need not supply all its own savings. Poor nations, for example, receive foreign aid. Far more savings cross national borders in the form of investment. Recent purchases of U.S. bonds by the Chinese and other foreigners are a case in point. However, most savings are generated and utilized domestically.

Capital does not accumulate if savings are simply stuffed in mattresses. To avoid this outcome, **financial intermediation**, which involves the channeling of savings provided by households to enterprises that invest in productive capacity, is needed. If banks and other providers of this service are inefficient, then capital formation falters and incomes suffer. But if financial intermediation is reliably efficient, then investment opportunities with attractive returns are consistently exploited, capital builds, and per capita gross domestic product (GDP) grows.

A sizable economic literature addresses the gains in living standards that are created as savings are generated and efficient investment takes place. Agriculture is a focus of this literature since that sector dominates many poor economies. Where markets are allowed to allocate resources efficiently, so that investment projects featuring the highest returns are undertaken, there is a tendency for **economic structure**, which describes the portions of GDP produced in various sectors, to diversify. In particular, agriculture's **GDP share** tends to decline as an economy develops—not because food production is falling, but rather because other sectors are growing faster. This **structural transformation** has much to do with Engel's Law (Chapter 2), which holds that the income elasticity of food demand declines as GDP per capita increases.

Structural transformation of this kind is only symptomatic of the fundamental development process, which it bears repeating is driven by savings and efficient investment. Nevertheless, ratcheting down agriculture's GDP share was a major goal of economic strategies pursued in various parts of the world during the twentieth century, as we document in this chapter. Economic diversification and growth were rarely achieved by distorting markets in favor of manufacturing and other non-agricultural sectors and at the expense of farmers. At an extreme, this approach created famine. To this day, sight is sometimes lost of one of the essential challenges of development, which is to establish the institutional and cultural antecedents needed for maximum well-being to be derived from available resources.

7.1 Economic Expansion and Structural Transformation

First investigated systematically during the 1950s, the economic diversification that accompanies development can be detected by examining a cross section of rich, poor, and in-between nations, by investigating structural change over time in a specific setting, or by doing some of both. Reported in Table 7.1 are data of the sort used in the combined approach.

Listed in that table are the same ten countries for which demographic data are reported in Chapter 2. Poverty is rife in Tanzania, as it is in most of Sub-Saharan Africa. Bangladesh is one of the poorest nations in Asia. Living standards are a little better in Senegal, yet still meager. An oil producer, Indonesia has joined the ranks of lower-middle-income nations. Syria, which unlike many of its Middle Eastern neighbors lacks petroleum deposits, is similarly categorized. Suffering from widespread poverty as recently as the 1970s, Thailand is now considered an upper-middle-income country; adjusted for inflation and purchasing

Table 7.1 Living standards and economic structure in 1990 and 2007, selected countries

	GDP per Capita (PPP adjusted, in 2005 U.S. dollars)		Agriculture's Value Added as % of GDP		Industry's Value Added as % of GDP		Services' Value Added as % of GDP	
	1990	2007	1990	2007	1990	2007	1990	2007
	(1)		(2)		(3)		(4)	
Tanzania	861	1,118	46	45	18	17	36	37
Bangladesh	680	1,178	30	19	21	28	48	52
Senegal	1,420	1,641	20	21	22	24	58	62
Indonesia	2,077	3,504	19	14	39	47	41	39
Syria	2,945	4,038	30	18	25	35	45	47
Thailand	3,769	6,983	12	12	37	46	50	43
Ecuador	5,501	7,023	13	7	38	36	49	57
Chile	6,589	13,087	9	4	41	47	50	49
Poland	8,164	15,638	8	4	50	31	42	65
Sweden	24,616	34,090	4	2	31	29	66	70

Source: World Bank (2009b).

power parity (PPP),[1] its GDP per capita was 85 percent higher in 2007 than it had been in 1990. Ecuador has experienced much less improvement, although it remains a little better off than Thailand. This reflects the fact that Latin American living standards are higher than those of the rest of the developing world, including Asia. Chile's GDP per capita has doubled during the past two decades, as has Poland's; if present trends continue, these two nations will become affluent in a few years. Sweden has long been one of the world's richest countries.

In the middle of the twentieth century, when investigation of economic diversification began in earnest, special attention was given to the tendency of industrial parts of the economy to grow faster than agriculture. In addition to reflecting Engel's Law, the relative importance of crop and livestock production has diminished because technological advances have allowed more food to be produced with fewer inputs. For example, Johnson (2000) stresses that farmers' use of mechanical reapers and other manufactured inputs in Europe and the United States during the 1800s and early 1900s released labor from rural areas precisely as industry's demand for this factor was expanding. In light of historical experience of this sort, development was viewed during the 1950s and afterward as going hand in hand with the expansion of manufacturing (Chenery and Syrquin, 1975; Kuznets, 1965). Johnson (2000) offers the broader observation that the Industrial Revolution would have proceeded much more slowly had it not been for new farming methods.

Going down the list of countries in Table 7.1, one can see that agriculture's GDP share (column 2) almost always decreases as living standards rise. One of the poorest countries in the world, Tanzania is a place where agriculture is larger than any other sector of the national economy and where most people raise crops and livestock. The ratio of value added[2] in agriculture to GDP is still nearly 50 percent. In Senegal, where average incomes are higher, agriculture's share has fallen nearly to one-fifth. The share is below 20 percent in every other country listed in Table 7.1.

The relationship between **industrialization** and average income is different. Industry's share of GDP (column 3) usually goes up as median earnings rise from a few hundred to several thousand dollars per annum. Clear examples of this change include Bangladesh, Indonesia, Syria, and Thailand. But as indicated in Table 7.1, the ratio of industrial value added to GDP tends to stabilize as average income reaches the range of $7,000 to

---

[1] PPP measures of variables such as GDP and GDP per capita reflect variations in prices from country to country (Chapter 2).
[2] A sector's value added comprises the difference between the value of its output and its input purchases from other sectors and is distributed between payments to labor and payments to capital (Chapter 3).

$12,000. Beyond that range, industry's share declines. This has happened in Poland, for instance. Note as well that the current share in Sweden is comparable to those of Bangladesh and Senegal.

In a number of countries, the service sector is the largest and fastest-growing part of the economy. If anything, services are more diverse than either agriculture or industry. Obviously, the sector's output comes in myriad forms. Some services require no skill and, quite often, back-breaking effort. Carrying goods to or from market, cleaning houses, or standing guard are all good examples, and activities like these employ many people in impoverished settings. Demand for the same services exists in wealthy places; for example, there are strong incentives for someone like Forrest Gump to be hired to mow Tiger Woods's lawn (Chapter 6). However, the rising preeminence of services in more developed parts of the world relates to the growth of activities requiring a lot of training and education. Not only is this true in affluent countries, like Sweden, but in nations that are becoming so as well, such as Chile and Poland. Financial intermediation, which requires substantial human capital, is an important part of the service sector and the entire economy in wealthy places. The same holds for schools and universities, which equip the workforce for skilled employment, as well as the health care system, which keeps people healthy and productive.

Just as industrialization in decades past provided the means to transform crop and livestock production and reallocate labor from farms to factories, the expansion of services has had profound and beneficial impacts on the rest of the economy, including crop and livestock production. Medical attention and pharmaceuticals, for example, reduce the chances that disease will impair the productivity of agricultural labor. The efficiency of farms in places like the United States can be traced directly to the marketing and other services supplied by agribusinesses (Chapter 3). Also, hybridization and the Green Revolution have been made possible by human capital formation and agricultural research and development underwritten by governments, private foundations, and international organizations.

The world's most successful economies are now dominated by services. In Sweden, the sector's value added exceeds seven-tenths of GDP. Three-fifths of the country's male workforce and nearly nine out of every ten Swedish women are employed in banks, educational institutions, hospitals and clinics, and other enterprises that are neither agricultural nor industrial. In the United States, where less than 2 percent of the labor force farms and another 21 percent is industrial, the ratio of services' value added to GDP is nearly 77 percent (World Bank, 2009b).

The Diversity of Structural Transformation

Conditioned as it is by geography, history, and a host of other factors, development never follows the same path in any pair of countries. In many settings, however, recent growth in living standards and structural transformation have been influenced greatly by evolution of the service sector.

Since many service jobs in impoverished settings require few skills and pay poorly, service employment represents an alternative for people who otherwise would engage in subsistence farming, which tends to be even less remunerative. This is the predicament of the majority of Tanzanians, who can neither read nor write and who must choose between eking out a living on the land and seeking menial work in Dar es Salaam or some other city. Poverty limits the size of the domestic market, so growth prospects for industry are limited. As already observed, most of the working population is employed in agriculture, which accounts for nearly half the country's GDP.

Most Bangladeshis are rural (Table 2.2), yet services now account for half the country's GDP. The value of the sector's output, not to mention the value of agricultural and industrial output, would be much greater if there were less illiteracy. Similar things can be said of Senegal, where more people have decamped to cities and towns but lack the skills needed for anything other than menial employment.

Thanks in part to petroleum development, industry accounts for nearly half of Indonesia's GDP. Structural transformation has been relatively modest in Syria since 1990, as has growth in GDP per capita. Thailand is an interesting case because incomes have shot up dramatically during the same period, yet agriculture has grown as fast as the rest of the economy. This reflects the country's strong comparative advantage in the production of rice and other crops.

In Ecuador, the service sector employs nearly the same share of the labor force as in Sweden: 70 percent versus 76 percent (World Bank, 2009b). However, low skills prevent many Ecuadorians from producing anything of great value, so living standards are modest. Due to the importance of copper mining and processing, industry occupies the same position in the Chilean economy that it occupies in the Indonesian economy. However, services are more important in the South American nation, which is much more prosperous.

Under communism, manufacturing was Poland's economic mainstay. A sure sign that a transition truly has happened is that services have emerged as the dominant sector. As in so many other places—affluent, impoverished, and otherwise—the service workforce is mainly female.

Two-thirds of the country's women are employed outside of agriculture and industry. For men, that portion is just 44 percent. About 15 percent of the workforce, male as well as female, labor on farms (World Bank, 2009b). Another sign that Poles have put communism behind them is that GDP per capita has nearly doubled since 1990.

As already indicated, Sweden is a wealthy place where the service sector accounts for most employment and output. It is representative of what one finds in Western Europe, North America, Japan, and other affluent parts of the world.

Living Standards and Income Distribution

Simon Kuznets, who along with Hollis Chenery pioneered the study of structural transformation in developing economies, hypothesized that **income inequality** tends to be modest in impoverished settings, rises as an emergence from penury occurs, and then declines after the average living standard passes a certain threshold (Kuznets, 1955). Like its environmental variant (Figure 5.4), the **Kuznets Curve** is shaped like an inverted U. The only difference between it and the Environmental Kuznets Curve (EKC) has to do with their respective vertical dimensions. While environmental damage is plotted up and down in Figure 5.4, income inequality is measured vertically in Figure 7.1.

The standard measure of income inequality is the **Gini coefficient**, or Gini concentration ratio (Perkins, Radelet, and Lindauer, 2006, p. 196). Its minimum and maximum values are, respectively, zero (reflecting

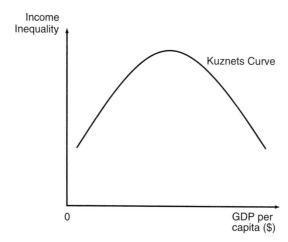

Figure 7.1 Living standards and income distribution

complete equality) and 100 (for the case of all income being captured by a single individual, with absolutely nothing left for anyone else). These extreme values are never observed in the real world, of course. As a rule, a Gini coefficient below 40 indicates modest inequality. In contrast, inequality is considered high if the concentration ratio is over 50.

When Kuznets offered his hypothesis about living standards and income inequality, data required for testing were not available. Subsequent empirical research failed to yield clear proof that a single Kuznets Curve exists, one that is applicable to the entire world (Perkins, Radelet, and Lindauer, 2006, pp. 201–202). Regional differences are an important reason for this. For example, Latin America and the Caribbean, where no country is rich but also where few are extremely poor, suffers from acute income inequality, due to ethnic and class divisions that date back centuries. In contrast, Gini coefficients are generally low in Eastern Europe, for reasons discussed below.

Regional differences are apparent in our sample of ten countries, although the Gini coefficients reported in Table 7.2 also hint at the existence of a Kuznets Curve. Tanzania, Senegal, and (especially) Bangladesh all have low living standards as well as modest income inequality. At the other end of the income spectrum, earnings are distributed quite evenly in Sweden. In between, GDP per capita has risen in recent years in Indonesia, yet its Gini coefficient is still under 40. Data required to estimate the coefficient are unavailable in Syria. Not far north of Indonesia, Thailand has much higher living standards and only a little more inequality.

Table 7.2 Average income and income inequality, selected countries

	2007 GDP per Capita, with PPP adjustment (1)	Gini Coefficient (survey year) (2)
Tanzania	$1,184	35 (2000)
Bangladesh	1,247	31 (2005)
Senegal	1,737	39 (2005)
Indonesia	3,711	39 (2005)
Syria	4,276	unavailable
Thailand	7,394	42 (2004)
Ecuador	7,437	54 (2007)
Chile	13,858	52 (2006)
Poland	16,089	35 (2005)
Sweden	36,712	25 (2000)

Source: World Bank (2009b).

Representative of their region, Ecuador and Chile have elevated Gini coefficients. Poland is also a good regional representative; its coefficient is identical to Tanzania's and not much above Bangladesh's.

Some of the economic realities underlying the Kuznets Curve are alluded to in the preceding discussion of income growth and structural change. In a place like Bangladesh, half the labor force farms and is poorly compensated for doing so. Another third is in the service sector, with most of this segment receiving a low wage for unskilled work. The same is true of many industrial laborers. Since the vast majority of Bangladeshis are poor, their country's Gini coefficient is low.

If development starts to take place under conditions such as these, the impoverished and unskilled majority benefit little. The poor may continue to farm. Alternatively, they may opt for unskilled service employment, as many Ecuadorians have done. Either way, their earnings rise modestly. In contrast, the minority of people with a high school education—or, better yet, a university degree—see demands for their skills go up substantially, and their incomes rise accordingly. This causes concentration ratios to climb into the middle 40s and beyond.

One place (which happens not to be listed in Table 7.2) where rapid economic expansion is undoubtedly coinciding with rising income inequality today is India. Global communications are allowing credit-card companies, airlines, and other businesses to transfer customer-service operations outside the United States. This outsourcing has created hundreds of thousands of well-paying jobs for educated Indians with good English (Chapter 11). Since there is no comparable improvement in earnings for the impoverished majority of their countrymen, Gini coefficients have risen.

The situation is different in Eastern Europe, mainly because schooling has been widely available there for many years. Indeed, this is the only part of the nonaffluent world where small elites do not parlay superior education and other advantages into earnings that greatly exceed those of their fellow citizens. Instead, a broad segment of the population is able to benefit from the demands for skilled workers and professionals that development creates. Thus, the distribution of income is not highly skewed in places like Poland.

Few nations can match Sweden's success in extending educational and other opportunities to the entire national population, which means that most Swedes are able to exploit the opportunities arising in an affluent, service-based economy. Also, heavy taxation pays for one of the world's best social safety nets. As a result, the country's Gini coefficient is lower than Bangladesh's.

7.2 Agriculture's Role in Economic Development

As a rule, agricultural value added as a share of GDP diminishes as living standards improve. This change is by no means undesirable, since it reflects increased specialization and trade and a general expansion of productive capacity—or, to be more precise, enhanced productive capacity for each member of the workforce. Put another way, agricultural development ought to be pursued in ways that cause the GDP shares of nonagricultural sectors to go up, to the benefit of living standards.

Johnston and Mellor (1961) were among the first economists to describe agriculture's contributions to economic growth and diversification in detail. In particular, they identified five such contributions.

1. *Increased food supplies for the domestic, nonagricultural economy.* If food supplies are unreliable, or can be obtained only at high prices, then wages are bid up and employment falls in the industrial and service sectors. This hampers economic diversification, which is also held back because households have less to spend on nonfood items. By the same token, savings are reduced, which diminishes the buildup of productive capacity required for GDP growth. A sustained increase in food supplies for the domestic economy has precisely the opposite effects. As food grows less scarce, upward pressure on wages abates and nonagricultural employment rises. Also, cheaper food allows households not only to eat better, but to diversify consumption expenditures and increase savings.

2. *More foreign exchange.* Many if not most developing countries have a comparative advantage in one or more commodities. As this advantage is exploited, foreign exchange is earned. Among other things, this allows for the importation of capital goods, which are an important component of productive capacity. But if comparative advantage is overwhelmed by currency overvaluation (Chapter 6) and other policy-induced distortions, then exports decline and less foreign exchange is earned. Imports may be maintained in the face of diminished exports by international borrowing, which of course leads to indebtedness.

3. *Labor reallocation.* It bears repeating that a major contribution of agricultural development to overall economic progress has been to release labor from the countryside, and to enable people to seek employment instead in industry or (more often than not these days) the service sector. Obviously, this transfer is impeded if a large segment of the workforce remains on farms because this is the only way for the entire population to be fed.

4. *Increased tax revenues and transfers of savings.* As it develops, the agricultural sector becomes a source of tax revenues that can pay for public goods—public goods not just for rural areas, but for the entire economy. In many parts of the United States, for example, taxes on farm real estate have been spent on education. Many of the young people benefiting from this investment in human capital have subsequently found productive work in manufacturing and the service sector. Another impact of increased agricultural productivity is that savings by farm households increase. If other parts of the economy are growing faster than agriculture and if financial intermediation is good, then some of the savings of rural households are reallocated to nonagricultural sectors, thereby accelerating growth and diversification. Of course, an agricultural economy consisting largely of subsistence farms has few financial resources to share.

5. *Enhanced demand in rural areas for nonfood products.* Likewise, an impoverished farm economy is a poor market for goods and services produced by nonagricultural sectors. But as farmers grow more prosperous, their increasing demand for these products becomes an important force for structural transformation and GDP growth.

The preceding observations about how agriculture can, and often does, contribute to economic progress should not be regarded as a precise blueprint for development, one that is universally applicable. Instead, prescriptions for development must be tailored to local circumstances. For example, Johnston and Mellor (1961) generally favor increasing the domestic supply of food, which diminishes its price, in order to accelerate economic growth and structural transformation. However, some nations have prospered by relying on imports instead. Lacking the natural resources needed to grow all their own food, Singapore and Hong Kong have specialized in nonagricultural activities, in which each country has a strong comparative advantage, and have bought farm goods from foreigners.

As Johnston and Mellor (1961) point out, complementarities and trade-offs exist among the contributions that agriculture makes to development. One complementarity arises because an increase in commodity exports, which enhances foreign exchange earnings, also causes rural incomes to rise. This allows for higher savings and increased purchases of nonfood items in rural areas. One example of a trade-off has to do with increased exports, which can cause prices paid by local consumers to rise as commodities are directed away from the domestic market or if resources that would otherwise produce output for that market are instead used to produce commodities for international buyers. Likewise, various trade-offs are associated with a government's attempts to extract tax revenues from the agricultural sector. If the burden on farmers grows too onerous, production suffers, for domestic as well as international markets. Savings of rural households also fall, as does their spending on nonfood items. Excessive taxation of agriculture may accelerate the exit of labor from farming. Otherwise, it diminishes the full range of contributions the sector can make to a larger and more diversified economy.

No pair of countries is likely to strike precisely the same balance among these and other trade-offs and complementarities. Something important to recognize, however, is that a good way to ameliorate trade-offs and to enhance complementarities is to make agriculture more productive. As farmers grow more efficient, domestic food supplies and exports can increase simultaneously and this expansion can coincide with the inter-sectoral reallocation of labor. Rural households are more prosperous, which enables them to pay more taxes, increase savings, and spend more on nonfood items—all at the same time. Rising agricultural productivity, then, can be a major driver of overall economic progress, especially in the many poor nations where farming remains an important part of the economy in terms of GDP and employment shares.

7.3 Trying to Develop at Agriculture's Expense

Lessons about agriculture's multiple contributions to development have not been learned easily. To the contrary, development strategies predicated on neglect or penalization of the sector were pursued in various parts of the world during the twentieth century. The record of this experimentation was invariably dreadful, not just economically but all too frequently in terms of lives lost as well.

Experiments in Communist Nations

In late 1917, Vladimir Lenin brought off a coup d'etat that allowed for the application of scientific socialism, as he preferred to call it, in the nation immediately rechristened the Union of Soviet Socialist Republics (USSR). He and other communist leaders were united in their enthusiasm for industrializing the country, which at the time was overwhelmingly agricultural. For obvious ideological reasons, there was little enthusiasm for private-sector participation in new or expanded manufacturing enterprises. Some in the new regime contemplated exploiting the Soviet Union's comparative advantage in agriculture, using the foreign exchange earned from crop exports to pay for imported technology and inputs so that industries owned by the state could expand. However, this option was ruled out once foreign trade was choked off due to the communists' repudiation of debts from the czarist era. The only way to finance socialist industrialization, as called for in the **New Economic Policy (NEP)**, was to tax peasants, who at the time comprised more than 80 percent of the population and who had ended up with most of the country's farmland after the breakup of large estates (Skidelsky, 1996, p. 50).[3]

By and large, taxation of the rural economy was accomplished indirectly, which was possible because of complete **socialization** of the service sector. In particular, all marketing was taken over by the government, which proceeded to pay low prices for farm products and to charge excessively for manufactured goods purchased by peasants. This manipulation of the domestic terms of trade (i.e., relative prices within the national economy) resulted in the extraction of wealth from the countryside, obviously not as voluntary savings but rather in the form of obligatory commerce with state-owned monopolies (Skidelsky, 1996, p. 51).

The peasants were not entirely without recourse. An obvious response to low official prices was to cut back on marketable output. Another was to feed more grain to livestock, which was privately owned. Yet another was to produce vodka and other liquor, which were fairly easy to sell through illegal channels. As a result of subterfuges such as these, revenues collected by the government through the indirect taxation of farmers proved disappointing. This put a brake on the expansion of state-owned industry.

Josef Stalin, who prevailed in the succession struggle that followed Lenin's death in 1924, had a brutally simple remedy for the shortcomings of the NEP. Soon after he consolidated power, virtually all agriculture was collectivized, with individual holdings—aside from parcels smaller than one hectare retained by individual families—incorporated into large-scale farms organized by the state. The **collectivization** of farming was accomplished with unspeakable cruelty. Between 1929, when Stalin became dictator, and 1941, when the Soviet Union was invaded by Nazi Germany, nearly 15 million people died, the majority rural folk. They either starved in famines engineered by communist authorities to drive recalcitrant farmers, denigrated at the time as *kulaks*,[4] off the land or

---

[3] State seizure of agricultural land right after the October 1917 "revolution" was not an option since "all land to the peasants" was one of the three major promises that Lenin made when he seized power. Of course, what the communist state gave peasants, in the form of parcels carved out of old estates, it took from them in the form of coercive food procurement at near-confiscatory prices.

[4] The literal meaning of *kulak* in Russian is fist. Communists applied the term to individual farmers so as to suggest that they, the farmers, were inordinately acquisitive, to the detriment of the wider good, and therefore deserving of extermination.

were worked to death as slave laborers in mines, factories, and infrastructure projects (Conquest, 1986, pp. 299–307). After the Second World War, other experiments with socialized agriculture behind the Iron Curtain had similarly disastrous results (Box 7.1).

> **Box 7.1 Communist crimes in the countryside**
>
> The Soviet Union was not the only nation where agricultural collectivization resulted in enormous loss of life. At least 25 million Chinese, mainly peasants, perished after Mao Tse-tung opted for the socialization of farming in the late 1950s, particularly after he decided on a crash program of industrialization (including in rural areas) that left agriculture starved of state support (Short, 1999, pp. 486–505). The consequences of imposing communism on the countryside have been no less appalling in smaller countries, most notably Cambodia under the Khmer Rouge and North Korea under the Kims, *père et fils*.

A few communist governments refrained from imitating the Stalinist approach to agriculture. Polish farms, for example, were not taken over by the state. This proved fortuitous after communism ended; the fact that items such as eggs and milk were readily available from private farmers allowed the market economy to gain a decisive foothold quickly (Yergin and Stanislaw, 1998, pp. 267–268). Elsewhere, collectivization was attempted, but was subsequently abandoned. After the death of Mao Zedong, rural communities in China began experimenting with family farming, individual marketing of livestock and produce, and other capitalistic practices in the late 1970s. Even though all these practices were illegal, communist authorities were not successful in clamping down entirely, as would have happened under Mao. To the contrary, elements of the **Household Responsibility System (HRS)**, as the community-instigated reforms have come to be known, began receiving state approval in the early 1980s (Gregory and Zhou, 2010). The HRS soon spread from the countryside to urban areas and is largely responsible for China's recent economic trajectory (Gregory and Zhou, 2010; Naughton, 1995, pp. 138–142).

Governmental Intervention in the Developing World

Except for communist states, most African, Asian, and Latin American nations have avoided the disaster of socialist agriculture, collectivized farming only having been dabbled with here and there. However, scores of governments in the developing world have succumbed to the temptation to interfere with market forces so as to accelerate structural transformation—exactly as was tried in the Soviet Union under Lenin. Especially during the first three or four decades after the Second World War, domestic terms of trade were manipulated to the detriment of agriculture so that manufacturing could grow faster. That is, government policies drove down prices of farm products relative to the prices of manufactured goods. This approach, called **Import-Substituting Industrialization (ISI)**, has proven no less disappointing than the NEP was.

As implied by the term, import-substituting, special emphasis is placed in ISI on national production of domestically consumed manufactured items. Various rationales have been offered for this emphasis over the years, including saving foreign exchange that otherwise would be spent on imported products and enhancing manufacturing employment. Another oft-repeated justification is that so-called **infant industries** should be given an opportunity to acquire comparative advantage in international markets through "learning by doing" in national markets, from which cheaper imports are kept out for a while (Perkins, Radelet, and Lindauer, 2006, p. 718).

Under ISI, the structural shift from agriculture to industry is accelerated both by pegging domestic prices of manufactured goods above international levels and by holding down the domestic prices of farm products. One way to accomplish the latter distortion is to impose controls on food prices, which is obviously welcomed by nonagricultural workers. Such controls also benefit firms employing these workers, since pressure to raise wages is reduced. To diminish the prices received by commodity producers serving foreign markets, exports can be taxed. In addition, ISI is promoted by currency overvaluation and its resulting import bias (Chapter 6). This bias is counteracted by trade barriers in the case of domestically produced manufactured goods, but not in the case of farm products and other primary commodities.

The economy suffers as domestic terms of trade are distorted in these and other ways. Consumers are denied access to manufactured imports that are cheaper and often of higher quality. Also, agriculture and other parts of the economy enjoying a comparative advantage in global markets are penalized since prices for these sectors' output are depressed and more must be paid for inputs purchased from protected sectors. For a few years after ISI is adopted, the sacrifices made by consumers and competitive sectors seem worthwhile because domestic industry expands rapidly as imported products are driven from national markets. But unless protected manufacturers truly learn by doing, which often does not happen, this period of rapid expansion comes to an end once they dominate domestic markets. With weak incentives to operate efficiently, protected industries rarely acquire comparative advantage, and hence are unable to compete internationally. Once they have taken over the domestic market, further industrial growth is constrained by expansion of the national economy (Perkins, Radelet, and Lindauer, 2006, p. 728), which often proves to be sluggish (Box 7.2).

> ### Box 7.2 ISI in Latin America
>
> Many countries, mainly the larger Latin American nations, experienced expansion in manufacturing immediately following the adoption of ISI, but later found economic growth and industrialization difficult to sustain. A general consequence of this stagnation was mounting **trade deficits**. Industries accustomed to protection rarely were competitive enough to export. Moreover, factories tended to rely on imported capital goods and raw materials, the prices of which were kept low by currency overvaluation. Meanwhile, exports from agriculture and other sectors with a comparative advantage were reduced by currency overvaluation and other distortions. Thus, the overall impact of ISI was a deteriorating balance of trade as imports run ahead of exports.
>
> During the 1970s, substantial foreign borrowing allowed trade deficits to be sustained throughout Latin America and other parts of the world where ISI had been adopted during the 1940s and 1950s. International debts also mounted because government spending consistently exceeded tax revenues. Rising indebtedness could not continue indefinitely and, immediately after Mexico declared a moratorium on interest payments in August 1982, Latin America was plunged into a severe and prolonged recession. At the end of the day, the countries that have put this crisis behind them have largely abandoned ISI and its costly distortions.

7.4 Agricultural Development for the Sake of Economic Growth and Diversification

In light of the economic dislocation suffered by countries that have pursued ISI, it is fortunate that there is an alternative path to structural transformation and economic growth. The **Outward-Looking Strategy** (Perkins,

Radelet, and Lindauer, 2006, pp. 729–730) was first pursued by a handful of places in East Asia, which opted to produce manufactured items for foreign markets precisely as Latin America and other parts of the developing world were shying away from international competition and instead providing incentives for domestic manufacturers to concentrate on national markets. The four East Asian "tigers"—Hong Kong, Singapore, South Korea, and Taiwan—chose export-oriented industrialization over ISI mainly for a simple, practical reason. Particularly during the 1950s and 1960s, their respective domestic markets were quite small, comprising in each case a few million consumers with low average earnings.

Regardless of the motivating forces, choosing the Outward-Looking Strategy has proven fortuitous. As a rule, domestic prices are distorted less than in countries practicing ISI. At times, the four tigers and other practitioners have undervalued their currencies, which has created an **export bias** (i.e., precisely the opposite of the import bias resulting from currency overvaluation). But chronic trade surpluses cannot be sustained indefinitely any more than chronic deficits can. This is because the accumulation of foreign money in a country with an undervalued currency eventually causes that currency to gain value, which in turn diminishes exports and stimulates imports.

Aside from distortions and imbalances of this sort, governments not absorbed with the task of market manipulation have tended to concentrate more on the buildup of productive capacity, broadly defined. Education has been supported. In addition, the legal guarantees that commerce requires—reliable and uniform enforcement of contracts, property rights, and other elements of the rule of law—have been established. As a result, the structural transformation and economic growth that have taken place in Hong Kong (Box 7.3), Singapore, South Korea, and Taiwan are the envy of the developing world.

Box 7.3 Outward-looking development in Hong Kong

Rather than pursuing ISI, Hong Kong achieved industrial development by focusing initially on lines of manufacturing, like textiles, that made heavy use of low-paid, unskilled labor, which was abundant during the 1950s and 1960s. Subsequently, a switch was made to electronic goods and other products requiring more skilled labor and capital. These days, the former British colony, like other rich places, specializes in banking and other services. Most industrial operations, the majority of which are owned by Hong Kong businessmen, have moved to China and other countries.

Positive elements of the Outward-Looking Strategy have been adopted elsewhere. Having cut taxes and played on the advantage of an educated workforce, Ireland—the "Celtic tiger"—enjoyed an economic boom in the late 1900s and first few years of the twenty-first century. As highlighted earlier in this chapter, income inequality remains high in Chile, which suggests that improvements in living standards have not been broad based. Nevertheless, it comes closer than any other country in Latin America to being the region's tiger. In addition, interference with market forces is fairly restrained in Malaysia and Thailand, which have invested substantially in human capital and other elements of productive capacity.

In settings such as Thailand that are well endowed with arable land and where agriculture continues to be an important sector, economic progress in the countryside has made important contributions to overall development. Much more than in Latin America and Sub-Saharan Africa, farmers in Asia have intensified their operations, making use of fertilizer and other purchased inputs to raise yields, especially during the Green Revolution (Chapter 3). Also, product composition has changed in response to evolving export opportunities as well as shifts in domestic demand resulting from higher incomes for consumers. Since total factor productivity (TFP) has gone up, the various benefits of agricultural development identified by Johnston and Mellor (1961) and surveyed in the second section of this chapter have actually materialized. Supplies of farm products have increased for domestic and international markets, labor has been released to nonagricultural sectors of the economy, and expenditures on nonfood items have gone up, as have savings (Mellor, 1995, pp. 10–16).

Given that the "expenditure multiplier" of increased farm earnings is sizable in countries with modest living standards, a virtuous cycle is set in motion by improvements in farm productivity where a comparative advantage in agriculture exists. As emphasized in this chapter and throughout this book, these improvements depend on undistorted prices as well as agricultural research and development, marketing infrastructure, and other public goods.

As illustrated in Figure 7.2, one link in the cycle driven by productivity growth has to do with its direct effects—to be specific, lower food prices and more exports. Also, spending on nonfood items increases as food becomes less scarce. The resulting economic diversification is further accelerated by the increased availability of workers formerly employed on farms. Diminished food scarcity also allows savings to increase. Higher farm earnings have the same effect. As savings and the foreign exchange generated by exports are invested in agriculture and other

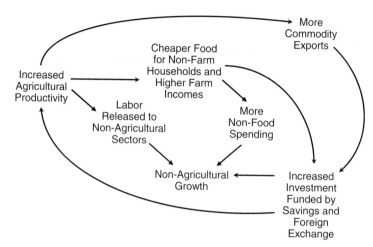

Figure 7.2 Productivity growth in agriculture and economic development

sectors, productivity is given a boost throughout the economy, which allows the entire virtuous cycle to continue.

7.5 Summary and Conclusions

The suppression of private farming by communist authorities not only took an enormous economic toll in the Soviet Union, Maoist China, and other nations, but cost tens of millions of lives. In Latin America and Africa, manipulation of market forces to agriculture's disadvantage is less egregious today than it was when ISI was being pursued vigorously. However, investment in public goods for crop and livestock production in these two regions still compares poorly with Asia's record (Mellor, 1995, pp. 307–329), even where there is a comparative advantage for agriculture. Consequently, opportunities for development, within the food economy and beyond it, continue to be missed.

In a fundamental sense, the greatest error committed in nations where communism or ISI was implemented has been to focus too much on the fates of particular sectors. As Tweeten and Brinkman (1976, p. 60) point out, an economy's prospects are based on three things that may have little to do with different sectors' GDP shares: the endowment of human and natural resources, institutions, and the prevailing culture. Up to a point, substitution among these three factors is possible. For example, natural resource scarcity can be overcome by investing in human capital or strengthening the rule of law, which is an institutional reform of critical importance in many places. However, cultural deficiency—for example, distrust that is acute and pervasive enough to discourage commerce and choke off investment—is something that must be addressed directly through education and by reinforcing the rule of law rather than being compensated for by the acquisition of additional resources.

In successful economies, resources (be these human, natural, or something else) are matched up, first, with institutions that are functional in the sense that the rule of law is secure and, second, a culture of trust made possible by functional institutions. Rarely if ever are price distortions debilitating in these economies. Neither are public goods severely undersupplied. Agriculture makes an appropriate contribution to economic progress and few people are food-insecure.

Unsuccessful economies, on the other hand, often possess abundant environmental wealth, and may even have an impressive endowment of human capital. The limiting factors, then, are institutional and cultural, with true rule of law only to be imagined and with everyone so distrustful that countless opportunities for trade and investment go unexploited. Under these circumstances, governments frequently try to stimulate development by meddling with prices, typically ignoring public goods as a result. The long-term outcome, in terms of agricultural development, the general trajectory of the economy, as well as food security, is invariably disappointing.

Key Words and Terms

collectivization
economic structure
export bias
financial intermediation
GDP share
Gini coefficient
Household Responsibility System (HRS)
Import-Substituting Industrialization (ISI)
income inequality

industrialization
infant industry
Kuznets Curve
New Economic Policy (NEP)
Outward-Looking Strategy
socialization
structural transformation
trade deficit

Study Questions

1. Aside from resources (human, natural, and otherwise), what does development require?

2. How does an economic structure normally change as an economy develops?
3. Are the richest countries of the world the most industrialized? Why or why not?
4. Explain the Kuznets Curve. How well does it describe patterns of income inequality?
5. Identify various contributions of agriculture to economic development as well as complementarities and trade-offs among these contributions.
6. Compare and contrast the New Economic Policy of the early Soviet Union and Import-Substituting Industrialization.
7. Why do trade deficits mount and growth flag the longer a country pursues Import-Substituting Industrialization?
8. Compare and contrast the Outward-Looking Strategy for economic development, as pursued by the four East Asian tigers for example, and Import-Substituting Industrialization.
9. How has agricultural development in Asian nations pursuing the Outward-Looking Strategy compared with agricultural development in Latin America and Sub-Saharan Africa?

8

Striving for Food Security

The world's food economy has accomplished wonders in recent decades, yet hundreds of millions still go hungry. Recognizing this, no one can dispute the shared judgment of the directors of the International Food Policy Research Institute (IFPRI) and Bread for the World that "the global community stands indicted for knowing much about how to reduce hunger, but not doing so" (Runge et al., 2003, p. xvii).

The goal of ending hunger has been debated and analyzed at length, although most agree that the critical factor in **food security** is **access** to adequate nutrition. Expressing the consensus is the declaration by the Food and Agriculture Organization (FAO) that food insecurity will have ended once "all people at all times have physical, social, and economic access to sufficient, safe, and nutritious food that meets their dietary needs and food preferences for an active and healthy life" (FAO, 2002b, p. 49).

Not addressed in this declaration is an important dimension of proper nourishment, which is the **utilization** of food once it has been consumed. Genetic factors (e.g., the predisposition of some people to certain allergies) can hinder absorption of particular vitamins or nutrients. So can improper food preparation, eating disorders, pathogens, and simple bad habits. A problem for countless Africans, Asians, and Latin Americans is limited access to clean water. These people often have intestinal parasites, which can cause **malnutrition** even among those who are actually eating enough.

Nutrition and health education, vitamin supplements, oral rehydration, treatment of parasites, and immunization are all needed to help households make better use of the resources available to them. Increased funding of these measures could eliminate a lot of malnutrition, especially among children. But while the advantages of an integrated approach are widely recognized, improved utilization is often not subsumed in food security initiatives.

Something else to understand is that these initiatives are normally predicated on food being available in sufficient quantities. True, **availability** is a critical issue if prolonged civil strife causes food-distribution networks to break down and farms to be abandoned. But as emphasized by Amartya Sen, a Nobel laureate who has investigated a number of famines, the food that hungry people need is often at hand, as was the case when Bengal suffered starvation in 1943 (Sen, 1981, pp. 57–63).

Global production in recent decades has consistently been enough to feed the entire human race (Chapter 3), so food security is mainly about access—access of a distinctly economic nature, to be precise. The skeletal inmates of refugee camps that have sprung up in places like the Horn of Africa as local strongmen and thugs have prosecuted their sordid, sanguinary rumbles are not really emblematic of global food insecurity. The fundamental problem is more mundane and widespread, though no less serious. Quite simply, the meager earnings of hungry people do not cover the expense of an adequate diet.

In addition to reporting how many people are not eating enough in various parts of the world, we examine **food aid** in this chapter. We also emphasize that steps taken to enhance food security are all but indistinguishable from the **standard model** required for broad-based economic progress.

8.1 Who and Where Are the Food-Insecure?

As much as one-third of the human race suffers from one or more forms of undernourishment. Approximately 2 billion people are anemic, because of iron-poor diets and other reasons. At least 90 percent of this group is in the developing world, where hundreds of millions also suffer from **micronutrient deficiencies** (Babinard and Pinstrup-Andersen, 2001).

The costs of dealing with problems of this sort are modest. For instance, iodine deficiency, which in extreme cases causes goiters, can be corrected simply by ingesting a little iodized salt at a per capita expense of a penny or two a week. Moreover, the expense of making sure that everyone gets enough micronutrients pales in comparison to the benefits of doing so. A far more ambitious undertaking is to eradicate food insecurity proper, which can be defined as the daily shortfall of 100 to 400 calories that hundreds of millions of human beings experience chronically or frequently. Nearly all this group is counted among the 1.2 billion people living in **extreme poverty**, in the sense that their earnings are less than or equal to $1.25/day (Box 8.1).

Box 8.1 Food insecurity in the United States?

Very few Americans subsist on a dollar or two a day, yet more than one in every ten people in the United States is reported to be food-insecure. By and large, the problem is not a chronic or frequent shortfall in the daily intake of calories. Instead, the incidence of food insecurity, which jumped from 11.1 percent in 2007 to 14.6 percent in 2008, is defined as the share of all households that "were food insecure at least some time during that year." This determination is based on periodic surveys in which participants report if during the past twelve months they ever:

- worried about their food running out before they were able to buy more;
- lacked money to purchase food after supplies had run out; or
- could not afford a balanced diet.

Additional details about food security and its measurement in the United States are available at ERS-USDA (2009b).

Reported in Table 8.1 are broad trends and regional differences in food insecurity. The number of people who usually do not consume enough of the dietary energy needed for normal activity and good health has fallen steadily since the late 1960s, from more than 900 million throughout Africa, Asia, and Latin America around 1970 (column 1) to 832 million in 2003–2005 (column 4). Relative to the total population of the developing world, the **incidence of food insecurity** halved during this period, from 35 percent of the total to 17 percent.

Food insecurity does not decline year in and year out. For example, the number of chronically hungry people climbed to 1.02 billion in early 2009, due to unemployment and lower incomes during the worldwide economic recession (FAO, 2009b). In addition, progress toward greater food security has not been uniform throughout the developing world.

Impressive gains have been registered in the Middle East and North Africa. Also, one in every ten East and Southeast Asians consumes too

Table 8.1 Undernourishment in the developing world

Region	1969–71 (1)	1979–81 (2)	1990–92 (3)	2003–05 (4)
East and Southeast Asia	475 million (41 percent)	378 million (27 percent)	279 million (17 percent)	219 million (11 percent)
South Asia	238 (33)	303 (34)	283 (26)	314 (22)
Latin America and the Caribbean	53 (19)	48 (14)	53 (10)	45 (9)
Middle East and North Africa	48 (27)	27 (12)	19 (4)	33 (4)
Sub-Saharan Africa	103 (38)	148 (41)	169 (31)	212 (29)
All Developing Countries	917 (35)	904 (28)	823 (20)	832 (17)

Sources: FAO (2009b) for 1990–92 and 2003–05 data; FAO (1997) for all other data.

few calories these days, which is a much smaller share of the population than it used to be largely because of great strides in China. Forty years ago, no other part of the world experienced more hunger. Today, the incidence of undernourishment in East and Southeast Asia is comparable to the incidence in Latin America and the Caribbean.

The two parts of the globe of greatest concern are South Asia and Sub-Saharan Africa. After growing faster than the total population during the 1970s, the food-insecure cohort in India and neighboring countries seems to be stabilizing. Numbering over 300 million today, this cohort is larger than what one finds in any other part of the world, although the incidence of food insecurity is lower than it used to be.

There is little positive news south of the Sahara, where the number of hungry people has more than doubled since the 1960s. If present trends continue, Sub-Saharan Africa soon will have not just the highest incidence of undernourishment, as has been the case for decades, but also more food-insecure people than any other part of the world, including South Asia.

Another perspective on food security is provided by the data in Table 8.2. An adult male with a moderate physical workload ought to consume about 2,800 calories/day, a woman engaged in moderate physical activity should take in 2,200 calories/day, and children (who are numerous in poor countries) need even fewer calories. Relative to these requirements, food supplies, including imports, are adequate as a rule. About 3,000 calories/day are available for every man, woman, and child

Table 8.2 Daily per capita calorie supplies during the 1990s

Region	1990–92 (1)	1997–99 (2)	Percentage Increase (3)
East and Southeast Asia	2,647	2,899	9.4
South Asia	2,330	2,400	3.0
Latin America and the Caribbean	2,710	2,830	4.4
Middle East and North Africa	3,010	3,010	0.0
Sub-Saharan Africa	2,120	2,190	3.3
All Developing Countries	2,540	2,680	5.5

Source: FAO (2002b).

in the Middle East and North Africa. During the 1990s, per capita daily supplies went from 2,710 to 2,830 calories in Latin America and the Caribbean and from 2,647 to 2,899 in East and Southeast Asia. Neither is there a shortfall in South Asia, where daily availability had risen to 2,400 calories per capita before the turn of the century.

Per capita supplies have gone up as well in Sub-Saharan Africa. However, much of this gain is accounted for by large relative increases in Angola, Mozambique, and Sierra Leone, where there has been a recovery from deep lows registered during civil wars. Food production per capita actually went down during the 1990s in 21 of the region's 38 nations. In several of these countries, local food availability is less than nutritional requirements. Where this shortfall occurs, warding off hunger requires food aid and commercial imports.

Just as food insecurity is more severe in some regions than in others, the incidence of undernourishment varies within a specific population or community. As a rule, women are more vulnerable than men. This is true even if females do most of the farming, as is the case in many settings. They are especially at risk during pregnancy or while nursing their offspring, when they need to eat more to maintain energy levels and avoid anemia.

Malnutrition is also severe among children, particularly if they were severely underweight at birth because their mothers were underfed during pregnancy. In addition, general poverty and the inequitable allocation of food within households cause many African, Asian, and Latin American youngsters to be malnourished. According to Smith and Haddad (2000), 167 million children under five years of age in developing countries—fully one-third of the age group—were undernourished in 1995. Also, there were 8.2 million more such children in 2000 than there had been ten years earlier (Runge et al., 2003, p. 19).

Finally, food insecurity tends to be more acute precisely where crops and livestock are raised, rather than in teeming cities. As Barraclough (1991, p. 42) and many others report, hunger is severe among landless peasants, smallholders, and hired agricultural workers. Of course, the problem for economically marginal people in the countryside, like the problem for the smaller share of the urban population that is undernourished, is incomes that are so miserably low that an adequate diet is unaffordable.

8.2 Achieving Food Security

From time to time, food security has been characterized as something to which each and every human being is entitled. Broad declarations that adequate nutrition is a basic human right (Box 8.2) are meant well. One wonders, though, what exactly is gained by declarations that, because of limited means or some other reason, no one is prepared to enforce. More important is what governments that make commitments actually accomplish in terms of food aid and alleviation of the poverty that puts adequate nourishment out of reach for many people.

Box 8.2 Food security: A human right

The entitlement to adequate nutrition was included in the International Declaration of Human Rights, which dates back to 1948, and reiterated in the International Covenant on Economic, Social, and Cultural Rights, which 145 governments signed in 1966. Adequate nourishment was also enshrined as a fundamental human right in the Rome Declaration on Food Security, which 182 countries endorsed during the World Food Summit of 1996. At the same meeting, world leaders pledged to reduce the number of undernourished people around the world to less than 400 million by 2015—a goal that currently looks out of reach.

Similarly difficult to achieve are the **Millennium Development Goals (MDGs)**, which 191 member states of the United Nations endorsed in September 2000. Along with seven other overarching ends, including the achievement of universal primary education and improvements in child and maternal health, the first MDG is to reduce the number of people living in extreme poverty and suffering from hunger to 50 percent of 1990 levels by 2020 (UNDP, 2009).

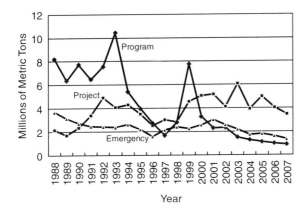

Figure 8.1 Program, project, and emergency food aid from 1988 to 2007
Source: World Food Program (2009).

Food Aid

When starving people flee to refugee camps, usually because of civil war or natural disaster, donated food arrives as well. This humanitarian response is a point of pride for those making the assistance possible. Americans, for instance, like to think of themselves as magnanimously caring for wretched cases on the far side of the world.

Never sizable relative to international trade in farm goods, food aid has declined in recent years (Figure 8.1). Cereal donations, which comprise most of the food given away at little or no cost, exceeded 12 million tons per annum through the middle 1990s, but subsequently have trailed off as yearly grain exports have climbed above 230 million tons.

Aside from being on a downward path, food aid fluctuates, which creates risks for recipient nations. Practically all donations originate in wealthy nations such as the United States,[1] with volumes in any given year depending on the gap between domestic production and consumption. A country with an elevated incidence of food security might receive above-normal donations when these are not needed, because of bumper crops elsewhere. Another possibility is for assistance to be curtailed when it is required most, perhaps because of inferior harvests in donor nations. Either way, the actual volume of donated food depends little on conditions in food-insecure places.

---

[1] Departing from standard practice, George W. Bush proposed in 2008 that up to one-fourth of U.S. food aid be purchased in or near recipient countries, although Congress rejected the idea (Eggen, 2008).

Something else to understand is that food aid has various purposes. So-called **program** donations are sold in local markets by recipient countries' governments, which then use the proceeds to underwrite agricultural research and extension and other activities. **Project** aid, which for a number of years has accounted for more than half of all donations, is used in food-for-work initiatives, school-feeding, and other development initiatives. Much of this category can be regarded as humanitarian assistance. So can the third category, which is **emergency** aid.

While the case for humanitarian assistance is unassailable, serious doubts have been expressed about the first category of food aid, which exceeded the other two categories through the early 1990s (Figure 8.1). To be specific, opponents have criticized program assistance for distorting international trade, discouraging domestic production in recipient countries, or both.

So that commodity donations do not crowd out trade, exporting nations, which include the United States and other major sources of food aid, agree among themselves that contributions will not happen if these impinge on imports by recipient countries. However, this leaves open the possibility that donated food will depress prices in recipient countries, thereby discouraging local agricultural development. Schultz (1960) highlighted this impact 50 years ago in a classic critique. The future Nobel laureate chided the United States and other donor nations for applying a humanitarian gloss to the international disposal of surpluses created by support prices and other policy-induced distortions. Food aid, he argued, was not the best way to alleviate hunger and underdevelopment.

Economic Growth and Lower Food Prices

Although it will continue to be offered, food aid cannot reach more than a few million people. If hundreds of millions of chronically underfed human beings who do not benefit from humanitarian assistance are to be adequately nourished, their earnings will have to rise, food will have to become cheaper (not because of governmental meddling with market forces, but rather because of improvements in agricultural productivity), or a combination of the two.

Barring substantial redistribution of income, of the sort that is seldom attempted and is even less frequently successful, lifting food-insecure people out of extreme poverty requires economic expansion. To examine the consequences of growth, Senauer and Sur (2001, cited in Runge et al.,

2003, pp. 28–29 and 209–213) have developed a model that combines statistical estimation of the linkage between per capita earnings and daily calorie consumption with projections of gross domestic product (GDP) and population in various parts of the world. Also incorporated in the model are interregional differences in minimal energy requirements, which have to do with variations in body stature. So are increases in these requirements that happen over time as demographic transition proceeds and, consequently, children become an ever smaller part of the population. A particular concern of the study is the portion of each region's population with earnings below the levels at which minimal energy requirements are barely met.[2]

Some of Senauer and Sur's (2001) results are presented in Table 8.3. If the GDP growth that the World Bank forecasts and the increases in population that the United Nations projects both materialize, the number of people whose diets do not meet minimal energy requirements will decline by nearly one-fourth between 2000 and 2025 (column 1). There will be more progress in Asia, where GDP per capita is increasing rapidly, as well as Eastern Europe and the Former Soviet Union, where the food-insecure cohort is relatively small. In contrast, the number of food-insecure people in Latin America and the Caribbean will only go down by 8 percent and there will be no absolute change in the Middle East and North Africa, where growth in living standards is slated to be modest at best. Even worse, there will be 52 percent more food-insecure people south of the Sahara, where economic expansion is slow and human numbers are going up rapidly.

The consequences of slower growth, brought about because GDP expansion is only half of what the World Bank projects, would be severe. Under this scenario, China and the Former Soviet Union and its erstwhile satellites would achieve modest progress toward food security, while there would be no change in the Indian subcontinent. Everywhere else, the ranks of the food-insecure would swell (Table 8.3, column 2).

Impoverished and food-insecure households spend half their earnings or more on edible goods, so lower food prices benefit them enormously. While Senauer and Sur (2001) refrain from forecasting prices, they

---

[2] Minimal requirements, which only allow for low levels of physical activity and which have been estimated by Senauer and Sur (2001), currently range from 1,790 calories/day in South Asia, where short, thin people predominate, to 1,900 calories/day in Eastern Europe and the Former Soviet Union, where people tend to be taller and heavier. Due to the aging of the population, minimal daily energy intakes will rise by 2050, to 1,900 calories in Sub-Saharan Africa, 1,940 calories in the Middle East and North Africa as well as Asia, and 2,000 calories everywhere else.

Table 8.3 Percentage changes between 2000 and 2025 in the number of people not consuming enough calories

Place	Current Trends in GDP per Capita Continue (1)	50 Percent Less GDP Growth (2)	Current Trends in GDP per Capita Continue, but with Lower Food Prices (3)
China	−71	−31	−81
Other East and Southeast Asia	−42	+16	−63
South Asia	−56	0	−71
Latin America and the Caribbean	−8	+52	−40
Middle East and North Africa	0	+72	−36
Eastern Europe and Former Soviet Union	−47	−27	−66
Sub-Saharan Africa	+52	+136	+8
Total	−23	+36	−47

Source: Senauer and Sur (2001).

investigate the consequences of a 10 percent increase in per capita calorie intake, as would result from price declines. Rather than falling by a little more than half, as would happen if recent trends in per capita GDP continue with no change in food prices, the number of food-insecure people in South Asia would be reduced by more than 70 percent. In Sub-Saharan Africa, the combination of rising living standards and cheaper food would still result in an 8 percent increase in the food-insecure segment of the population: much less than the 52 percent growth observed if current trends hold (Table 8.3, column 3).

8.3 The Food Security Synthesis and Economic Development

Focusing on the challenge of improving the economic access of impoverished people to available supplies of food, Tweeten (1999; 2007) has distilled a **food security synthesis**. Made up of a seven-step logical framework, this synthesis addresses fundamental causes of undernourishment as well as cures for the problem.

1. *Transitory and chronic food insecurity is caused mainly by poverty.* As already indicated, nearly all the underfed Africans, Asians, and Latin Americans identified in Table 8.1 have daily incomes of $1.25 or less, and are therefore extremely poor. These people lack the buying power needed to overcome the frictions of time (e.g., unpredictable and unstable harvests from year to year) and space (e.g., local food shortages) that cause them to be food-insecure.

2. *Poverty is best alleviated through economic development that is broad-based and sustainable.* Altruism is commendable and plays a critical role in feeding members of a close-knit group. But in communities with little or nothing to share, there is little scope for altruism and the issue of redistribution is moot. Certainly, the linkage between economic expansion and progress toward greater food security is recognized by the FAO.

> The need is for policy measures that address all aspects of food insecurity with a view to providing safety nets for the vulnerable and to creating the conditions that can lead to an eradication of endemic hunger. *This has to mean economic growth* [emphasis added]. . . . Improving the equitableness of the income distribution can only achieve so much (in countries with low and falling income), and, as seen time and again, will be strongly resisted by the potential losers. So growth is necessary, and against a background of economic growth, experience shows that it is easier, although never easy, to implement measures that increase equity, particularly if the growth is broadly based to include the agricultural sector (FAO, 1997, p. 3).

3. *The best way to pursue broad-based economic development is to follow the standard model.* Described below, this approach can be applied anywhere and provides a workable prescription for economic progress, thereby ensuring enough buying power for food self-reliance (Chapter 6) and food security. The standard model is not one-size-fits-all, in the sense that each and every one of its elements must be implemented thoroughly. However, some of its key features are essential for a sound economy and have been embraced by virtually every country that has experienced broad-based development.

4. *Political failure* explains why some countries do not adopt enough components of the standard model to end poverty and food insecurity. Sometimes, economic illiteracy helps to explain why the standard model is not followed. However, political (or government) failure is usually the

main problem. Individuals and groups with power and authority frequently gain from policies that egregiously compromise the public interest. As these individuals and groups resist application of the standard model, direct costs (e.g., the inefficiencies created by state-owned enterprises and policy-induced distortions as well as the transaction costs associated with bureaucratic bloat) are created. So are indirect costs, which relate to the effort devoted to winning favors from the public sector.

5. *Political failure is inseparable from institutional failure.* The ability of political leaders and bureaucrats to pursue inefficient policies, which harm overall well-being, has much to do with institutions of government that are weak, mismanaged, and corrupt, and consequently incapable of enforcing the rule of law and avoiding domination by special interests. Since markets require sound institutions, institutional reform is a key element of the standard model.

6. *Poorly structured and inadequate institutions often trace to cultural factors.* Only a handful of economists have examined ties between culture and economic performance systematically (Landes, 1998; Sen, 2000), yet these ties are of basic importance. For example, institutional failure is hard to avoid if few citizens are involved in political processes and if people resign themselves to government that is unrepresentative, corrupt, and incompetent. No less than anyone else, leaders are products of a nation's culture. If they view their position as a chance for personal aggrandizement, rather than an opportunity to serve the public, the rule of law is bound to be weak. Also, tribal animosities and other cultural characteristics militate against sociocultural change, of the sort required for economic progress.

7. *The core challenge is sociocultural change.* Achieving food security ultimately rests on the adoption of sociocultural norms that are conducive to economic progress. Two examples of such norms are the Protestant ethic (Tawney, 1966; Weber, 1930) and Confucian values (Johnston, 1966), both of which put a premium on honesty in interpersonal dealings, hard work, and enterprise. Likewise, much of Japan's economic success has been credited to "social and political attitudes that made economic development a priority bordering on an obsession" (Sowell, 1998, p. 345).

Sociocultural norms having an impact on development have been a recurring concern of economists, sociologists, and others. Safe to say, this subject will continue to command the attention of the finest minds in the social sciences for a long time to come.

The Standard Model

As observed in the preceding chapter, economic progress requires that human, natural, and other resources be complemented by a culture of trust sustained by institutions that enforce the rule of law. Once the cultural and institutional underpinnings of broad-based development are satisfied, economically sound prescriptions for public policy tend to be followed. To describe these prescriptions, Williamson (1990) coined what is now a widely used term, **Washington Consensus**. As the same individual subsequently admitted, the term has the unfortunate connotation of a program imposed by the **International Monetary Fund (IMF)** and World Bank, both of which are headquartered in the U.S. capital (Williamson, 2000).

The standard model, which Tweeten (1999) proposes to describe a broader set of prerequisites for economic progress, does not share this disadvantage. Numerous studies provide statistical evidence of the model's success (Tweeten, 2007). Anecdotal evidence is also compelling. For example, the East Asian "tigers"—Japan, Singapore, South Korea, Taiwan, and China (including Hong Kong, which is a long-time member of this group)—have enjoyed impressive economic gains by adopting critical elements of the approach.

An undergirding principle of the standard model is that things that the market does best should be left to markets—in particular, the competitive allocation of most goods and services—and that government should focus on doing what the marketplace neglects or is incapable of performing. One fundamental task for the state is to provide public goods (Chapter 3), which the market economy does not supply if left to its own devices. Something else the marketplace cannot do for itself is to establish the institutional framework that is needed for the market economy to exist. If the government performs the latter function and supplies public goods, then competitive markets can be counted on to determine what, where, and how to produce the vast majority of goods and services. Not merely unnecessary, governmental manipulation of prices or output levels in markets is usually undesirable (Tweeten, 1999).

A critical part of the standard model is sound public administration, which reinforces the institutional framework for efficient markets.

- *Security, stability, and order*. Courts must interpret laws and administer justice evenhandedly. If the rule of law is secure, contracts can be made

and acted on, which allows commercial exchange and investment to take place.
- *Property rights.* Ownership that is exclusive as well as tradable is another prerequisite for the market economy. Private investment, for example, is all but impossible if property rights are weak, in part because the collateral needed for loans does not exist in a legal sense. Also, property rights encourage investment because the value of improvements accrues to owners (or their heirs) when assets are sold. These rights need to be enforced consistently by courts, of course. A system for delineating and registering properties is required as well (Chapter 5).
- *Competition.* The full benefits of market exchange are realized when there are many buyers and many sellers. At the least, government must avoid coddling monopolies, by protecting them from foreign competition. More than that, competition has to be promoted, through the application of antitrust laws. Where a natural monopoly exists, because the presence of two or more firms would cause economies of scale to be lost, regulation is appropriate.
- *Honesty and competence in government.* Some economists used to be complacent about corruption, regarding the impacts of bribery as little different from the consequences of taxation. However, the toll on economic progress associated with peculation in the public sector is now widely appreciated (Sachs, 1997). To contain the problem, transparency and accountability in government dealings are required. So are merit hiring, competitive pay, and civil-service protection, which are also needed for a competent and politically neutral bureaucratic workforce. A free press, which is a core element of democracy, also serves as a brake on governmental malfeasance.

Along with providing a market economy's institutional underpinnings, the state needs to avoid macroeconomic policies that distort incentives and hinder private enterprise.

- *Fiscal responsibility.* Chronic deficits, which arise because government does not live within its means, can be sustained for a while by borrowing: rising public indebtedness, in other words. But a point is often reached at which the state prints money to settle its bills, which ignites inflation. Inflation erodes the value of private savings, thereby reducing investment. To be sure, borrowing that permits governmental spending in excess of tax revenues might be justified if this is required for investments in public goods that accelerate economic expansion. A case can also be made for countercyclical fiscal policy: that is, deficits during recessions and surpluses at other times. But chronic deficits, which are the hallmark of fiscal irresponsibility, are by all means to be avoided.

- *Monetary restraint.* Sound fiscal policy eases the task of sound monetary policy, as needed for price stability. The total absence of inflation is not really ideal, since deflation (i.e., a prolonged decline in the general price level) is almost always accompanied by a slowdown in economic activity.[3] Instead, a general price-rise in the low single digits has merit. This is normally accomplished by increasing the money supply more than growth in GDP. Monetary authorities who are largely independent of immediate political concerns are most likely to follow this guideline, although their ability to do so is compromised if government is fiscally irresponsible.
- *Appropriate taxation.* One of the greatest challenges facing governments in the developing world is the collection of taxes required for essential public services. The economic disincentives of taxation are reduced if marginal rates are low and the tax base encompasses a large part of the economy. But it is also true that progressive taxation minimizes the burden on the poor. Successful governments tax "bads" (e.g., tobacco, alcohol, and emissions) instead of "goods," like investment and exports. Also, user fees charged for electricity, irrigation water, and other things provided by the public sector cover the costs of these goods and services. Taxes on sales, value-added, and property distort the economy less than taxes on exports and corporate profits.

Just as sound macroeconomic policy creates the price stability that causes markets to flourish, a liberal trade policy promotes efficiency by encouraging everyone to specialize according to comparative advantage.

- *Openness to trade and investment.* Although infant industries occasionally have acquired comparative advantage after being protected from foreign competition, this strategy has serious pitfalls (Chapter 7). Rather than taking advantage of barriers to imports to bring costs down below international prices, firms often grow indolently dependent on protection,

---

[3] As prices fall, consumers postpone purchases whenever they can, in the hope of getting a better deal in the future. Also, business profits are squeezed, insofar as labor contracts keep wages from falling along with prices. Firms are likely to respond to this by laying off workers. Slumping sales and unemployment occurred on a large scale during the Great Depression, when deflation persisted for a number of years.

and therefore lobby for its continuation. As they do so, protection originally proposed as temporary is apt to become permanent, thereby locking a country into inefficient industrialization and arrested growth.
- *A realistic foreign exchange rate.* Efficient transmission of international prices to domestic markets hinges on a proper rate of exchange between the national currency and the currencies of trading partners. Currency overvaluation, which creates an import bias, has occurred frequently in the developing world (Chapter 6). To deal with this problem, Fischer (2001) recommends two solutions. One is floating (or flexible) exchange rates, with currency values determined in the marketplace. The other, which particularly suits a small country that trades mainly with a larger nation, is adoption by the latter's currency. Illustrative of the latter approach is dollarization, which Panama underwent in 1904 and which Ecuador and El Salvador accomplished more recently.

Rather than trying to develop by manipulating exchange rates and other market forces, developing countries need to exploit the opportunities proliferating in a globalizing world by raising productivity. Granted that productivity gains in the private sector are impressive, the government's role of providing public goods is also important.

- *Infrastructure.* Commercial activity is facilitated by all-weather roads, communications networks, and so forth. Likewise, seaports, airports, and the like are needed for international trade. While private firms can administer many such facilities efficiently, much of this infrastructure is best classified as public goods, which means that governmental financing is unavoidable.
- *Agricultural research and development.* Support for the scientific and technological base for crop and livestock production creates high returns (Chapter 3). However, this investment is deficient in the developing world. Whereas spending on agricultural research and development amounts to 2 to 3 percent of the value of agricultural output in affluent nations, the ratio is closer to 1/2 percent in Africa (Pinstrup-Andersen, 2002). If this imbalance is not redressed, African farmers, who comprise the majority of the labor force and produce half the GDP of many countries south of the Sahara, will not benefit from genetic engineering and other biotechnologies, which show promise of raising yields and reducing pressure on threatened ecosystems.

- *Education.* Universal access to schools is also essential. Denying access to women or ethnic minorities, for example, is enormously costly, in terms of production and incomes. Since the social payoff of elementary education is sizable, universal primary schooling, financed by government, is a priority for food security and development.
- *Public sanitation and health.* Along with education, public investment in human capital involves attention to water and wastes. Otherwise, parasites and bacteria impair food utilization and sap vitality, even interfering with the physical and mental development of the very young. Human capital is also built up through networks of health clinics, at which immunization, vitamin supplements, knowledge about the prevention of acquired immune deficiency syndrome (AIDS) and other diseases, family planning services, and guidance for pre- and postnatal care are provided.
- *Environmental quality.* Yet another public good is the quality of air, water, and other natural resources. Correcting market failure requires intervention by the state (Chapter 5). However, this intervention is inappropriate if its administrative costs exceed the economic harm of the market failure that is being corrected. Also, research and development allows for agricultural commodities and other goods and services to be produced with less environmental damage. For example, achieving higher yields on land well suited to agriculture reduces the incentives to cultivate fragile lands.
- *Food and income safety nets.* In just about any part of the world, emergency food aid is made available in the wake of environmental disasters and civil disturbances. Also, many governments offer food and other sustenance to those constituents who are either unable to provide for themselves or who lack support from family and other private sources.

The standard model and its various elements do not fit easily into any single ideological category. It certainly is not state-centered, at least as a Marxist would define the term. If public administration is generally sound and if appropriate policies are in place to deal with problems like environmental market failure, competitive markets can and ought to be treated as if they are on autopilot. But neither is the standard model libertarian, in the sense that government is expected to wither away as its functions are taken over by private entrepreneurs. To the contrary, the state is uniquely suited to administering the marketplace and supplying public goods.

Implementation Issues

While the preceding list of prescriptions defines proper roles for government as well as markets, it does not really amount to a comprehensive and precise blueprint for the sort of economic progress needed for everyone to be food-secure. Experience shows that the exact balance struck between the state and the marketplace in one country can differ from the balance chosen somewhere else, even though development is happening in both places. Also, converting the list of prescriptions into a blueprint would require that questions of sequencing be addressed.

In the world's poorest nations, putting a high priority on primary education and roads makes sense. The same is true of agricultural investment. Ninety percent of the food consumed in developing nations is produced domestically. Tweeten and McClelland (1997, pp. 1–31) reviewed a large number of studies finding high returns on resources of science and technology to improve agricultural productivity. But in spite of these realities, the neglect of agriculture, which Schultz (1964) criticized more than four decades ago, continues.

Sequencing is a major concern for development agencies. In its **structural adjustment** initiatives, the IMF offers loans to countries in financial distress in exchange for the trimming of fiscal deficits, adoption of realistic exchange rates, and related changes in public policy. Likewise, the World Bank, which specializes in the financing of development projects with a longer-term payoff, sometimes attempts to tie its lending to policy reform. But, enforcing this sort of **conditionality** can be difficult (Easterly, 2001, pp. 115–120).

It is important to keep in mind that financial considerations, of the sort addressed with the help of multilateral donors (Box 8.3), do not apply to many elements of the standard model. Trade liberalization and other

---

Box 8.3 Foreign aid's contributions to development

When spent in the right ways, foreign assistance can help nations break out of **poverty traps**. These traps exist when incomes are too low to pay for investments in infrastructure, technology transfer, and education, which are needed to raise living standards. Consider, for example, a country where a large segment of the population engages in subsistence farming and is unable to pay for agricultural research and development. External monies may be the only way for the country to deal with the poverty trap related to chronically low agricultural productivity.

But just as there are beneficial and important uses of foreign aid, there are many inappropriate ways that this assistance has been used in the past, including the underwriting of subsidies and other policies for distorting markets. This is one reason why the contributions that foreign aid has made to development often have been disappointing (Easterly, 2001, pp. 35–44; Moyo, 2009, pp. 3–68).

Although it has its place, foreign assistance rarely has been the primary driver of economic progress in Africa, Asia, and Latin America. In part, this is because the amounts available are modest. Currently, approximately $100 billion are provided by affluent nations and multilateral institutions, such as the World Bank. This is a fraction of the net inflow of **foreign direct investment (FDI)** to the developing world, which in 2006 amounted to $1.6 trillion (UNCTAD, 2008). The best way to stimulate FDI is to follow the standard model, as China has done in large measure but too many poor countries fail to do.

reforms in economic policy do not require public expenditures. Initiatives like strengthening the judiciary and related agencies, so that enforcement of contracts and property rights becomes more reliable, are not free. However, the expenses are modest in absolute terms and, of even greater importance, exceeded by the benefits of strengthening the marketplace's institutional framework.

Finally, the most effective contribution that wealthy countries can make to economic progress and food security in Africa, Asia, and Latin America has nothing to do with foreign aid. As emphasized in Chapter 6, trade restrictions in the former countries impinge significantly on exports from less prosperous parts of the world. Removing these restrictions would create a substantial impetus for adoption of the standard model and, hence, improvements in well-being around the globe.

8.4 The Standard Model and Communitarian Values

In light of the great debate over the relative merits of capitalism and socialism that dominated intellectual life for much of the twentieth century, the coalescence that has happened in recent decades about sound economic policy as well as the prosperity it creates is impressive. Call this coalescence the Washington Consensus, the standard model, or anything else, the range of opinions about how to pursue development is narrower than it used to be (Yergin and Stanislaw, 1998). Reflecting this coalescence is the accord that now exists concerning the problem of food insecurity. In particular, there is agreement that the policy reforms and other initiatives required for economic progress are also needed to alleviate hunger and vice versa. Regardless of which broad aim is chosen, the same standard model applies (Tweeten, 1999).

As reflected in the latter steps of Tweeten's (1999) food security synthesis, cultural norms ultimately determine whether or not the standard model is applied. Among these norms are enterprise, thrift, hard work, and honesty in personal dealings (Johnston, 1966; Tawney, 1966; Weber, 1930). However, there are other cultural norms, including what one might call **communitarian values**—a sense of community, or social capital. These values, which influence our thinking about food security, determine how the standard model is adopted. Commendably, a sense of community at the national level can minimize divisive tribalism and can instead focus attention on nationwide goals such as economic progress and the alleviation of hunger. At another extreme, however, communitarian impulses can block the standard model's application.

By no means is a sense of community economically irrational, either for individuals or for entire groups and populations. In many parts of the world, the natural environment is harsh and unpredictable and, consequently, human survival has been tenuous. In such settings, sharing is a well-entrenched tradition. If one family's harvest fails this year, it can count on getting something to eat from neighbors or distant relatives, provided they have enough to spare. This option, which has obvious value, is kept alive by sharing what one has when other folks who are down on their luck come calling. Barring a catastrophe, which leaves everyone hungry at the same time, individuals survive from one year to the next and the community remains intact.

Regardless of environmental and other conditions that created habits of sharing in the first place, communitarian impulses can have a down side in terms of economic growth—especially when these impulses apply to financial resources, not just food. For example, someone with savings might be viewed mainly as a source of largesse by traditionally minded members of his or her family and community. The thrifty person then faces a choice. To continue amassing funds, he or she must strain, perhaps sever, his or her closest human ties. The alternative is to scale back on capital accumulation. Inasmuch as many people choose the latter alternative, material standards of living suffer.

A sense of community is alive and well even in affluent places with a long history of capitalism. It is even reasonable to suppose that prosperity, which capitalist development has made possible, makes the application of communitarian values easier, although acting on communitarian impulses often has a negative impact on economic growth. Sweden is a good example. Like many of its neighbors, Scandinavia's leading nation has a generous and comprehensive social safety net, which helps to explain its low Gini coefficient. Also, consumption is taxed heavily to generate the resources that pay for human capital formation, which raises living standards and reduces inequality simultaneously.

While potential trade-offs between growth and equity must be kept in mind, it is also true that the distribution of productive assets can be so skewed that economic expansion is depressed (Deininger and Squire, 1997). In particular, redistribution of farm real estate sometimes "becomes the only option for improving rural livelihoods rapidly and substantially," even though governments usually are reluctant to undertake land redistribution because it is socially divisive and often provokes violence (Barraclough, 1991, p. 130). A better way to reduce asset inequality is to build up other sorts of productive wealth for the poor, especially their human capital. This approach makes sense in Sub-Saharan Africa, for example, which is impoverished and agriculturally dependent. Even in that part of the world, human capital comprises three-fifths of all assets, while agricultural land and other environmental wealth make up just 10 to 20 percent (Dixon and Hamilton, 1996).

As indicated in the preceding section, human capital formation and other productivity-enhancing investments are key components of the standard model for economic progress and food security, not just a good way to reduce asset inequality. So is appropriate taxation, which is needed to pay for these investments but which does not impinge on economic expansion.

Communitarian values and desires to reduce inequality are strong, so advocates of the standard model are likely to end up frustrated if they simply ignore these values and desires. Runge et al. (2003, p. 6) are not off the mark when they characterize the alleviation of poverty and hunger as a public good, in the sense that failure to do so constitutes an indictment of the prevailing economic order in the eyes of many and therefore dims the prospects for policies that facilitate economic expansion. The challenge, then, is one of balance, with communitarian impulses being harnessed to maintain safety nets that are both adequate and do not impinge on the broad-based development needed to raise the living standards of the food-insecure.

Key Words and Terms

access to food	International Monetary Fund (IMF)
availability of food	malnutrition
communitarian value	micronutrient deficiency
conditionality	Millennium Development Goal (MDG)
emergency food aid	poverty trap
extreme poverty	program food aid
food aid	project food aid
food security	standard model
food security synthesis	structural adjustment
foreign direct investment (FDI)	utilization of food
incidence of food insecurity	Washington Consensus

Study Questions

1. The FAO's definition of food security emphasizes economic access to an adequate diet. Explain this emphasis, taking into account other dimensions of human nourishment.
2. In what parts of the developing world has food security improved and in what regions is food insecurity still a severe problem?
3. Distinguish between program and other categories of food aid and assess the likely impacts of each on imports and domestic production in recipient countries.
4. What combination of economic expansion and lower food prices would be needed to prevent the number of food-insecure people from rising appreciably in Sub-Saharan Africa?
5. Is there any substantial conflict between trying to achieve broad-based economic development and pursuing wider food security?
6. Compare and contrast the elements of the Washington Consensus and the elements of the standard model.
7. What are the potential inconsistencies between the standard model and communitarian values?

9

Interregional Differences and Similarities

Up to this point, the book's content has been thematic, relating to the overall demand and supply of food, agriculture's environmental impacts, the sector's role in globalization and economic development, and food security. Differences among various parts of the world have been noted. Malnourishment, for example, has been reduced substantially in the emerging economies of East and Southeast Asia. In contrast, the incidence of food insecurity remains elevated in South Asia and Sub-Saharan Africa. Also, forests are being lost at a faster pace in the latter region than anywhere else, whether rich, poor, or in between.

From this chapter on, the book's organization is geographic. Our purpose here is to highlight the various challenges facing the food economy, such as conserving the natural resources on which agriculture depends while consistently providing adequate nourishment for everyone.

This chapter sets the stage for the next six, which are about the world's prosperous nations, Asia, Latin America and the Caribbean, the Middle East and North Africa, Eastern Europe and the Former Soviet Union, and Sub-Saharan Africa, respectively. Comparing and contrasting broad regional characteristics and trends, we begin this chapter and the six that follow with an analysis of recent economic performance. Next comes a survey of recent demographic developments: increased life spans, diminished human fertility, and reduced population growth, to be specific. We then turn to agricultural development, describing how mixes of inputs have changed as farmers in different parts of the world have responded to growing demands for food.

9.1 Economic Growth and Income Distribution

Clearly apparent in Table 9.1 is the economic expansion that various parts of the world experienced during the last third of the twentieth century. Not limited by any means to affluent settings, gross domestic

Table 9.1 Economic trends from 1965 to 2007, current per-capita incomes, and recent measures of income inequality

Region	GDP growth (average annual percent) 1965–1999 (1)	Population growth (average annual percent) 1965–1999 (2)	Growth in GDP per capita (average annual percent) 1965–1999 (3)	Growth in GDP per capita (average annual percent) 2000–2007 (4)
1. High-Income Nations (OECD)	3.3	0.7	2.6	1.7
2. East and Southeast Asia	7.4	1.8	5.5	8.0
3. South Asia	4.6	2.2	2.4	5.5
4. Latin America and Caribbean	3.9	2.1	1.8	2.1
5. Middle East and North Africa**	4.6	2.7	1.9	3.2
6. Eastern Europe and Former USSR	—	—	—	6.1
7. Sub-Saharan Africa	2.9	2.9	0.0	2.8

Region	Gross fixed capital formation (average annual percent) 1965–1999 (5)	Export growth (average annual percent) 1965–1999 (6)	Income per capita (PPP $) 2007 (7)	Income inequality (Gini coefficient) late 1990s early 2000s (8)
1. High-Income Nations (OECD)	3.7	6.3	36341	—
2. East and Southeast Asia	8.9	8.2*	4969	38
3. South Asia	5.2	7.6	2532	34
4. Latin America and Caribbean	3.7	5.7	9678	50
5. Middle East and North Africa**	4.9	3.7	7402	36
6. Eastern Europe and Former USSR	—	—	11262	33
7. Sub-Saharan Africa	—	—	1870	47

Source: World Bank (2009b). *(1978–1999 for column 6); **(1966–1999 for columns 1, 2, 3, 5, and 6).

product (GDP) increases (column 1) have had much to do with increased specialization and trade, as reflected in export trends (column 6). Furthermore, economic expansion has gone hand in hand with the buildup of productive capacity, including fixed capital (column 5).

Regional Trends in GDP per Capita

Nowhere have investment, increases in trade, and economic growth coincided more than in East and Southeast Asia. Overall output has risen much faster than human numbers (Table 9.1, column 2) in China, which is the largest of the four **BRIC** economies (Box 9.1) as well as the dominant country in the region, and most of its neighbors. From 1965 through 1999, annual increases in GDP per capita averaged 5.5 percent (column 3), which caused living standards to double every 13 years.[1] Economic expansion in East and Southeast Asia has flagged at times, such as during the financial crisis of 1997. However, the upward trajectory continues, as indicated by annual growth of 8.0 percent in GDP per capita from 2000 through 2007 (column 4).

Productive capacity has been accumulating and exports have been increasing in India (another BRIC economy) and neighboring countries as well. With economic expansion accelerating in South Asia after 1990, living standards have improved noticeably, including since the turn of the twenty-first century.

The rest of the developing world fared less well for many years. In Latin America and the Caribbean as well as the Middle East and North Africa, there was less capital formation (Table 9.1, column 5) and export growth (column 6) than in Asia, so improvements in GDP per capita lagged (column 3). Data on investment and growth are spotty south of the Sahara, although long-term trends were undoubtedly weak. Population growth (column 2) was generally in line with economic expansion (column 1) during the last third of the twentieth century, so GDP per capita in 1999 was little better than what it had been in the middle 1960s. Long-term trends for the same period cannot be calculated for the Former Soviet Union and Eastern Europe. The region's difficulties during the 1990s are amply documented, particularly in Russia, yet there is no doubt that economic deterioration had set in earlier. After all, any improvement

---

[1] As noted at the beginning of Chapter 2, doubling time of a population, an economy, or anything else that happens to grow at a constant exponential rate is found by applying the Rule of 72—that is, dividing 72 by the annual percentage growth rate.

> **Box 9.1 The emerging BRICs**
>
> As highlighted in Table 9.1, living standards have grown faster this century outside the prosperous **Global North** than among members of the OECD. Epitomizing this new trend are four economies now taking a prominent place on the global stage: Brazil, Russia, India, and China.
>
> Each of the so-called BRICs is large and has a sizable population—although Russia, with fewer than 150 million people and a national territory of continental dimensions, is underpopulated. Brazil differs from the other three in that its shares of global output and military expenditures have declined during the past 15 years, as the portions of the other three nations have increased. However, each country's share of global exports has risen steadily, thereby signaling a willingness to engage (not shy away from) international markets (World Bank, 2009b).
>
> Brazil, China, India, and Russia are not a cohesive block, even though leaders of the four nations have gathered on three occasions—most recently in Yekaterinburg, Russia, in June 2008. Commercial ties sometimes are more than comfortable: when China purchases minerals and agricultural commodities from Brazil, for instance. But rivalries also exist, in the competition over oil and other resources in Africa, for example. Moreover, the political systems of the four BRICs are by no means harmonized. While Brazil and India are democratic, China and Russia are not.
>
> Each BRIC dominates its respective region. In addition, the course of economic events this century will be determined in large part in the four national capitals: Brasilia, Moscow, New Delhi, and, above all, Beijing.

before the fall of communism would have been trumpeted by Soviet authorities and their allies and apologists.

At times, economic progress has been held back in a number of countries, and even entire regions, due to chronic currency overvaluation, brought about whenever exports of oil or some other commodity have provided the means to keep exchange rates stable in spite of domestic inflation (Chapter 6). Often called **Dutch Disease**, this condition usually has deleterious impacts, such as elevated unemployment, outside the industry or sector that is the source of commodity exports (Box 9.2).

Development strategies predicated on import-substituting industrialization (ISI) also has had debilitating long-term consequences in Latin

> **Box 9.2 Dutch Disease**
>
> Dutch Disease (Perkins, Radelet, and Lindauer, 2006, pp. 675–682) arises in an **enclave industry**, which by definition employs limited labor and other local factors of production, that is the source of substantial exports. As the foreign exchange paid for these exports is repatriated and traded for the domestic currency, that currency gains value. This diminishes exports and increases competition from imports in other industries and economic sectors.
>
> The same industries and sectors also lose jobs, which can reduce overall employment if displaced laborers are unable to find work in the enclave industry that is the source of Dutch Disease. Examples of this outcome include Iran, Nigeria, and Venezuela, where agriculture and rural employment have suffered because of the currency appreciation resulting from petroleum development. Argentina has experienced a different strain of Dutch Disease, one arising in agriculture and adversely affecting other parts of the economy (Chapter 12).

America and the Caribbean and other developing regions (Chapter 7). So have elevated military expenditures in the Middle East and North Africa (Chapter 13). Eastern Europe and the Former Soviet Union suffered for decades under communism (Aslund, 2002, pp. 20–69) and then underwent a transition that was mishandled in many countries (Chapter 14). During the Cold War, proxy conflicts pitted communist forces against allies of the West across Sub-Saharan Africa, where development also has been stifled due to government policies that throttled agriculture (Chapter 15).

Before the turn of the twenty-first century, some of these impediments to development were being dismantled. Better macroeconomic management lowered inflation in Latin America and the Caribbean, for example, and proxy conflicts in Africa wound down with the fall of communism. Additionally, exporters of oil and other commodities benefited from higher prices after 2000, which have been driven largely by the rapid expansion of China, India, and other emerging economies in Asia. From 2000 through 2007, annual increases in GDP per capita averaged 3.2 percent in the Middle East and North Africa and 6.1 percent in Eastern Europe and the Former Soviet Union (Table 9.1, column 4). This was an improvement on previous trends. Growth also accelerated somewhat in Latin America and the Caribbean as well as south of the Sahara. In contrast, GDP per capita has grown more slowly since 2000 in affluent members of the **Organization for Economic Cooperation and Development (OECD)** than in the rest of the world.

Income Distribution Differences and Economic Convergence

Just as average incomes (Table 9.1, column 7) and trends in GDP per capita vary among different parts of the world, income inequality, as indicated by Gini coefficients (Chapter 7), is not the same everywhere (column 8). Inequality is modest in most wealthy countries (although the United States is a partial exception) as well as in South Asia, where poverty is widespread. Gini coefficients are also low in the Middle East and North Africa, although data on income distribution are unavailable for Iraq during Saddam Hussein's reign and for several oil-exporting monarchies, and in Eastern Europe and the Former Soviet Union, although the distribution of income has become much more uneven in Russia during the past couple of decades (Chapter 14).

Nowhere is income inequality more acute than in Latin America and the Caribbean, where class and ethnic cleavages dating back to the era of Iberian colonialism persist to the present day. Even more miserable, however, is Sub-Saharan Africa, where a regional Gini coefficient nearly as high as that of Latin America and the Caribbean coincides with average living standards below those of every other part of the Global South.

On numerous occasions, economists have tested an hypothesis, called **convergence**, about trends in GDP per capita. The hypothesis states that living standards in various countries come together—not because average earnings deteriorate in better-off places, but rather as a consequence of faster income growth in poorer settings (Perkins, Radelet, and Lindauer, 2006, pp. 93–97). Evidence in support of convergence is far from overpowering. However, the authors of one empirical study do not reject the hypothesis mainly because of rapid improvements in Asia, which is more populous than all other parts of the world combined and which was impoverished not so long ago (Bourguignon and Morrisson, 2002).

As mentioned above, average annual growth in GDP per capita in the OECD has fallen behind rates of increase for the rest of the world since 2000 (Table 9.1, column 7), which is consistent with economic convergence.

9.2 Population Dynamics

When obliged by the facts to admit that GDP per capita has gone up, critics of the economic perspective on human progress often fall back on the argument that positive trends in conventional economic indicators do not really reflect an improvement in the well-being of people. But when they make this contention, the critics are on unfirm ground. As pointed out by Fogel (2004a and 2004b) and Johnson (2000), many noneconomic indicators of well-being, such as average food consumption and life expectancy, have improved markedly in poor countries, often without comparable growth in monetary earnings (Chapter 2). Since noneconomic indicators are of fundamental importance, it is in a sense comforting that the convergence hypothesis applies at least as much to these variables as to standard economic measures, such as GDP per capita.

Increased Human Longevity

Convergence of a very encouraging sort is happening for life spans in most of the world. Thanks mainly to diminished mortality among infants and small children (Chapter 2), life expectancy at birth went up by seven years in East and Southeast Asia, eight years in Latin America and the Caribbean, and fully 11 years in South Asia and the Middle East and North Africa between 1980 and 2007 (Table 9.2, columns 1 and 2). Each of these increases exceeded the five-year gain in affluent nations. In relative terms, lifespan increases in the four developing regions have been particularly impressive.

Although newborns in Asia, Latin America and the Caribbean, and the Middle East and North Africa can now expect to live 65 to 73 years, scope for improvement remains in these places. In India and surrounding nations, for example, 44 percent of all children under five years of age are malnourished (Table 9.2, column 3). Also, living conditions are crowded and impoverished, as indicated by the high incidence of tuberculosis (column 4). As deficiencies of this sort are remedied, more infants and small children will survive and end up living longer than their ancestors did.

The trend toward longer life spans has been weaker in two parts of the world. In Eastern Europe and the Former Soviet Union, life expectancy at birth was 68 years in 1980 and had inched up to 70 years as of 2007. Within the region, greater improvements have occurred in Eastern Europe. On the other hand, there has been a steep decline, especially for men, in Russia and the Ukraine, due to widespread alcoholism, the increased incidence of communicable and noncommunicable illness, and other factors (Chapter 14).

The other part of the world where life expectancy at birth has not increased much is Sub-Saharan Africa. Infectious diseases that have long plagued humankind, such as tuberculosis (Table 9.2, column 4) and malaria, still take a heavy toll in the region. Likewise, the incidence of the human immunodeficiency virus (HIV), which is spread almost entirely by sexual contact in Africa and which often results in death due to

Table 9.2 Regional indicators of mortality, health, and poverty

Region	Life expectancy at birth (years)		Childhood malnutrition (percentage of 5-year-olds who are severely underweight) 2007	Incidence of tuberculosis (per 100,000) 2007
	1980 (1)	2007 (2)	(3)	(4)
1. High-Income Nations (OECD)	74	79	—	16
2. East and Southeast Asia	65	72	19	136
3. South Asia	54	65	44	174
4. Latin America and Caribbean	65	73	10	50
5. Middle East and North Africa	59	70	—	41
6. Eastern Europe and Former USSR	68	70	—	84
7. Sub-Saharan Africa	48	51	35	369

Region	Prevalence of HIV (percentage of adults) 2007 (5)	Incidence of extreme poverty (percentage of population living on $1.25/day or less)	
		1987 (6)	2007 (7)
1. High-Income Nations (OECD)	0.3	—	—
2. East and Southeast Asia	0.2	54	15
3. South Asia	0.3	54	40
4. Latin America and Caribbean	0.5	14	16
5. Middle East and North Africa	0.1	6	2
6. Eastern Europe and Former USSR	0.6	1	5
7. Sub-Saharan Africa	5.0	55	46

Sources: World Bank (2009b).

acquired immune deficiency syndrome (AIDS), substantially exceeds incidences in other parts of the world (column 5).

The devastation resulting from HIV/AIDS is profound. Health services, which are barely able to deal with diseases long endemic to the region, are overwhelmed, even when cheap generic drugs that control the symptoms of AIDS are available. Economies shrink as the workforce is decimated. Even the passing of skills and knowledge from one generation to the next is interrupted, as children are orphaned and as their teachers die.

Reduced Human Fertility

Falling death rates set in motion a demographic transition (Chapter 2). To begin, growth in human numbers accelerates, since the decline in mortality is not matched immediately by a fall in the number of births. But sooner or later, people recognize that the threat of mortal disease has receded, including for the very young, and respond by having fewer children. The resulting decline in the total fertility rate (TFR) eventually causes the population to stop increasing.

Although the avoidance of premature mortality triggers demographic transition, which is not entirely over until distribution of the population among age cohorts has stabilized (Lee, 2003), it is not the only cause of diminished fertility. The alleviation of poverty is also a causal factor. Changes in the **incidence of extreme poverty** (Table 9.2, columns 6 and 7) are generally consistent with trends in GDP per capita. The percentage of the people in East and Southeast Asia living on $1.25/day or less fell by nearly 70 percent between 1987 and 2007. There was a decline in South Asia as well. The incidence of extreme poverty also declined modestly in Sub-Saharan Africa and rose slightly in Latin America and the Caribbean. The Middle East and North Africa, where the incidence was barely 6 percent in the late 1980s, managed to reduce extreme poverty during the next two decades, even though average living standards were not going up rapidly. In contrast, the rising incidence of extreme poverty in Eastern Europe and the Former Soviet Union signals that the postcommunist transition has been deeply troubled.

As living standards have risen and many people have emerged from poverty, more resources have been devoted to the formation of human capital. Especially when females benefit from this investment, TFRs move toward or even below the replacement level, of 2.1 births per woman (Chapter 2). As reported in Table 9.3, literacy, which is a fundamental indicator of human capital, increased during the last decade of the twentieth century throughout the developing world, even south of the Sahara.

Table 9.3 Regional indicators of human capital

Region	Males		Females	
	1990 (1)	2007 (2)	1990 (3)	2007 (4)
(A) Adult illiteracy rates (percentage aged 15 years and older)				
1. High-Income Nations (OECD)	1	1	1	1
2. East and Southeast Asia	12	4	28	10
3. South Asia	40	26	67	48
4. Latin America and Caribbean	11	8	13	10
5. Middle East and North Africa	30	18	53	36
6. Eastern Europe and Former USSR	2	1	6	4
7. Sub-Saharan Africa	36	29	55	46

Region	Primary		Secondary		Tertiary	
	1980 (5)	2007 (6)	1980 (7)	2007 (8)	1980 (9)	2007 (10)
(B) Gross enrollment ratios (percentage of relevant age groups)						
1. High-Income Nations (OECD)	102	101	87	101	36	70
2. East and Southeast Asia	111	111	44	76	4	22
3. South Asia	77	108	27	50	5	10*
4. Latin America and Caribbean	105	117	42	88	14	34
5. Middle East and North Africa	87	106	42	71*	11	26
6. Eastern Europe and Former USSR	99	98	86	89	31	54
7. Sub-Saharan Africa	81	98	15	33	1	6

Source: World Bank (2009b).
*for 2006, not 2007.

However, the numbers reported in the table's first four columns reflect pronounced inequalities between males and females. Latin America and the Caribbean are striking exceptions, as are Eastern Europe and the Former Soviet Union. Also, the gap between female and male illiteracy in East and

Southeast Asia is falling, although it is still sizable. Elsewhere, a wide gap remains. Illiteracy rates for women are double the rates for men in the Middle East and North Africa and nearly so in Sub-Saharan Africa and South Asia. More schooling for women enhances their economic empowerment, which in turn reduces human reproduction. Improved female literacy also accelerates expansion of the economy as a whole, which benefits by tapping into the full talents and capabilities of the entire population.

The future state of human capital in different parts of the world is indicated by data reported in columns 5 through 10 of Table 9.3. Outside of Sub-Saharan Africa, gross **enrollment rates**, which equal the number of students attending primary, secondary, or tertiary (i.e., university-level) school divided by the relevant age group,[2] have risen—dramatically so in some places. It is becoming normal for children to attend secondary school, and even to graduate from high school, in Latin America and the Caribbean as well as the Middle East and North Africa, which are both highly urbanized. The same holds for East and Southeast Asia, even though much of that region is not prosperous and a large portion of the population is rural. South Asia, which in spite of recent growth still has low living standards, has made impressive strides, with gross enrollment rates at the primary and secondary levels reaching 108 percent and 50 percent, respectively, in 2007. In contrast, there has been less progress south of the Sahara. Secondary enrollment rates, for example, had risen only to 33 percent as of 2007.

Economic expansion is difficult in the face of persisting and high illiteracy, yet human capital formation does not guarantee economic progress (Easterly, 2001, pp. 71–84). In Eastern Europe and the Former Soviet Union, educational spending was considerable, including at the tertiary level, under the communists. However, this investment did not result in the region's living standards converging toward Western levels, which means that the economic returns to education were negligible. The absence of market incentives kept GDP per capita stagnant, even though most adults had many years of schooling.

If higher female literacy truly reflects greater economic empowerment for women, the chances are excellent that human fertility will fall. Already low in 1980, TFRs in wealthy nations have changed little (Table 9.4, columns 1 and 2). Since the average Chinese woman

---

[2] Since students not in the normal age group can attend school, gross enrollment rates can exceed 100 percent. An alternative measure is the net enrollment rate, which equals the enrolled share of the relevant age cohort and which therefore cannot exceed 100 percent. While gross enrollment rates have been estimated for practically every nation, net rates are unavailable for a number of countries.

Table 9.4 Regional population trends from 1980 to 2007

Region	Total fertility rate (births per woman lifetime)		Adolescent fertility rate (births per 1,000 women aged 15–19 years) 2007	Contraceptive prevalence rate (percent of women aged 15-49 years) 2007
	1980 (1)	2007 (2)	(3)	(4)
1. High-Income Nations (OECD)	1.9	1.8	22	—
2. East and Southeast Asia	3.1	1.9	17	78
3. South Asia	5.2	2.9	67	53
4. Latin America and Caribbean	4.2	2.4	77	—
5. Middle East and North Africa	6.2	2.8	30	62
6. Eastern Europe and Former USSR	2.5	1.7	29	—
7. Sub-Saharan Africa	6.7	5.1	118	22

Region	Population growth (average annual percent)		Crude death rate (per 1,000 population) 2007	Crude birth rate (per 1,000 population) 2007	Age dependency ratio (dependents/working-age population)	
	1980–1990 (5)	2000–2007 (6)	(7)	(8)	1980 (9)	2007 (10)
1. High-Income Nations (OECD)	0.6	0.8	8	12	0.5	0.5
2. East and Southeast Asia	1.6	1.0	7	14	0.7	0.4
3. South Asia	2.2	1.9	8	25	0.8	0.6
4. Latin America and Caribbean	2.1	1.5	6	20	0.8	0.6
5. Middle East and North Africa	3.0	2.1	6	24	0.9	0.6
6. Eastern Europe and Former USSR	1.0	0.1	12	14	0.6	0.4
7. Sub-Saharan Africa	3.0	2.9	15	39	0.9	0.9

Source: World Bank (2009b).

delivers fewer than two children, the TFR for East and Southeast Asia is a little less than what is required for the next generation to be as large as the current generation. The same threshold is being approached quickly in Latin America and the Caribbean. In spite of low living standards, human fertility in India and the surrounding region has plummeted since 1980. The same has happened in the Middle East and North Africa, even though the socioeconomic status of women compares poorly with the status of men in many parts of the region. A decline even has occurred in Sub-Saharan Africa, although the regional TFR remains above five births per woman.

In places like Sub-Saharan Africa, elevated human fertility, in general, coincides with a high fertility rate for adolescents (Table 9.4, column 3), in particular. As a rule, this reflects a low average age of marriage and family formation. The situation is different in East and Southeast Asia, where women tend to marry and start having babies at an older age; as a result, the region's adolescent fertility rate is about equal to the rate for wealthy countries. Of course, women who desire small families because they are urbanized, have put poverty behind them, can take advantage of post-primary education, and regard infant mortality as a remote risk act on that desire by using contraceptives. In contrast, a large portion of the women who live south of the Sahara are poor and uneducated and many of them reside in rural areas, where a large family can have benefits. Under these circumstances, contraceptive prevalence is low (column 4).

Natural Increase

Human numbers do not stop growing once mortal disease has been contained and women are having just enough children to replace themselves and their male partners. Instead, population growth continues because of demographic momentum (Chapter 2). However, the revolution in human fertility that has taken place in much of the world since 1980 has led to a sharp deceleration of natural increase, which is the difference between the crude birth rate (CBR) and the crude death rate (CDR) and which comprises all but a tiny fraction of population growth in most regions (Table 9.4, columns 5 through 8).

In two sets of countries, the difference between birth and death rates is negligible. One of these is the group of high-income nations, where TFRs have been below the replacement level for more than a generation. With CBRs barely above CDRs, immigration from poorer settings accounts for roughly half of all population growth. Without this immigration, which mainly adds to the working-age cohort, the **age dependency ratio** (Table 9.4, columns 9 and 10), which equals the number of people who are either below or above normal working age divided by the remainder of the population, would be even higher than it now is. The other part of the world where natural increase is coming to a halt is Eastern Europe and the Former Soviet Union, in which human fertility is well under 2.1 births per woman and life spans have declined for some parts of the population.

In most of the Global South, population growth, though still greater than 1.0 percent per annum, is clearly diminishing. In East and Southeast Asia, where the average woman bears few children, human numbers continue to go up because of demographic momentum (Chapter 2). During the next few decades, the population of China and nearby countries will age, which will result in a lower CBR and a slight increase in the CDR. The age dependency ratio of East and Southeast Asia already looks like that of an affluent place, although children are a slightly larger segment of the dependent group. Similarly, stabilization of human numbers is not too far off for Latin America and the Caribbean. In South Asia and the Middle East and North Africa, family size has fallen noticeably in the last three decades. However, elevated TFRs a generation ago have created substantial demographic momentum, which will take several years to dissipate.

As in so many other demographic and economic categories, Sub-Saharan Africa stands apart. Declines in human fertility have been modest and the toll taken by HIV/AIDS and other diseases has been staggering, so natural increase has only eased off slightly since 1980 and annual population growth is only a little below 3 percent. Also, the age dependency ratio, which is the highest in the world, indicates that demographic momentum will be substantial for many years to come—for the simple reason that children, not the elderly, comprise the vast majority of the dependant population. HIV/AIDS might cut back on demographic expansion. Nevertheless, natural increase will continue for many more decades.

9.3 Agriculture's Response to Demand Growth

Growth in the demand for food almost always used to lead to a geographic expansion of farming. But during the last century or two, land has become a progressively less important factor of agricultural production. In recent decades, increases in demand, which clearly are influenced these days by income trends and not just population growth, have caused farmers to increase their use of fertilizer, irrigation, machinery, and other non-land inputs.

Changes in the Mix of Inputs

According to Hayami and Ruttan's (1970) hypothesis of induced innovation, changes in agricultural inputs made for the sake of raising farm output depend on the relative scarcities of different factors of production (Chapter 3). Where labor is scarcer than land, mechanization (i.e., increased use of tractors and other equipment, along with the fuel to drive this machinery) occurs, thereby increasing what each agricultural worker produces. But where land is scarcer than labor, irrigation is encouraged, as is the increased use of fertilizer, improved seeds, and other purchased inputs. Yields (i.e., output per hectare) go up as a result.

These two paths to increased production can be discerned from regional patterns of factor use. In prosperous nations as well as Eastern Europe and the Former Soviet Union, labor is scarcer than land, as reflected in low rural population densities (Table 9.5, column 1). So mechanization, which is measured by the number of tractors per square kilometer of arable land (columns 2 and 3), is far advanced. Though still high in Russia and neighboring lands, tractor use has fallen as communist-era subsidies for machinery have been cut and increased exposure to market forces has obliged farmers to use inputs more efficiently. Outside of the Caribbean and Central America, rural population densities are not high south of the U.S. border, so there is intensive use of machinery in many countries. One consequence of mechanization in Latin America has been rapid deforestation (column 4), because an individual farmer with a tractor and other equipment can clear land faster than a peasant using a machete and fire.

Elsewhere in the developing world, rural population densities are higher, which implies that land is relatively scarcer. As the hypothesis of induced innovation suggests, incentives are strong to irrigate, fertilize, and use high-yielding varieties. Sure enough, nearly 40 percent of the arable land in South Asia is irrigated (Table 9.5, column 5). Also, fertilizer use has risen well above 100 kilograms/hectare in China, India, and neighboring countries (columns 6 and 7). Indeed, with fertilizer applications having eased off in affluent nations and fallen dramatically in Eastern Europe and the Former Soviet Union, East and Southeast Asia now have the highest application rate in the world. Between widespread irrigation and intensive use of chemical inputs, the region's average cereal yield is high, exceeded only by that of high-income countries (column 8). Except for a few places, virtually all land that is well suited to crop production is being farmed and, thanks to the intensification of agriculture, deforestation is abating in Asia.

Input use in the Middle East and North Africa, where no important opportunities exist for agricultural extensification, is a little anomalous. Rural

Table 9.5 Regional trends in the use of land and other agricultural inputs

Region	Rural population density (rural inhabitants per sq. km of arable land) 2005 (1)	Agricultural Machinery (tractors per sq. km of arable land)		Change in forest area (annual percent) 2000–2005 (4)
		1979–1981 (2)	2003–2005 (3)	
1. High-Income Nations (OECD)	323	3.9	4.3	0.1
2. East and Southeast Asia	547	0.6	1.0	0.2
3. South Asia	617	0.3	1.5	−0.1
4. Latin America and Caribbean	232	1.0	1.2	−0.5
5. Middle East and North Africa	665	0.6	1.5	0.3
6. Eastern Europe and Former USSR	129	2.2	1.9	0.0
7. Sub-Saharan Africa	351	0.2	0.1	−0.6

Region	Irrigated percentage of cropland 2001–2003 (5)	Fertilizer consumption (kg per ha of arable land)		Cereal yield (kg per ha) 2003–2005 (8)
		1979–1981 (6)	2003–2005 (7)	
1. High-Income Nations (OECD)	11	131	147	5,079
2. East and Southeast Asia	37	116	263	4,499
3. South Asia	39	36	117	2,477
4. Latin America and Caribbean	12	59	111	3,209
5. Middle East and North Africa	34	42	106	2,269
6. Eastern Europe and Former USSR	11	145	36	2,209
7. Sub-Saharan Africa	4	16	12	1,162

Source: World Bank (2009b).

population density there is nearly identical to South Asia's and not much lower than that of East and Southeast Asia, which suggests that yield-enhancing technology should predominate. But while the irrigated portion of arable land in the region is high, fertilizer application rates are not. Furthermore, there are more tractors per square kilometer of farmland in the Middle East than in any other group of countries, save Eastern Europe and the Former Soviet Union as well as affluent nations. This is explained by extraordinarily high-population densities and correspondingly low mechanization in some places (e.g., the Nile River Valley) and highly mechanized (and subsidized) farming in oil exporters, like Saudi Arabia.

In agriculture, as in economic performance and the demographic transition, Sub-Saharan Africa lags the rest of the developing world. Although its rural areas are not as densely populated as the Asian and Middle Eastern countryside, machinery use, which was limited in the early 1980s, is on the decline. Irrigation development remains at an incipient stage, and fertilizer applications are below levels required to maintain soil fertility. Many impoverished farmers south of the Sahara subsist by mining soil nutrients in one setting, clearing bush and forests by hand somewhere else, and then repeating the cycle (Chapter 5). Rapid deforestation in the region (Table 9.5, column 4) is symptomatic of this way of life.

Production Trends

Farmland in the temperate latitudes is well-suited to the production of corn, wheat, and other grains, which directly or indirectly comprise most of the human diet (Chapter 3). In the world's high-income nations, which are mainly in the same latitudes, mechanization (to economize on the employment of scarce labor), fertilization, and (where appropriate) irrigation have been drawn on to raise yields to levels unmatched in any other part of the world. In places like Australia and North America, growth in human numbers (Table 9.6, column 1) has been outstripped by increases in food supplies (column 2), generally, and livestock output (column 3), specifically.

Eastern Europe and parts of the Former Soviet Union also enjoy the agricultural advantages of a temperate environment. However, these advantages were negated by the imposition of collective farming—an imposition from which the region is still recovering (Chapter 14). Production per hectare is little more than half the yields achieved in prosperous nations. While there is considerable potential for agricultural recovery in places like Ukraine, the urgency of this task is reduced by the absence of demographic expansion.

Having benefited substantially from the Green Revolution, Asia is another setting where food supplies have increased faster than the

Table 9.6 Regional trends in agricultural output during the 1990s

Region	Population growth (%) (1)	Food production growth (%) (2)	Livestock production growth (%) (3)
1. High-Income Nations	6	13	10
2. East and Southeast Asia	14	56	98
3. South Asia	21	26	36
4. Latin America and Caribbean	19	31	32
5. Middle East and North Africa	25	34	37
6. Eastern Europe and Former USSR	2	—	—
7. Sub-Saharan Africa	30	25	14

Source: World Bank (2009b).

population. In addition, farmers in China, India, and other nations have responded to growing demands for livestock products, resulting from growth in GDP per capita. Aside from a few impoverished and densely populated places in the Caribbean Basin, hunger in the Western Hemisphere has little to do with agricultural limitations. The resources available for crop production are considerable, though not fully exploited. Also, intensification has occurred, which explains why yields are not much below those of East and Southeast Asia and increases in food and livestock output comfortably exceed population growth. Trends have been positive as well in the Middle East and North Africa, which is another region where the Green Revolution has had an impact.

The situation in the world's second-largest continent remains discouraging. Human numbers continue to shoot up south of the Sahara. There is little irrigation and fertilization, so yields are abysmal. Not coincidentally, farmers also are encroaching rapidly on forests and other habitats. Sub-Saharan Africa is the only part of the world where the population has grown faster than food supplies. Furthermore, GDP per capita has been stagnating or declining in all but a few places. Hence, incentives to increase livestock production have been weak.

9.4 Summary

The food economies of various parts of the world are linked, primarily through international trade, yet demand and supply trends in each region are distinctive in at least some respects. In places like the United States and

Western Europe, where living standards are high and natural increase is winding down, overeating is a more serious problem than hunger. Agriculture, which employs a small share of the labor force because of mechanization and attractive employment options off the farm, is highly productive in many settings, thanks to temperate conditions and the use of improved seeds and other purchased inputs. Prosperous countries are the source of most grain exports and all but a tiny share of food aid.

The developing world's stellar performer has been East and Southeast Asia. Rates of savings and investment are high in the area, and an export-oriented approach to industrialization and development has been pursued (Chapter 7). Very low in the middle 1960s, regional GDP per capita has risen at a fast clip since then, which has had a substantial impact on the consumption of food, generally, and of livestock products, specifically. Human capital formation has accelerated and increases in population have decelerated. Growth of nonagricultural sectors has driven up the opportunity costs of labor, capital, and other factors of production used on farms. Agriculture has responded by intensifying, although it is becoming internationally uncompetitive in some of the region's wealthier countries.

Although South Asia is poorer than East and Southeast Asia, trends in the former region are encouraging. Thanks mainly to a marked decline in TFRs, human numbers are not going up nearly as quickly as they were a generation ago. Nowadays, regional growth in GDP per capita is exceeded only by that of East and Southeast Asia. In addition, agricultural intensification has occurred, during the Green Revolution and since then, thereby alleviating food insecurity.

In many ways, the economic performance of Latin America and the Caribbean has been disappointing. When compared to what other parts of the developing world possess, the region's natural resources are abundant. Illiteracy is limited and population growth has slowed. Moreover, living standards south of the U.S. border in the middle of the twentieth century were the envy of Africa and Asia and are still higher today. The failure of Latin America and Caribbean to match the growth in GDP per capita that Asia has experienced is largely a consequence of profligate monetary and fiscal policies, protectionism, low savings rates, and pronounced inequality. Macroeconomic management has improved and trade has been liberalized, much to the benefit of agriculture—a sector with excellent prospects thanks to favorable environmental conditions.

In the Middle East and North Africa as well as Eastern Europe and the Former Soviet Union, living standards improved little for a generation or more. In the former region, rapid population growth has put a brake on increases in GDP per capita. But in both places, economic expansion has been held back by sizable military expenditures and the distortion of incentives by autocratic states reluctant (to say the least) to give market forces free rein. Something positive to say about the two regions is that extreme poverty has been largely avoided and, in Eastern Europe, income inequality has been and remains modest because education is widely available. Also, the former USSR and its satellites have considerable potential for agricultural recovery.

As emphasized throughout this chapter, trends are less encouraging in Sub-Saharan Africa, where living standards have slipped below those of South Asia. Population growth is unparalleled, as is deforestation. In contrast, agricultural development is stymied in many parts of the continent and it is difficult to envision an end to widespread food insecurity, poverty, and illness. On top of everything else, HIV/AIDS is devastating many parts of the region.

Key Words and Terms

age dependency ratio
BRIC
convergence
Dutch Disease
enclave industry
enrollment rate
Global North
incidence of extreme poverty
Organization for Economic Cooperation and Development (OECD)

Study Questions

1. Are trends in GDP per capita in different parts of the world consistent with the hypothesis of economic convergence?
2. Is there any evidence that life expectancies at birth in various parts of the world are converging?
3. Relate regional differences in female illiteracy to interregional differences in human fertility.
4. Explain interregional differences in mechanization and intensification in terms of varying factor endowments.

10

Affluent Nations

The classic novel, *Anna Karenina*, begins with the observation that happy families are all alike, but every discontented household is distressed in its own way (Tolstoy, 1992, p. 1). The same can be said of entire countries. Those with widespread poverty are heterogeneous. While most are tropical, a few are in the temperate latitudes. Culture and historical experience vary widely in the developing world. Of particular concern to this book, living standards and food security are improving in some places, while others remain mired in poverty, with much of the population frequently or chronically hungry.

In contrast, nations where most people are comfortably middle class and few lack food closely resemble one another. Most goods and services are allocated in markets and democracy and individual liberties are secure. A few rich places are close to the equator: Hong Kong and Singapore, which are free-trading city-states with highly educated populations, as well as small realms blessed with oil or natural gas, such as Brunei, Kuwait, and Qatar. But aside from these exceptions, affluent countries are outside the tropics.[1] The majority are Western as well. Japan is obviously a counter-example, although the cultural and institutional antecedents for commerce are no less entrenched there than in Western Europe and North America.[2]

Per capita gross domestic product (GDP) continues to rise in affluent parts of the world, although not very rapidly in recent years. Also, fertility rates have been low for decades, thereby causing human numbers to stabilize or even shrink. Food shortages are a remote concern, either because domestic output exceeds domestic consumption or because food imports

---

[1] Temperate zones have various advantages over tropical settings, including a lower incidence of communicable diseases and a winter for cutting into populations of agricultural pests (Chapter 3). Admonishing development specialists not to forget that climate and other geographical factors are important, Landes (1999, pp. 17–22) underscores the myriad environmental advantages enjoyed by Western Europe, where temperate isotherms extend farther from the equator than in any other part of the world.

[2] The same observations apply to Hong Kong, Singapore, South Korea, and Taiwan, which are discussed in the next chapter.

are easily paid for with nonagricultural exports. Instead of worrying about undernourishment, most people have reason to fret about eating too much. In addition, government policies create incentives for excessive agricultural output, which weighs on global commodity markets (Chapter 6).

10.1 Standards of Living

Wide gaps exist between prosperous nations belonging to the Organization for Economic Cooperation and Development (OECD) and the **Third World**, as the Global South used to be called. Moreover, GDP per capita has gone up steadily in those nations, more so between 1965 and 1999 than contemporaneous improvements in Eastern Europe and the Former Soviet Union, Latin America and the Caribbean, the Middle East, and Africa. Only since the turn of the twenty-first century have growth rates in these regions pulled ahead of rates in the OECD and started to approach those of Asia, which long has experienced the fastest economic expansion (Chapter 9).

As indicated in Table 10.1, annual increases of 1.0 to 2.5 percent in real GDP per capita have been normal in the developed world. Three exceptions during the 1980s were Japan, the United Kingdom, and New Zealand, with growth rates of 3.4, 3.0, and 0.7 percent, respectively (column 1). In Great Britain, Prime Minister Margaret Thatcher ended inflationary monetary management and other policies harmful to private enterprise. Gross investment (column 4) accelerated as a result, to the benefit of living standards. In contrast, reform did not take place in New Zealand until the middle 1980s, as indicated by the country's having the highest rate of inflation (column 3) in the OECD. Since the risks and uncertainties created by inflation were discouraging capital formation and since export growth (column 2) was disappointing, economic expansion was anemic.

A quarter century ago, Japan seemed to be yielding new lessons about growth and development. Its **industrial policy**, with the **Ministry of International Trade and Investment (MITI)** guiding financial resources to firms and sectors supposedly poised for takeoff, was extolled as a model for the United States (Thurow, 1992, p. 37), where emphasis was placed on bringing down tax rates and easing the regulatory burden on business during Ronald Reagan's presidency. However, the Japanese economic paradigm subsequently lost its luster. After soaring during the 1980s, prices for real estate and other assets collapsed during the last decade of the twentieth century. This crippled a number of large banks, which were heavily burdened with nonperforming loans in part because industrial policy had tempted

Table 10.1 Economic trends for selected OECD members from 1980 to 2007

Country (ranked by column 1)	A. Average annual rate of growth, 1980–90 (%)				Country (ranked by column 5)	B. Average annual rate of growth, 1990–99 (%)				Country (ranked by column 9)	C. Average annual rate of growth, 2000–07 (%)		
	GDP per capita (1)	Exports (2)	Inflation (3)	Gross invest. (4)		GDP per capita (5)	Exports (6)	Inflation (7)	Gross invest. (8)		GDP per capita (9)	Exports (10)	Inflation (11)
Japan	3.4	4.5	1.7	5.7	Australia	2.6	7.9	1.6	6.1	Sweden	2.0	6.4	1.6
United Kingdom	3.0	3.9	5.7	6.4	Denmark	2.4	3.8	1.7	4.8	United Kingdom	1.8	5.4	2.4
Spain	2.6	5.7	9.3	5.7	United States	2.4	9.3	1.8	7.0	New Zealand	1.8	4.2	2.3
Denmark	2.3	4.3	5.6	3.7	Netherlands	2.1	4.8	2.1	1.5	Spain	1.6	4.9	3.7
Italy	2.3	4.1	10.0	2.0	Spain	2.0	10.9	4.0	−0.5	Australia	1.6	3.0	3.2
Canada	2.1	6.3	4.5	4.9	United Kingdom	1.9	6.0	2.9	1.8	Canada	1.4	3.2	1.8
Germany	2.1	—	—	2.4	New Zealand	1.7	5.4	1.5	8.1	Japan	1.3	6.5	−1.2
United States	2.1	4.7	4.2	4.4	Canada	1.2	8.8	1.3	2.6	United States	1.2	3.2	2.4
Sweden	2.0	4.3	7.4	4.2	France	1.2	4.9	1.5	−3.2	Netherlands	1.2	5.9	2.4
Australia	1.9	6.9	7.3	3.0	Germany	1.1	4.1	2.0	0.5	Denmark	1.2	6.2	2.1
France	1.8	3.7	6.0	2.6	Japan	1.1	5.1	0.1	1.1	Germany	1.0	7.8	0.8
Netherlands	1.7	4.5	1.6	2.3	Sweden	1.1	8.3	2.1	−2.2	France	1.0	4.0	1.7
New Zealand	0.7	4.0	10.8	4.4	Italy	1.0	7.2	4.1	−1.0	Italy	0.4	1.2	2.5

Sources: World Bank (2001), pp. 278–279 and 294–295; World Bank (2009b).

politicians to meddle with the financial sector (Neely, 1999). Due to Japan's financial crisis, capital formation was disappointing during the 1990s (column 8), even though there was no inflation (column 7) and even though Japanese households consistently save large shares of total earnings.

A few other countries aside from Japan moved up or down noticeably in growth-rate rankings after the 1980s. Thanks to lower inflation and accelerated investment, Australia made a positive change. Meanwhile, economic expansion flagged in some of the leading economies of continental Europe, including France, Germany, and Italy. Spain's experience was different. After joining the **European Common Market**—the precursor to the European Union (EU)—in 1986, the country experienced faster growth than most of its neighbors for more than two decades, including the first few years of the twenty-first century (Table 10.1, column 9).

More recently, Spain's boom has turned to bust, and for the same reasons underlying downturns in Ireland, the United Kingdom, and even the United States. Each of these countries had experienced soaring real-estate prices, driven by cheap credit, as well as excessive borrowing, most notably by people who in other times never would have qualified for loans. In the United States, excessive borrowing also had much to do with implicit guarantees of mortgages by the government, arising out of politicians' conviction that more people should own their own homes. In any event, the real-estate boom ended once interest rates rose and distressed borrowers were obliged to sell their properties, often at a loss (Reid, 2008). Through 2010, economic growth remained weak and unemployment was still elevated in these same economies.

Long before the global economy went into recession in 2007, growth in GDP per capita was decelerating gradually in the OECD. As indicated in Table 10.1, rates of expansion during the 1980s (column 1) were mostly higher than those of the 1990s (column 5), which were in turn superior to average rates between 2000 and 2007 (column 9). Growth might have slowed because of technological and natural resource constraints. Alternatively, affluent economies might have opted to trade off some expansion for income redistribution or for improvements in environmental quality, which are not captured in GDP.

Since above-average earnings are needed to reward risk-taking and self-improvement, some income inequality is essential for growth. However, inequality also arises for other reasons. Consider the United States, which has a higher Gini coefficient than many other OECD members (World Bank, 2009b). The country is fairly open to immigrants, many of whom lack English fluency and other skills required for remunerative employment. In addition, its Gini coefficient is high because of bad schools in inner cities and other impoverished settings. Dealing effectively with problems of the latter sort would undoubtedly diminish inequality and enhance **social mobility** (related to the prospects that someone who is poor early in life can rise into or beyond the middle class). Improving educational opportunities for people who currently lack them ought to enhance overall output as well.

One additional observation to make here relates to downturns in the business cycle, which have happened every eight to ten years during the past half century. Formerly, OECD members bounced back quickly after these events. During the early 1980s, for example, monetary contraction in the United States, Great Britain, and other leading nations led to a severe recession. But as the numbers in column 1 of Table 10.1 indicate, OECD members did not take long to recover, largely because monetary contraction put an end to the inflation that had undercut economic performance during the 1970s. In contrast, the consequences in Latin America and the Caribbean were more severe and long-lasting (Chapter 12).

A departure from this old pattern has happened since the most recent dip in the business cycle, which by many measures has been the worst since the Great Depression. Emerging economies such as China and Brazil were growing much faster than the OECD by the end of 2009. A weak recovery in the United States and other places was largely explained by excesses during the real-estate boom earlier in the decade (see above): exaggerated risk taking by financial institutions deemed "too big to fail" by governments, overexpansion in the housing market thanks to buyers who could not afford mortgages (including loans guaranteed by the public sector) that were bundled into securities that puzzled most investors, the rating of these securities by trusted agencies as safe, and so forth (Ferguson, 2008, pp. 241–274). Sorting out the resulting mess, as Japan had to do during the 1990s, was always going to be long and costly.

Despite persistent attempts to end business cycles, economic upswings continue to be followed by downturns. When and where recessions take a toll on employment and earnings, food insecurity is aggravated. This is true even in prosperous nations, especially those that recover slowly from periodic crises.

10.2 Population Dynamics

Living standards that are the envy of the rest of the world are matched in the OECD by long life spans and the rarity of infant deaths. More than 99 percent of all the babies born in affluent nations live beyond their first birthday, and life expectancies at birth are around 75 years for males and 80 years or more for females (World Bank, 2009b). Instead of succumbing to infectious diseases, as do many Africans for example, most people in places like the United States die of cardiovascular disease and cancer. Much of this mortality is an inevitable consequence of old age, yet some of it is linked to bad habits: smoking, of course, as well as overeating.

Thanks to the progress made against mortal illness, the demographic transition in OECD members is far advanced. By 1980, human fertility (Table 10.2, column 2) already had fallen below the level at which one generation is as large as the next, Spain and a few other nations being minor exceptions. Since then, total fertility rates (TFRs) have risen a little in the United States (where an influx of immigrants inclined to have larger families has had an impact), Denmark, France, Sweden, and the Netherlands (Box 10.1). In all other countries listed in Table 10.2, family size in 2007 (column 3) was the same or lower than it had been a quarter century earlier. In Spain, the TFR plummeted after 1980, reaching 1.4 births per woman in 2007.

Table 10.2 Human fertility and its determinants in selected OECD countries from 1980 to 2007

Country (ranked by column 1)	2007 GDP per capita, PPP (constant 2005 international $) (1)	Total fertility rate		Urban % of total population in 2007 (4)	Female % of labor force	
		1980 (2)	2007 (3)		1980 (5)	2007 (6)
United States	43,055	1.8	2.1	81	41	46
Netherlands	36,956	1.6	1.7	81	37	45
Canada	36,260	1.7	1.5	80	40	47
Denmark	34,905	1.5	1.8	86	46	47
Sweden	34,090	1.7	1.8	84	47	47
United Kingdom	33,717	1.9	1.8	90	41	46
Germany	33,181	1.4	1.3	74	39	45
Australia	32,735	1.9	1.8	89	37	45
Japan	31,689	1.8	1.3	66	39	41
France	31,625	1.9	2.0	77	40	46
Italy	28,682	1.6	1.4	68	33	40
Spain	28,536	2.2	1.4	77	28	41
New Zealand	25,281	2.0	2.0	86	40	46

Source: World Bank (2009b).

Box 10.1 Rebounding fertility in affluent settings

An interesting exception to the general tendency of human fertility to decline as living standards rise is highlighted in recent contributions to the demographic literature.

To be specific, Myrskylä, Kohler, and Billbari (2009) have investigated the relationship between fertility and the **human development index (HDI)**, a widely used measure of well-being that reflects life expectancy, educational attainment, as well as GDP per capita (Perkins, Radelet, and Lindauer, 2006, pp. 45–46). Throughout the developing world, the HDI and the number of births per woman move in opposite directions. Consistent with this pattern are low levels of human fertility in Canada, Japan, and a few other places with high HDIs.

But in most of the countries at the top of the HDI rankings, TFRs are recovering. This is partly a consequence of the arrival of immigrants from poorer countries, who tend to have more children. Also having an influence, though, are workplace flexibility, day care that is subsidized by governments and private employers, and other features of life in the world's best-off places.

Human fertility that is low, declining, or both is to be expected where infant mortality is limited, living standards are high (Table 10.2, column 1), and a majority of the population is urban (column 4). More than anything else, small family size is an outcome of economic opportunities for women that are generally comparable to those enjoyed by men. As reported in columns 5 and 6, the female portion of the OECD workforce is rising and currently approaches 50 percent in a number of countries. In addition to driving total and per capita GDP to levels that could not be achieved if women were denied equal economic status, female empowerment raises the opportunity cost of bearing and raising children. Recognizing these costs, the vast majority of women of childbearing age use contraceptives, regardless of the instruction provided by some religious leaders. Indeed, contraceptive use is so prevalent in the OECD that it is not regularly documented by standard international sources, such as the World Bank.

Due to lingering demographic momentum, the crude birth rate (CBR) still exceeds the crude death rate (CDR) (Table 10.3, columns 3 and 2, respectively) in most nations where GDP per capita is over $25,000. Australia, New Zealand, and the United States have the highest rates of

Table 10.3 Population trends in selected OECD countries

Country (ranked by column 1)	2007 GDP per capita, PPP (constant 2005 international $) (1)	Crude death rate in 2007 (per 1,000 population) (2)	Crude birth rate in 2007 (per 1,000 population) (3)	Average annual population growth 2006-07 (%) (4)	International migrants ÷ total population (%)	
					1980 (5)	2005 (6)
United States	43,055	8	14	0.8	7	13
Netherlands	36,956	8	11	0.4	3	11
Canada	36,260	7	11	0.9	15	20
Denmark	34,905	10	12	0.3	3	8
Sweden	34,090	10	12	0.4	8	12
United Kingdom	33,717	9	13	0.4	6	10
Germany	33,181	10	8	0.0	6	13
Australia	32,735	7	14	1.2	20	21
Japan	31,689	9	9	0.1	1	2
France	31,625	8	13	0.6	11	11
Italy	28,682	10	9	0.5	2	5
Spain	28,536	8	11	1.4	2	11
New Zealand	25,281	7	15	1.2	15	21

Source: World Bank (2009b).

natural increase. In contrast, a generation or more of TFRs consistently below 2.1 births per woman have been enough to cause birth rates to slip below death rates in Germany and Italy. Other countries are about to join this group. One of these is Spain, where (as in Italy) most people are identified as Roman Catholics. Denmark and Sweden are also on the verge of demographic contraction. So is Japan.

With human numbers stabilizing in the developed world, its share of the global population is falling. Equal to 15 percent in 2000, this share is projected to reach 11 percent of the global total (9 billion or so) midway through the twenty-first century (UNPD, 2003). Furthermore, as natural increase declines toward (or perhaps below) zero, the demographic transition in wealthy nations will manifest itself especially in a shifting age structure of the population, with the elderly making up an ever larger share of the total. To be specific, low TFRs combined with longer life spans are expected to raise the ratio of dependents (young as well as old) to the total population from 0.5 in 2000 to 0.7 in 2050 (UNPD, 2003).

The demographic flux unleashed if human fertility is low for a generation or more leaves no socioeconomic reality unaffected. More is spent on health care as the population ages. Also, the trade balance deteriorates if retired people use pension income and accumulated savings to buy imported goods. In addition, high age dependency ratios (and thus relatively few workers) coupled with high incomes make rich countries a magnet for immigrants from poor countries, who can work and pay the taxes needed to support pensions for aging natives. High immigration coupled with natural increase accounts for population growth of 1 percent or more (Table, 10.3, column 4) in Australia, New Zealand, and Spain. Migrants now comprise more than a fifth of the population in Australia, Canada, and New Zealand (column 5).

Immigration is readily accomplished in North America and Australia, which are accustomed to assimilating newcomers from various parts of the world. The same cannot be said, though, of many parts of Europe. There, the arrival of foreigners, which inevitably alters the national identity, can create social tensions. To this day, there are few immigrants in Japan, where national identity and ethnicity are tightly intertwined.

10.3 The Food Economy

As with every other sector, the food economy is profoundly altered by the achievement of high living standards and the maturing of the population. For one thing, demand for environmental quality is positively tied to average incomes. Accordingly, society insists on limiting the discharge of agricultural wastes into rivers, lakes, and streams. Similarly, natural landscapes that provide scenic views and other amenities are less likely to be converted to urban housing or cropland. For that matter, **multifunctionality**—a term meaning that rural holdings are not simply a source of commodities, but of pleasing scenery and other environmental values as well—becomes a part of the farm policy lexicon.

No less than land and other natural resources used by agriculture, farm labor is affected by affluence and changes in the age structure of the population. The opportunity cost of agricultural employment has been driven up in places like the United States because of high earnings in the rest of the economy (Chapter 3). Responding to these earnings, many farmers choose not to work full-time producing crops and livestock. Part-time farming has become so common in the OECD that nonagricultural wages and salaries comprise the majority of earnings for a large number of farm households.

Lucrative opportunities outside of agriculture also strengthen incentives for mechanization, which in turn diminishes the demand for farm operators and workers. As producers react to these incentives, the average size of agricultural holdings increases, which drives up capital requirements for individual farmers—in particular, the investment in real estate and other assets needed to produce commodities at a low cost while simultaneously supporting an operator and his or her family. By reducing the hard, physical labor that used to characterize farming, mechanization and related technological change also increase the involvement of women in commodity production.

In addition to being driven by supply-side factors, the transformation of OECD food economies is a consequence of shifting patterns of food consumption. Income growth has a positive impact on sales of animal products and fruits and vegetables (Chapter 2). Demand also increases for ready-to-eat items. Along with depending on the opportunity cost of time, which clearly is elevated in affluent settings, consumption of processed and prepared goods requires the use of microwaves and other appliances, which poor households in the developing world cannot afford but that just about every OECD household possesses. Another feature of consumption in a rich country relates to food safety. Where most people are poor, safety concerns are by no means absent, although these tend to be muted. In contrast, middle-class consumers have little patience for products that do not meet exacting standards of quality.

Some consumers think they can avoid contamination by eating organic products, although Rutgers University food scientist Joseph D. Rosen points out that there is no scientific basis for this belief (Truth about Trade and Technology, 2004). A more common safeguard is to purchase name brands offered by companies that go to great lengths to maintain the quality (and reputations) of their products. Expenditures are required on these companies' part as edible goods are processed and move through marketing channels. Due to these costs, brand-name products usually fetch higher prices than substitutes. Organic foods also cost more on average than conventionally produced foods. This is because modern inputs that enhance productivity, such as transgenic seeds and synthetic fertilizers, are not used (Knutson et al., 1989).

As the demands of affluent and maturing customers for processed and prepared foods of assured high quality are catered to, agribusinesses gain importance. In the United States, for example, farmers comprise a tenth of the food economy's workforce. Another tenth supplies fertilizer, seeds, machinery, fuel, and other inputs. Eighty percent is employed beyond the farm gate, storing, processing, transporting, and retailing food.

Some populists regard agribusinesses' dominance of the food economy as exploitative. However, claims that these firms incur unnecessary costs and absorb food receipts that rightfully belong to growers are mostly unfounded. Agribusinesses' earnings reflect the value added that they create in supplying food in the form, time, and location that consumers desire. The rising importance of these firms also reflects specialization and trade—which benefit everyone, farmers and consuming households alike. It is no accident that weak development of this sort coincides in much of the African, Asian, and Latin American countryside with agricultural self-sufficiency, poverty, and hunger.

Agricultural Subsidies and Protectionism

For good reason, Gardner (2002, pp. 271–277) emphasizes the linkage between high farm incomes, on the one hand, and elevated compensation outside of agriculture, on the other. This linkage, which applies throughout the OECD, explains part-time farming, mechanization, and other supply-side adaptations. Another consequence of good wages and salaries outside of agriculture is that governments often prop up farmers' earnings.

Moving commodities from one country to another is not particularly expensive, so attempts to enhance growers' incomes with policies such as support and target prices (Chapter 4) almost always require protection from imports. As is easily appreciated, pegging prices in an individual country above values in international markets would unleash a torrent of imports if there were no tariffs or quantitative restrictions. It is hardly surprising, then, that agriculture's **producer subsidy equivalent (PSE)**, which measures the portion of total farm receipts attributable to all forms of governmental support (including price premiums associated with trade restrictions and governmental regulation of commodity markets as well as various kinds of payments), is correlated with the sector's **nominal protection coefficient (NPC)**, which is found by dividing domestic prices by international prices (Table 10.4).

Table 10.4 Trends in state support of farm incomes and agricultural protectionism

Country or Region	Units	1986–1988	2005–2007
Australia			
Total producer support	$U.S. million	1,014	1,584
Producer subsidy equivalent	Percent	7	5
Nominal protection coefficient	Ratio	1.04	1.00
New Zealand			
Total producer support	$U.S. million	432	105
Producer subsidy equivalent	Percent	10	1
Nominal protection coefficient	Ratio	1.02	1.01
Canada			
Total producer support	$U.S. million	6,048	6,913
Producer subsidy equivalent	Percent	36	21
Nominal protection coefficient	Ratio	1.39	1.33
United States			
Total producer support	$U.S. million	36,782	34,849
Producer subsidy equivalent	Percent	22	12
Nominal protection coefficient	Ratio	1.14	1.05
European Union*			
Total producer support	$U.S. million	98,585	131,921
Producer subsidy equivalent	Percent	40	29
Nominal protection coefficient	Ratio	1.76	1.19
Norway			
Total producer support	$U.S. million	2,800	2,953
Producer subsidy equivalent	Percent	70	62
Nominal protection coefficient	Ratio	4.15	2.12
Switzerland			
Total producer support	$U.S. million	5,385	4,884
Producer subsidy equivalent	Percent	77	60
Nominal protection coefficient	Ratio	4.80	1.89

(*continued*)

Table 10.4 (*Continued*)

Country or Region	Units	1986–1988	2005–2007
Japan			
Total producer support	$U.S. million	49,535	39,682
Producer subsidy equivalent	Percent	64	50
Nominal protection coefficient	Ratio	2.63	1.94
Poland			
Total producer support	$U.S. million	12,055	24,757
Producer subsidy equivalent	Percent	70	62
Nominal protection coefficient	Ratio	3.32	2.46
Turkey			
Total producer support	$U.S. million	3,118	12,288
Producer subsidy equivalent	Percent	16	22
Nominal protection coefficient	Ratio	1.17	1.23
OECD			
Total producer support	$U.S. million	239,269	262,533
Producer subsidy equivalent	Percent	37	26
Nominal protection coefficient	Ratio	1.5	1.2

Source: OECD (2008), pp. 88 and 89.
*The EU had 12 members in 1986–88, 25 in 2005–06, and 27 in 2007.

A few OECD members meddle little with farm earnings and agricultural trade. One of these is Australia, which has a strong comparative advantage in various commodities and which chooses not to undermine its advocacy of free trade in the World Trade Organization (WTO) by distorting its own food economy. Equal to 1 percent in the late 1980s, the country's PSE stood at 5 percent from 2005 through 2007. During this same period, the NPC fell from 1.04 to 1.00. New Zealand, which along with Australia belongs to the Cairns Group (Chapter 6), has a low PSE and NPC as well (Table 10.4).

Canada and the United States also have a comparative advantage in agriculture, especially in cereals, yet each of these countries distorts prices and supports farmers more than the antipodal nations. Like its southern neighbor, Canada has lowered its PSE and NPC in recent years. From 2005 through 2007, Canadian prices were 33 percent above world levels and government support amounted to 21 percent of farm receipts. Gradual reform of the U.S. **Farm Bill** (Box 10.2), which is the most important piece of agricultural legislation in the United States, has caused the PSE in the world's largest grain exporter to decline from 22 to 12 percent since the late 1980s, as its NPC has fallen from 1.14 to 1.05 (Table 10.4).

Box 10.2 The U.S. Farm Bill and the EU's CAP

Intending to raise commodity prices and enhance farm incomes during the Great Depression, the United States enacted the **Agricultural Adjustment Act (AAA)** in 1933. This legislation, which featured a combination of price supports and reductions in the area planted to major crops, was supplemented in 1954 by a food-aid program, which among other things raised domestic prices by diverting food supplies to poor nations. Two years later, the **Conservation Reserve Program (CRP)** was instituted, to reduce agricultural land use, contain soil erosion, and increase wildlife habitat. A shift to voluntary participation occurred during the 1960s, with price supports and reductions in farmed area applying to a particular crop if most of the producers of that crop voted affirmatively in a referendum (Peterson, 2009, pp. 132–135).

The implementation of deficiency payments, which depend on the difference between market prices and government-guaranteed targets (Chapter 4), in 1973 marked a major departure in U.S. agricultural policy. Substantial changes also were made in the 1996 Farm Bill, which eliminated production controls and emphasized direct payments to farmers that were supposed to be decoupled from output. Although this new form of support was to be phased out gradually, it was modified and renewed in the Farm Bills of 2002 and 2008 (Peterson, 2009, pp. 135–137).

Europe underwent major change after the **Treaty of Rome**, which Belgium, France, Germany, Italy, Luxembourg, and the Netherlands signed in 1957. Protectionist measures that various nations had employed to benefit their farmers were replaced with a shared set of farm policies called the CAP. Members of what eventually became the EU, which 27 countries currently belong to, have committed to uniform food prices, which have been maintained by reducing agricultural land use and by restricting imports from outside the union. Price supports and payments to farmers have been paid for in part by tariffs on imported commodities. In addition, individual governments contribute financially to the CAP, which absorbs a sizable portion of the EU budget (Peterson, 2009, pp. 160–164).

Gradual reform of the EU's **Common Agricultural Policy (CAP)** has occurred as well (Box 10.2), although the PSE was still 29 percent and the NPC remained at 1.19 from 2005 through 2007. Protection is even greater in wealthy European nations outside the EU that, in the absence of protection, would import nearly all their food. One such nation is Norway, where over three-fifths of farm receipts come from the state (which takes in billions of dollars annually thanks to offshore oil production) and domestic food prices are 112 percent above world levels. Another is Switzerland, with a PSE of 60 percent and NPC of 1.89. Both these countries are in the same position as Japan, which is an efficient producer of various nonfarm goods and services but has little comparative advantage in agriculture. Rice producers and other growers in East Asia's wealthiest economy stay in business only because half their receipts come in one way or another from the state and because farm commodity prices in domestic markets are nearly double world prices (Table 10.4).

To put PSEs and NPCs in wealthy nations into perspective, consider Turkey, which aspires to join the EU. Governmental support of farm earnings and the gap that exists in the country between domestic and global prices have converged toward EU levels, which means that Turkish farmers are increasingly sheltered from international market forces.

Outside of Japan, protectionism and subsidization are declining (Anderson, 2009). Further liberalization of the global food economy would create substantial benefits (Chapter 6). Producers in less fortunate parts of the world would gain from higher international prices as well as better access to European, Japanese, and North American markets. The greatest gains, however, would be for OECD consumers, who would pay less for commodities purchased from more efficient producers in other parts of the world, and OECD taxpayers, who as indicated at the bottom of Table 10.4 provide over a quarter of a trillion dollars annually to bolster farmers' earnings.

Production Technology and Output Trends

The generous subsidies received by producers in affluent nations have not shielded them from forces of change. During the twentieth century, OECD agriculture became industrialized: technology driven, capital and management intensive, and with standardized production procedures. In recent years, it also has acquired postindustrial features as individual farmers have taken on service activities. On many farms, marketing, utilization of information systems, and asset management in the home office create as much profit as work done in the field, shop, or barn.

Table 10.5 Agricultural inputs in selected OECD countries, 1979 to 2005

Country (ranked by GDP per capita in 2007)	Rural population density (rural inhabitants per sq. km of arable land) 2005	Fertilizer use (kg per ha of arable land)		Irrigated land (% of cropland)		Agricultural machinery (tractors per sq. km of arable land)	
		1979–81	2003–05	1979–81	2001–03	1979–81	2003–05
	(1)	(2)	(3)	(4)	(5)	(6)	(7)
United States	33	109	156	10.8	12.5	2.5	2.7
Netherlands	356	862	913	58.5	60.5	22.4	16.5
Canada	14	42	58	1.2	4.5	1.5	1.6
Denmark	34	245	116	14.5	9.7	7.1	5.1
Sweden	52	165	105	2.4	4.3	6.2	6.2
Un. Kingdom	108	319	302	2.0	3.0	7.4	8.7
Germany	184	425	215	3.7	4.0	13.4	7.9
Australia	5	27	49	3.5	4.9	0.7	0.7
Japan	996	413	411	56.0	35.8	27.2	44.2
France	77	326	208	7.2	13.3	8.4	6.5
Italy	245	229	173	19.3	8.6	11.2	23.1
Spain	74	101	160	14.8	20.6	3.3	7.1
New Zealand	38	188	674	5.2	3.8	3.5	5.0

Source: World Bank (2009b).

In addition, agriculture has been affected by environmental concerns. Instructive in this regard are patterns of fertilizer use in Western Europe. Among the EU members listed in Table 10.5, application rates (columns 2 and 3) went up during the 1980s and 1990s only in the Netherlands and Spain, the latter from a rate that was low by regional standards three decades ago. Elsewhere in the region, there was a noticeable decline—significantly, a decline that coincided with higher yields. Environmental regulation, aimed at protecting water quality, had an effect in some places, although a falling NPC also was a driving force. Demanded by taxpayers and consumers, diminished incentives for excessive farm output had an environmental payoff.

Application rates are unchanged in Japan, where farmers started using more fertilizer for the sake of higher yields during the nineteenth century (Chapter 3). In contrast, there have been sizable increases in Australia and Canada, which had the lowest application rates in the OECD from 1979 to 1981. The fastest growth has occurred in New Zealand, which, as already indicated, has a comparative advantage in agriculture and neither protects nor subsidizes its farmers.

While agriculture is capital-intensive throughout the OECD, the specific form of investment differs from place to place. Dry conditions in some parts of Italy and Spain explain irrigated percentages of cropland (Table 10.5, columns 4 and 5) that are above OECD norms, though well below what one sees in dry parts of the developing world. The Netherlands is not an arid nation. However, three-fifths of its arable land is irrigated because on-farm water management, undertaken for proper drainage, is intense. Japan's long history of agricultural intensification explains why more than a third of its farmland is irrigated.

Although agriculture is highly mechanized throughout the affluent world, the ratio of tractors to agricultural land (Table 10.5, columns 6 and 7) varies considerably, mainly because of different natural resource endowments. At one extreme, land suited to agriculture is scarce, yields are high, and off-farm jobs plentiful in Japan and the Netherlands. Consequently, average farm size is modest and tractors and other machinery are small, though numerous. At the other extreme are Australia, Canada, and the United States, where land is abundant and thus agricultural equipment tends to be large and tractor-to-land ratios are lower.

Starting in the late 1800s, use of machinery, improved seeds, fertilizer, pesticides, and other inputs aside from labor and land increased substantially in Europe, Japan, and North America. Since agricultural land use remained stable, production growth resulted entirely from this intensification (Chapter 3). In spite of stable or diminished fertilization in many countries, per hectare output has continued to rise (Table 10.6, columns 1 and 2). New ways are still being found to produce more output from land and human resources.

Of course, yields vary, depending on environmental conditions, and it is a mistake to suppose that a nation can compete internationally only if its agriculture is intensive. Consider Australia and Canada, both of which have a comparative advantage in cereal production. In each country, average per hectare output that is low by OECD standards is more than offset because land is not very scarce. Topography strongly favors mechanization, and economies of size are considerable. In contrast, Japan is not internationally competitive because, even though yields are high, the opportunity cost of land and other farm inputs is not fully covered.

Japan's lack of comparative advantage explains why food production has fallen in the country (Table 10.6, columns 3 and 4). A decline also has occurred in Sweden, where farmers find competing with foreigners difficult. But for other countries listed in Table 10.6, the index of food

Table 10.6 Trends in cereal yields, food production, and value added per agricultural worker

Country (Ranked by GDP per capita in 2007)	Cereal yield (kg per ha)		Food production index (1999–2001 = 100)		Value added per agricultural worker (2000 $)	
	1979–81 (1)	2005–07 (2)	1979–81 (3)	2004–06 (4)	1979–81 (5)	2003–05 (6)
United States	4,152	6,512	77.0	105.7	13,743	42,745
Netherlands	5,689	7,753	88.0	92.3	15,836	42,049
Canada	2,173	3,083	65.1	10.8	5,053	44,133
Denmark	4,040	5,863	82.1	100.5	8,649	38,441
Sweden	3,595	4,791	104.1	97.3	13,997	35,378
United Kingdom	4,792	7,081	96.9	98.7	14,345	26,942
Germany	4,166	6,538	91.6	98.6	6,889	25,657
Australia	1,321	1,411	61.4	93.7	18,041	29,908
Japan	5,252	6,013	102.2	100.5	14,032	35,668
France	4,700	6,731	91.3	96.3	12,569	44,080
Italy	3,548	5,311	98.3	99.1	6,555	23,967
Spain	1,986	3,050	72.3	102.7	4,947	18,619
New Zealand	4,089	7,071	67.5	115.7	15,195	27,189

Sources: World Bank (2009b).

production has increased. As was true earlier in the twentieth century, productivity-driven growth in food supplies has outstripped increases in demand. The latter increases have been modest both because human numbers are not going up rapidly and because the income elasticity of demand for food is generally low where GDP per capita is elevated (Chapter 2).

Finally, productivity growth in agriculture has driven up value added per worker in the sector (Table 10.6, columns 5 and 6). Since farming must compete for labor with other sectors, GDP per capita, which is a good indicator of the opportunity cost of labor dedicated to crop and livestock production, is positively related to value added per agricultural worker. Note, for example, the relatively low numbers in columns 5 and 6 for New Zealand and Spain, where living standards are lower than in other countries listed in the table, and the relatively high numbers for Canada, France, the Netherlands, and the United States. Observe as well that the gap between value added per agricultural worker in Japan (Table 10.6, column 6) and that country's GDP per capita (Table 10.3, column 1) is not as great as what one observes elsewhere in the OECD, although this is largely due to subsidies and trade barriers (see above).

10.4 Dietary Change and Consumption Trends

Transformed in various ways because of elevated living standards, technological improvement, and capital intensity, OECD agriculture also has been greatly influenced by the large appetite of affluent people for livestock products. This impact arises because multiple plant-derived calories must be fed to cattle, chickens, and other domesticated animals for each calorie of meat, eggs, and the like eaten by people. The former inputs include grain, soybean products, and other feed components. Forage, produced on rangeland and pasture, is another source of plant-derived calories for livestock, especially cattle and sheep. But regardless of how domesticated animals are fed, use of land and other resources is affected if demand for livestock products is elevated, as it is in the world's wealthy countries.

Rask has developed a methodology for describing dietary composition, one involving the conversion of data from the U.N. Food and Agriculture Organization (FAO) on food and beverage consumption into cereal equivalents (Box 10.3). Applying that methodology to three time periods, 1979 to

---

Box 10.3 A methodology for dietary assessment

In his description of dietary composition, Rask (1991) does not refer to prices because of the distortions caused by protectionism and other policies. Instead, every calorie contained in various products is converted into cereal-equivalent calories as follows.

Cereals (excluding beer)	1.0000	Bovine meat	19.8000
Starchy roots	0.2648	Mutton and goat meat	19.0000
Pulses	1.0998	Pig meat	8.5000
Oil crops	0.9070	Poultry meat	4.7000
Vegetable oils	2.7479	Other meats	12.0000
Fruits (excluding wine)	0.1475	Animal fats	3.1000
Vegetables	0.0766	Milk (excluding butter)	1.2000
Treenuts	0.7955	Butter and ghee	21.0000
Sugar and sweeteners	1.1161	Eggs	3.8000
Alcoholic beverages	0.2450		

(continued)

> (*continued*)
>
> This methodology, though it involves a large number of calculations, is conceptually straightforward. However, it has a pair of limitations. One is the implicit assumption that all plant-derived calories eaten by livestock are equally scarce, even though the forage that cattle and other domesticated animals graze is not as scarce as the corn and other grain provided at feed lots. The other limitation is that the plant-derived calories required to produce one calorie's worth of livestock products are assumed to be invariable. Since forage production and the feeding efficiency of livestock both improve over time, Rask's (1991) methodology tends to overstate agricultural use of natural resources.

1981, 1989 to 1991, and 1997 to 1999, we find that the United States is unusual in that intakes of plant as well as animal products (Table 10.7, columns 1 and 2) have gone up, albeit slightly. In other countries where per capita consumption (column 3) has increased, most of the change has related to animal products. Illustrative in this regard is Denmark, where the average person was eating about the same amount of plant products in the late 1990s as he or she did two decades earlier. Meanwhile, per capita consumption of animal products rose from 1.40 to 1.59 cereal-equivalent tons per annum. Similar changes happened in the Netherlands, Italy, and Spain. Relative increases in the consumption of eggs, meat, and milk were even more pronounced in Japan, although this consumption continues to make up a smaller portion of the human diet than in other countries with comparable living standards.

Changes in per capita intake of animal products are likewise the larger part of consumption trends where growth has occurred. For example, lower consumption of meat, milk, and so forth has more than offset Canadians' small increases in consumption of plant products. The same is true of France, Germany, Sweden, the United Kingdom, Australia, and New Zealand. The latter two nations are highly efficient producers of grain and livestock, which keeps prices low and encourages eating.

In most of the OECD, food self-sufficiency—defined simply in Table 10.7 (column 4) as domestic production (expressed in cereal equivalents) divided by consumption—has been achieved, is increasing, or both. Japan is an exception, as are Italy and Sweden. None of these nations has any reason to worry about food supplies, since, as emphasized in this chapter, commodity imports are easily paid for by selling nonagricultural products to foreigners.

Table 10.7 Food consumption trends in selected OECD countries, 1979 to 1999

Country	Years	Plant Product Consumption per Capita (in cereal equiv. tons) (1)	Animal Product Consumption per Capita (in cereal equiv. tons) (2)	Total Food Consumption per Capita (in cereal equiv. tons) (3)	Total Food Self-Sufficiency* (4)
United States	1979–1981	0.26	1.84	2.11	0.99
	1989–1991	0.29	1.81	2.10	0.99
	1997–1999	0.31	1.84	2.15	1.02
Denmark	1979–1981	0.23	1.40	1.62	2.39
	1989–1991	0.23	1.63	1.86	2.34
	1997–1999	0.25	1.59	1.84	2.36
Canada	1979–1981	0.22	1.71	1.93	1.19
	1989–1991	0.24	1.60	1.84	1.07
	1997–1999	0.26	1.53	1.79	1.15
Netherlands	1979–1981	0.24	1.45	1.69	1.57
	1989–1991	0.26	1.46	1.72	1.78
	1997–1999	0.24	1.58	1.82	1.92
Australia	1979–1981	0.23	2.35	2.58	1.74
	1989–1991	0.24	2.24	2.49	1.66
	1997–1999	0.25	1.98	2.23	2.07
Germany	1979–1981	0.26	1.67	1.92	0.97
	1989–1991	0.26	1.65	1.91	1.01
	1997–1999	0.27	1.45	1.72	0.98
Japan	1979–1981	0.26	0.68	0.93	0.73
	1989–1991	0.26	0.82	1.08	0.67
	1997–1999	0.26	0.85	1.10	0.56
France	1979–1981	0.24	2.05	2.29	1.05
	1989–1991	0.25	2.02	2.27	1.03
	1997–1999	0.25	1.92	2.17	1.08
Italy	1979–1981	0.32	1.38	1.70	0.78
	1989–1991	0.31	1.54	1.85	0.76
	1997–1999	0.31	1.54	1.85	0.78
United Kingdom	1979–1981	0.22	1.48	1.71	0.76
	1989–1991	0.25	1.38	1.62	0.81
	1997–1999	0.26	1.29	1.55	0.81
Sweden	1979–1981	0.22	1.51	1.73	0.99
	1989–1991	0.22	1.42	1.65	0.99
	1997–1999	0.24	1.46	1.70	0.94
New Zealand	1979–1981	0.21	2.57	2.78	3.34
	1989–1991	0.23	2.21	2.44	3.44
	1997–1999	0.24	2.07	2.32	3.60

(*continued*)

Table 10.7 (*Continued*)

Country	Years	Plant Product Consumption per Capita (in cereal equiv. tons) (1)	Animal Product Consumption per Capita (in cereal equiv. tons) (2)	Total Food Consumption per Capita (in cereal equiv. tons) (3)	Total Food Self-Sufficiency* (4)
Spain	1979–1981	0.27	1.12	1.38	0.96
	1989–1991	0.28	1.35	1.63	0.95
	1997–1999	0.28	1.53	1.81	1.00

*Total food self-sufficiency equals total domestic food production divided by total domestic food consumption.

Indeed, a more serious imbalance in much of the OECD is a self-sufficiency ratio that is too high. An excessive ratio may be above or below unity. Regardless, the gains from efficient specialization and trade are being foregone wherever domestic farmers are producing too much, invariably because of governmental meddling with market forces.

The Obesity Problem

Domestic farm output is not the only thing that is too high in many rich countries. Aside from Japan, where per capita consumption of plant and animal products is just 1.10 cereal-equivalent tons per annum (or 3.01 kilograms a day), daily food intake is between 4.25 and 6.36 cereal-equivalent kilograms in the OECD members listed in Table 10.7. Needless to say, consumption levels in this range are well above what the average African, Asian, or Latin American eats during a typical day. Moreover, these levels are excessive relative to nutritional requirements, which causes a large part of the population to be overweight.

Since excessive weight is a worldwide problem, not just an issue in rich countries, a full discussion of causes, impacts, and solutions is deferred until the last chapter of this book, which addresses the future of the food economy and in which the scope of discussion is global rather than regional. Only noted here are the problem's dimensions in the OECD.

One perspective on **obesity** is to count the number of people whose body mass index (BMI), which equals a person's weight in kilograms divided by the square of his or her height in meters,[3] exceeds 30. The

---

[3] To calculate the BMI using English units of measurement, weight in pounds is multiplied by 704.5 and then divided by the square of height in inches. The ideal BMI for adults of both sexes is 20 to 22. Anyone with a BMI over 25 is categorized as overweight and, as already noted, the cutoff for obesity is 30.

portion of Americans over the age of 18 categorized as obese according to this criterion has gone up in recent decades from 13 percent in the early 1960s, to 14 percent in the late 1970s, to 22 percent in the early 1990s, and on to 30 percent at the turn of the twenty-first century. Even more worrying is the rising incidence of children with an elevated BMI, who tend to struggle their entire lives with weight. Whereas 4 percent of the U.S. cohort between the ages of 2 and 17 was obese in the middle and late 1960s, 14 percent was so categorized in 1999 and 2000 (Rashad and Grossman, 2004). To put these numbers, and the obesity problem, into perspective, keep in mind that 14.6 percent of U.S. households were food-insecure at one time or another in 2008 (ERS-USDA, 2009b), when the country suffered a severe economic downturn.

Another view of weight problems is provided by data on mortality and health care costs. According to the U.S. **Centers for Disease Control and Prevention (CDC)**, as many as 400,000 Americans died in 2000 because of poor diets and lack of exercise; this was only 35,000 less than the number of their countrymen whose lives were claimed the same year by tobacco-related illness (Song, 2004). Future health care expenditures traced to obesity will exceed expenditures traced to smoking. The reversal will occur in part because the incidence of obesity is rising as the proportion of the adult population that smokes is falling. Overeating will also become the most costly U.S. health problem because people whose excess weight makes them ill tend to cling to life longer, and hence require health care for more years, than those who die from cancer, cardiovascular failure, emphysema, and other diseases related to tobacco use.

In light of the mounting human and economic toll of overeating, proposals for governmental action to deal with the problem are being debated actively. Improved food labeling, to help consumers who need to watch their weight, arouses little controversy, as does education about the health consequences of weighing too much. Sweden has banned food advertising during television programs aimed at youngsters; nevertheless, the incidence of childhood obesity is not appreciably different from the incidence in places similar to Sweden where such advertising is not restricted (Duncan, 2003). No country has attempted to deal with the market failure that arises because the medical expenses resulting from obesity are neglected entirely by farmers, agribusinesses, restaurants, and other food providers and not fully internalized by heavy people themselves. In the United States, for example, a sizable portion of these expenses is paid by the government's Medicare and Medicaid programs. An obvious corrective intervention would be similar to a tax on pollution (Chapter 5). For example, a charge on foods such as soft drinks and snacks laden with fat and sugar could slow or reverse the long-term weight gain trend in the United States (Cutler, Glaeser, and Shapiro, 2003; Lakdawalla and Philipson, 2002).

Safe to say, the issue is far from settled. Indeed, it is not far-fetched to envision a time when overeating might command more attention of policymakers than food insecurity—not just in the OECD, but around the world.

10.5 Summary

With due allowance for differences in geography and (especially in Japan's case) cultural antecedents, a number of generalizations apply to the world's rich nations. Not just elevated, standards of living are historically unprecedented. Moreover, additional improvement will follow as trade continues to expand and savings are invested in human, material, and technological capital. Natural population increase is becoming a thing of the past. Throughout the OECD, there is little immediate prospect that human fertility will rise above the replacement level, mainly because of the economic opportunities available to women in an affluent society. Since families with one or two children are the rule, demographic contraction and aging of the population can be forestalled only by accelerated migration from poorer countries, which Japan, for example, is reluctant to accept.

No less than any other sector, the food economy is greatly affected by generalized affluence and the rising predominance of the middle-aged and elderly. High earnings outside of agriculture create overpowering incentives for mechanization, thereby causing farm employment to decline and the average size of holdings to increase. The gradual demise of traditional family farms is indicated not just by reductions in the agricultural workforce, but by its maturation as well. The average age of U.S. farm operators was 55 years in 2002 and has yet to peak. Since an individual operator must harness assets worth millions of dollars to produce commodities like corn and soybeans efficiently, there will be fewer commercial farms in the future and many of these will owe their existence to generous parents who have passed the family homestead to a son or daughter, the proprietor's engagement of capital markets (even including the use of equity financing), or both (Tweeten and Hopkins, 2003). Since human fertility has declined in rural areas no less than in cities, much of the workforce aside from farm operators themselves will comprise younger immigrants from poorer nations.

Throughout the OECD, farming has been protected from international competition and subsidized. As reported in this chapter, PSEs and NPCs remain high in Japan and a couple of rich European nations outside the

EU. However, protectionism has all but disappeared in Australia and New Zealand and seems to be following a downward trend in the EU and North America. Further modification of government policies that distort the food economy is likely, both because freer trade serves broad national interests and because of mounting budgetary pressures. The impulse for reform also has to do with environmentalism. Where the population is stable (or declining) and growing more prosperous, demand for clean waterways is bound to outweigh the desire to stimulate commodity output with price supports, import quotas, and the like. When the choice between the two is stark, the former demand ought to trump the latter desire—even when large producers, who capture most of the gains from subsidies and protection, lobby hard for the largesse coming their way.

Due to policy-induced distortions, exports have been excessive from some parts of the developed world, most notably Western Europe, and imports have been choked off in Japan and elsewhere. To be sure, reform will correct inefficiencies. Nevertheless, reform may not greatly alter patterns of agricultural trade between the OECD and the rest of the world. With demographic stabilization and low income elasticities for food, demand for imported food is unlikely to grow appreciably in wealthy nations. Meanwhile, the continuation of technological progress in agriculture, including in biotechnology, will drive up supply. A significant portion of the additional production is bound to be exported, especially from those OECD members with a comparative advantage in agriculture.

Thanks to specialization and trade among farmers and agribusinesses, which are facilitated by the infrastructure, scientific advances, and other public goods made possible in large part by government, the food economy supplies abundant output to consumers at a low cost. With food insecurity minimal in the OECD, other issues command more attention. One of these is fine-tuning of the safety net, so that work incentives are maintained as people are fed who otherwise would go hungry. Another issue is environmental protection, which has considerable appeal for people who are secure about the basics of life. In the twenty-first century, the primary concern related to the food economy is excessive consumption. No consensus has been reached yet about dealing with obesity, although this is certainly a less painful problem than hunger.

In the twenty-first century, demands on agriculture in prosperous countries will be unprecedented. Domestic appetites for food will rise slowly and rich countries could be called upon to provide food to millions of the poor and hungry in developing countries troubled by poor harvests due to global warming. Millions of hectares of cropland in the United States and elsewhere might be used to produce feedstock for the biofuel industry or plant trees to sequester carbon. Meanwhile, advocates of alternative agriculture will seek organic products, lots of space for farm animals and wildlife, and other concomitants of the good life. In the United States, agriculture as a sector with comparative advantage will fill a critical role in correcting a long neglected, unsustainable trade deficit and servicing massive debt to foreign nations. If crop yields rise slowly, these demands could drive up prices, thus favoring farmers but burdening consumers.

Key Words and Terms

Agricultural Adjustment
   Act (AAA)
Common Agricultural
   Policy (CAP)
Conservation Reserve
   Program (CRP)
Centers for Disease Control
   and Prevention (CDC)
European Common Market
Farm Bill
human development index (HDI)

industrial policy
Japanese Ministry of International
   Trade and Investment (MITI)
multifunctionality
nominal protection coefficient (NPC)
obesity
producer subsidy equivalent (PSE)
social mobility
Third World
Treaty of Rome

Study Questions

1. Why is GDP per capita growing more slowly now in the OECD than in much of the developing world?
2. How do per capita GDP trends and income distribution in the United States compare with income trends and distribution in other OECD nations?
3. What are the chances that population contraction will be avoided in Western Europe and Japan? In the United States?
4. Are agricultural subsidies and trade barriers growing or diminishing in the OECD?
5. Explain the linkage between agricultural mechanization and per capita GDP in the OECD.
6. What is the general trend in the consumption of livestock products in affluent nations?
7. What are the trends in the incidence of obesity in the United States, both for children and for the entire population?

11

Asia

Defined in this chapter and throughout the book not to include Japan (which has been counted among the world's prosperous nations for decades) or any part of the Middle East or the Former Soviet Union, Asia has been the setting for intensive farming and great civilizations for millennia. It is also home to half the human population. The world's two largest nations are Asian: China, where more than 1.3 billion people live, and India, with a population exceeding 1.1 billion. So are the fourth, sixth, and seventh most populous countries: Indonesia (228 million), Pakistan (166 million), and Bangladesh (160 million), respectively (World Bank, 2009b).

Domination by occidental empires having come to an end in Asia by the middle of the twentieth century, two of the largest conflicts between communists and their opponents were fought in Korea and Vietnam after the Second World War. During this period, average incomes were low and poverty widespread throughout the continent. But just as the urgency of development was universally recognized, more than a few observers were pessimistic. Paul Ehrlich, for one, held out no hope, even predicting that India and many of its neighbors would succumb to famines of historically unprecedented proportions (Ehrlich, 1968, pp. 38–41).

As dire warnings of this sort were being issued, the Green Revolution was under way. Thanks to the introduction of high-yielding varieties of rice and other crops, food production has more than kept pace with consumption in Asia since the 1960s. With the specter of famine receding, accelerated development has occurred—so much so that the global center of economic gravity, which moved across the Atlantic from Europe to the North America during the twentieth century, is now shifting across the Pacific.

Yet a number of problems must still be resolved. Being so populous, Asia has more poor and hungry people than the rest of the world combined, Sub-Saharan Africa included (Chapter 8). Moreover, the Green Revolution is proving difficult to sustain, at least in its historical form. This is mainly because the advances of the 1960s, 1970s, and 1980s relied heavily on hydrologic resources, which are growing scarcer with each passing year (Chapter 3). Further progress toward food security will require new sorts of technological improvement—including agricultural biotechnology, which has yet to be universally accepted in Asia.

11.1 Trends in GDP per Capita

Since the late 1970s, Asia has set a standard for economic growth matched by no other part of the world (Chapter 9). Living standards have improved at a fast pace in China and South Korea as well as in Indonesia, Singapore, and Thailand. The World Bank and other international agencies ceased reporting data for Taiwan after the Republic of China (with its capital in Taipei) was obliged in 1972 to surrender its U.N. seat to the People's Republic of China (with its capital in Beijing). Be that as it may, the Taiwanese economy has been one of Asia's best performers, with average incomes rising by 5 to 6 percent per annum during the 1980s and 1990s (Council for Economic Planning and Development, 1999, p. 1).

An important driver of economic expansion has been the spread of labor-intensive manufacturing, which among other impacts has accelerated rural-to-urban migration. However, agricultural development, stemming from technological advances that raised land productivity during and since the Green Revolution (Chapter 3), have contributed even more to Asia's overall progress, including a 50 percent reduction since 1981 in the number of people living in extreme poverty (Anderson and Martin, 2006, p. 7).

In light of recessionary conditions throughout Latin America and the Caribbean, the Middle East, and Africa during the 1980s, the improvement in per capita gross domestic product (GDP) that occurred the same decade in most of Asia (Table 11.1, column 1) was impressive. The reasons for this success are indicated in columns 2 through 4. Exports shot up quickly. Rates of capital formation were also high—in part because inflation was held in check. Exports and investment drove living standards higher in the 1990s as well (column 5). Likewise, improvements have been sustained since the turn of the twenty-first century (column 9).

Increases in GDP per capita have varied over time, with some nations falling back relative to their neighbors between the 1980s and 1990s and others improving. The uncertainty associated with the 1997 transfer from British to Chinese rule caused growth to decelerate in Hong Kong. Also, South Korea, Thailand, and Indonesia slipped in the regional rankings as the twentieth century was drawing to a close. This was largely the result of a major financial crisis that erupted in 1997, when investors' loss of confidence in the

Table 11.1 Economic trends in Asia from 1980 to 2007

Country (ranked by column 1)	A. Average annual rate of growth, 1980–1990 (%)				Country (ranked by column 5)	B. Average annual rate of growth, 1990–1999 (%)				Country (ranked by column 9)	C. Average annual rate of growth, 2000–2007 (%)			Remittances ÷ exports (%), 2004–2007	Freedom status in 2008**
	GDP per capita	Exports*	Inflation*	Gross invest.		GDP per capita	Exports	Inflation	Gross invest.		GDP per capita	Exports*	Inflation*		
	(1)	(2)	(3)	(4)		(5)	(6)	(7)	(8)		(9)	(10)	(11)	(12)	(13)
China	8.6	19.3	5.9	11.0	China	9.6	13.0	8.2	12.8	China	8.4	24.0	3.7	0.8	NF
South Korea	8.2	12.0	6.1	11.9	Vietnam	6.3	27.7	16.8	25.5	Cambodia	6.8	18.0	3.3	3.4	NF
Thailand	5.9	14.1	3.9	9.4	Singapore	6.1	—	1.6	8.5	Myanmar ‡	6.6	20.6	18.6	—	NF
Hong Kong	5.7	14.4	7.7	4.0	Myanmar	5.1	7.5	25.9	14.7	Vietnam	5.5	19.5	5.9	10.4	NF
Singapore	5.0	—	1.9	3.7	South Korea	4.7	15.6	5.8	1.6	Mongolia	5.3	—	14.9	8.8	F
Indonesia	4.3	2.9	8.5	7.0	India	4.3	11.3	8.6	7.4	India	5.3	15.3	4.4	12.4	F
India	3.7	5.9	8.0	6.5	Sri Lanka	4.1	8.4	9.7	6.2	Laos	4.4	—	12.2	0.1	NF
Pakistan	3.6	8.4	6.7	5.9	Malaysia	3.8	11.0	5.0	6.2	Hong Kong	3.7	9.9	−2.0	0.1	PF
Malaysia	2.5	10.9	1.7	2.6	Laos	3.8	—	22.9	—	South Korea	3.7	10.7	10.3	0.0	F
Sri Lanka	2.6	4.9	11.0	0.6	Thailand	3.5	9.4	4.6	−2.9	Thailand	3.7	7.7	2.0	0.9	PF
Vietnam	2.5	—	210.8	—	Bangladesh	3.2	13.2	3.9	7.0	Sri Lanka	3.4	—	10.2	26.4	PF
Nepal	2.0	3.9	11.1	1.8	Indonesia	3.0	9.2	14.4	5.1	Indonesia	3.3	10.1	11.6	4.7	F
Bangladesh	1.9	7.7	9.5	1.4	Nepal	2.4	14.3	8.6	5.7	Bangladesh	3.3	12.8	4.1	48.4	PF
Mongolia	2.5	—	−1.6	1.7	Cambodia	2.0	—	28.7	—	Singapore	2.9	—	1.5	—	PF
Myanmar	−1.2	1.9	12.2	—	Hong Kong	1.8	8.4	5.2	6.3	Malaysia	2.8	6.8	4.1	0.9	PF
Philippines	−1.6	3.5	14.9	−2.1	Pakistan	1.5	2.7	10.7	2.1	Pakistan	2.5	10.7	9.0	29.7	PF
					Mongolia	−1.2	—	66.5	—	Philippines	2.5	7.6	5.1	21.6	PF
					Philippines	0.9	9.6	8.4	4.1	Nepal	1.1	—	5.6	122.2	PF

Sources: World Bank (2001), pp. 278–279 and 294–295; World Bank (2009b); and Freedom House (2008).

*Trends in exports and inflation are for 2000–2004.

**F = free; PF = partly free; and NF = not free.

financial markets of these and other countries led to widespread capital flight and severe currency devaluations (Perkins, Radelet, and Lindauer, 2006, pp. 586–590). In contrast, Vietnam fared better after the 1980s as it deemphasized central planning and invested in human capital and physical infrastructure. On the export side, the country has become one of the largest suppliers of rice in the international market and has moved past Colombia to become the world's second largest source of coffee.

Vietnam's expansion was still robust after 2000. Likewise, China has continued to experience rapid improvement in living standards. The other three countries with the fastest growth in GDP per capita from 2000 to 2007 (Table 11.1, column 9) have been: Cambodia, where exports of textiles and clothing have burgeoned thanks to the availability of cheap labor; Myanmar (formerly Burma), which has benefited from rising prices of timber, mining products, and other commodities it exports; and Mongolia, which had a difficult transition from communism during the 1990s but since then has benefited from macroeconomic stability and market-friendly policies. India, where the pace of improvement has picked up steadily since the 1970s, had the sixth-fastest growth rate.

Remittances home from Indians living overseas are sizable relative to exports (Table 11.1, column 10). Most other nations where these transfers are an important source of foreign exchange have experienced less economic growth. Other factors associated with sizable remittances include an elevated incidence of poverty (e.g., in Bangladesh and Nepal) and ready access to places where labor is scarce (especially in the Persian Gulf). Many Philippine migrants enjoy the added advantage of being well-educated and speaking good English.

According to Freedom House, political and civil liberties are in place in just four of the countries listed in Table 11.1 (column 11). Leading this group is India, which for decades has had the honor of being the world's most populous democracy. Also categorized as free is Mongolia, which is the only formerly communist nation in the region to adopt democratic institutions. Five other nations, including China, are categorized as non-free. Moreover, annual increases in GDP per capita averaged 5.5 percent or more in four of these since 2000 (column 9), which tempts some to think that authoritarianism somehow promotes economic expansion. However, continued growth over many years is by no means guaranteed in at least two of these countries: Cambodia, which is highly specialized in clothing and textiles, and Myanmar, which has tied its fortunes to commodity exports. Insights of greater importance about economic development and the future of Asia are to be gained by examining its two largest economies.

China Ascendant

In a transformation triggered by agricultural reform, the world's largest country emerged from economic dormancy 30 years ago. After the death of Mao Zedong, local (and illegal) experimentation with capitalistic practices in the Chinese countryside began in the late 1970s, which put an end to 20 disastrous years of collective agriculture (Chapter 7). This liberalization, called the Household Responsibility System (HRS), caused farm output to shoot up, more than meeting national food requirements (Naughton, 1995, pp. 138–142). Within a few years, market reforms were spreading to other parts of the economy (Gregory and Zhou, 2010), with rapid industrial expansion in enterprise zones along the coast. Besides textiles and food products, in which an emerging economy usually has a comparative advantage, electronic goods and machinery of various sorts are now being produced.

Remarkable in its own right and clearly superior to what has been achieved in the Former Soviet Union (Box 11.1), recent economic progress in China also has created problems. Rapid industrial expansion has created glaring regional disparities, with living standards in the rural interior comparing poorly with those of coastal cities. In addition, there is tremendous inefficiency among **state-owned enterprises (SOEs)**, which dominate the steel industry, ship-building, mining, energy, and transport. The mounting losses of these firms are being covered by loans from government banks; this has, in turn, diminished the solvency of the banking sector, with nonperforming loans amounting to about half of all debt held by the sector (Kynge, 2002; Righter, 2002). The weakness of financial institutions bodes ill for future capital formation. So does the lack of a legal system adapted to a market economy (Ahmad, 2004).

Box 11.1 China versus Russia

China's economic accomplishments are put into perspective by the dismal record of Russia and other remnants of the old USSR.

One disadvantage of the Former Soviet Union has been the entrenched legacy of socialist agriculture. Farming was collectivized, with atrocious loss of life, soon after Stalin seized power in the late 1920s (Chapter 7). Consequently, the rural population today has had little or no experience with free markets and individual decision making, so agriculture has been and continues to be a disappointing feature of the economic landscape (Chapter 14).

(continued)

> (*continued*)
>
> In contrast, Chinese peasants endured collective agriculture for about two decades, which was not enough to eradicate their familiarity with markets and making decisions on their own. Having been coerced into joining village-level communes around 1960, peasants quickly exploited openings for capitalistic behavior after Mao Zedong died in 1976 (Gregory and Zhou, 2010).
>
> Another clear advantage of China not shared with the Former Soviet Union is the large and affluent population of ethnic Chinese living in Taiwan, Hong Kong, Singapore, and other parts of Southeast Asia that is willing to invest in the ancestral homeland. In no small measure, the tens of billions of dollars made available by these people were an important factor in the initial drive for industrial development, export growth, technological improvement, and solidification of the private sector in China (Ahmad, 2004).

As in other parts of the world, recession took a toll in 2008. However, China has made a speedy recovery, thanks in part to quick implementation of a stimulus package of $586 billion in new government spending as well as a tsunami of lending (Naughton, 2009, pp. 278–279). Annual GDP expansion during the first half of 2009 was less than the 10 to 12 percent registered from 2004 to 2007. But at 6 to 8 percent, this expansion outstrips the slow growth (at best) that Europe, Japan, and North America have registered.

Imbalances remain. Small and medium-sized private firms have had limited access to loans from government banks. At the global level, the pattern of recent years of massive lending to the United States, which has kept China's currency cheap relative to the dollar, has caused Chinese exports to run consistently ahead of its imports and hundreds of billions of dollars to end up in Beijing. This pattern cannot be sustained indefinitely, particularly as sentiment for protectionism grows in the United States and other countries that have chronic trade deficits with China and as the accumulation of U.S. currency in foreign lands adds to fears of a run on the dollar (Ferguson, 2008, pp. 283–340).

Other challenges facing China garner less international attention, in spite of their severity. One of these is land dispossession in rural areas, often instigated as corrupt local officials pursue pet projects, which triggered at least 80,000 serious incidents in 2005 (Walder, 2009, p. 262) and more than 125,000 such in 2008 (Merkell-Hess, 2009, p. 291). Another challenge is to deal with the aging and impending contraction of the population (see below).

The Other Asian Colossus

The other Asian nation of continental dimensions, India has accomplished a rare feat in the developing world by maintaining a functional democracy since achieving independence in 1947. For most of this period, the former crown jewel of the British Empire embraced socialist policies that the Labor Party championed in the old colonial power after the Second World War. Protectionist barriers and an intrusive regulatory regime were established and banks and major industries were nationalized. Not surprisingly, GDP grew slowly up to 1980 (Ahuluwalia, 2002).

Even though performance of the national economy as a whole was middling during the 1950s, 1960s, and 1970s, India was able to achieve a breakthrough in agricultural productivity. It was the first developing country to experience the Green Revolution, with impressive increases in rice and wheat yields on farms of all sizes and benefits for non-farming households in the form of employment growth and cheaper food (Chapter 3). Agricultural development contributed directly to declines in the incidence of extreme poverty from 1970 onwards, especially in rural areas (Datt and Ravallion, 2002).

During the 1980s, gradual liberalization of India's closed, regulated economy was occurring, which led to the growth performance documented in the left-hand panel of Table 11.1. Reform accelerated during the last decade of the twentieth century. For example, average import duties declined from 72 percent in 1991 and 1992 to 25 percent in 2001, when all remaining quantitative restrictions on trade were removed (Ahuluwalia, 2002). In addition, some SOEs were privatized and licensing requirements, which formerly stifled the private sector, were relaxed (Long, 2004). Consequently, GDP per capita rose by 4.3 percent per annum during the 1990s and by 5 percent or more after 2000 (Table 11.1, columns 5 and 9).

Liberalization is still a work in progress in India. Still among the highest in the developing world, the country's tariffs are nearly double the levels in China and Southeast Asia. Also, privatization has been slow; SOEs still account for 35 percent of industrial value added. In addition, deficient transportation infrastructure and modest incentives for foreign investment further limit exports and growth (Ahuluwalia, 2002), including in labor-intensive industries capable of employing the tens of millions of rural and urban Indians who work part-time or not at all at present. A more encouraging trend has been the increase of relatively well-paid jobs in information technology and remote business services outsourced from wealthy places like the United States. Here, a comparative advantage has been created because tens of thousands of young Indians have the education and excellent English that these jobs require (Long, 2004), although they have faced more competition in recent years from well-educated Anglophones in the Philippines.

India has demonstrated that a large and pluralistic democracy can open its economy to world trade, with positive effects on living standards and the incidence of poverty. Continued growth depends on not abandoning the path of economic reform on which the country clearly has embarked. Likewise, the progress that liberalization has created must not be jeopardized by protectionism, either in India or in other parts of the world.

11.2 Population Dynamics

Before 1980, a large segment of the Asian population was destitute. But thanks to economic progress in China, India, and a number of other countries, hundreds of millions were subsequently lifted out of poverty. Global inequality declined during the last two decades of the twentieth century mainly because of the region's economic expansion (Chapter 9). This expansion also has facilitated the rapid adoption of health and sanitation technology, which along with improved nutrition has diminished the threat posed by mortal disease.

Reduced Human Fertility

The demographic change set in motion as life expectancies go up is far enough along in many Asian nations that human fertility is either approaching or has fallen to or below the replacement level (Table 11.2, columns 2 and 3). As in other prosperous settings, total fertility rates (TFRs) are minimal in Hong Kong, Singapore, and South Korea. If current rates are sustained in China, where urban couples with more than one child have been penalized, and other countries, where no such penalties have been applied, demographic stabilization will be achieved within a few decades.

As in other parts of the world, GDP per capita (Table 11.2, column 1) obviously has an impact on human fertility. Births per woman have fallen to or below 2.1 in every country with a GDP per capita above $3,000 (column 3), with Malaysia and the Philippines being modest exceptions. There

Table 11.2 Human fertility and its determinants in Asia from 1980 to 2007

Country (ranked by column 1)	2007 GDP per capita, PPP (constant 2005 international $) (1)	Total fertility rate		Urban % of total population in 2007 (4)	Adult illiteracy rate 2007 (%)		Female % of labor force		Contraceptive prevalence rate in 2000–2007 (% women 15–49) (9)
		1980 (2)	2007 (3)		Males (5)	Females (6)	1980 (7)	2007 (8)	
Singapore	46,939	1.7	1.3	100	3	8	34.8	41.1	—
Hong Kong	39,953	2.0	1.0	100			33.8	45.3	—
South Korea	23,399	2.8	1.3	81			36.7	40.8	—
Malaysia	12,766	4.2	2.6	69	6	10	35.2	35.2	—
Thailand	7,682	3.2	1.9	33	4	7	46.8	46.8	79
China	5,084	2.5	1.7	42	4	10	43.4	45.7	87
Sri Lanka	4,007	3.5	1.9	15	8	11	31.6	32.3	70
Indonesia	3,506	4.4	2.2	50	5	11	35.0	36.9	60
Philippines	3,217	5.2	3.2	64	7	6	38.5	38.3	47
India	2,600	5.0	2.7	29	23	45	27.6	28.2	47
Vietnam	2,455	5.0	2.1	27			48.0	47.9	74
Pakistan	2,357	7.0	3.9	36	32	60	22.4	18.7	28
Laos	2,044	6.4	3.2	30	23	39	40.5	50.5	32
Cambodia	1,702	5.8	3.2	21	14	32	53.9	48.7	24
Bangladesh	1,172	5.4	2.8	27	42	52	40.3	39.2	54
Nepal	991	5.6	3.0	17	30	56	34.4	45.2	37
Myanmar	854	4.5	2.1	32	6	14	44.6	45.3	37

Source: World Bank (2009b).

is good reason to expect that, as economies continue to expand, fertility will be in this range in poorer countries, which actually experienced the greatest declines in TFRs between 1980 and 2007. In India, where average income is still under the $3,000 threshold, the number of births per woman went from 5.0 to 2.7 during these 27 years. Even though GDP per capita is no higher than $2,500 in Vietnam, the country's TFR fell by more than one-half in the same period and is currently under the replacement level.

Economic expansion tends to go hand in hand with urbanization (Table 11.2, column 4) and the latter change correlates with diminished human fertility. These linkages certainly hold for Asia. Something interesting about the region, though, is that living standards have risen and TFRs have fallen dramatically in at least a few countries where most of the population is still rural. Sri Lanka, Thailand, and Vietnam are cases in point. So are China and India. A lesson here is that, where development has benefited the countryside and not just cities, rural families have responded by having fewer children, just as urban households have done.

As emphasized throughout this book, female economic empowerment, as reflected by literacy and remunerative employment, is a major influence on human fertility. Due to a history of sex discrimination, male illiteracy is appreciably lower than the rate for women throughout Asia, the Philippines being the sole exception (Table 11.2, columns 5 and 6). The gap between the sexes is pronounced and total illiteracy is especially high in South Asia (aside from Sri Lanka) and Cambodia and Laos, both of which are poverty-stricken. In contrast, female illiteracy is not particularly high in countries where GDP per capita has risen above $3,000 and the female portion of the labor force has been rising, especially in nations that are urbanized and relatively prosperous (columns 7 and 8). This has driven up the opportunity cost of raising children, thereby causing more women to do what is required to limit fertility. Their choices are reflected in rates of contraceptive prevalence (column 9) in places like Sri Lanka and Thailand, not just China, that are comparable to what one observes in prosperous places.

Natural Increase

Now that TFRs are less than or equal to 2.1 births per woman in so much of Asia, natural increase in the region is being driven largely by demographic momentum. Death rates (Table 11.3, column 2) will probably go down in the next few decades in India and a handful of the poorest countries: Cambodia, Laos, Myanmar, and Nepal. But elsewhere, the crude death rate (CDR) is at or close to its nadir, so will probably creep up gradually as the population ages.

Table 11.3 Population trends in Asia

Country (ranked by column 1)	2007 GDP per capita, PPP (constant 2005 international $) (1)	Crude death rate in 2007 (per 1,000 population) (2)	Crude birth rate in 2007 (per 1,000 population) (3)	Average annual population growth in 2006–07 (%) (4)
Singapore	46,939	4	10	2.1
Hong Kong	39,953	6	10	0.5
South Korea	23,399	5	10	0.2
Malaysia	12,766	4	21	0.8
Thailand	7,682	8	15	0.3
China	5,084	7	12	0.3
Sri Lanka	4,007	6	19	0.3
Indonesia	3,506	6	19	0.6
Philippines	3,217	5	26	0.9
Mongolia	3,056	6	22	0.5
India	2,600	8	24	0.7
Vietnam	2,455	5	19	0.6
Pakistan	2,357	7	27	1.1
Laos	2,044	7	27	0.9
Cambodia	1,702	9	26	0.9
Bangladesh	1,172	8	25	0.8
Nepal	991	8	28	0.8
Myanmar	854	10	18	—

Source: World Bank (2009b).

By regional standards, the gap between death and birth rates (Table 11.3, columns 2 and 3) is elevated in Malaysia and the Philippines. As already observed, these are the only two nations where GDP per capita has risen above $3,000 and yet human fertility remains above 2.1 births per woman. In all other countries that have broken through this income threshold, low TFRs have caused natural increase to diminish, to 13 per thousand (equal to the birth rate of 19 per thousand minus the death rate of 6 per thousand) in Indonesia and Sri Lanka and below 10 per thousand elsewhere. Among poorer nations, Nepal and Pakistan stand out for being the only two where the gap between birth and death rates is as high as 20 per thousand.

With natural increase decelerating in Asia, the region's population will undoubtedly peak during the next few decades. This is the main reason to anticipate that human numbers as a whole will stabilize this century (Chapter 2). However, the ending of population growth does not mean that demographic issues will disappear. To the contrary, countries such as China and Thailand must soon face the same problem that currently confronts Japan and other wealthy nations—paying pensions and otherwise tending to an aged cohort that makes up a sizable and growing part of the total population. Dealing with this problem will not be easy in places where there is little prospect of living standards reaching Japanese levels and where **population policies** have created gender imbalances and other adverse consequences (Box 11.2).

Box 11.2 The consequences of low human fertility in China

Just as China's economic expansion during the past three decades has been unprecedented, few countries ever have experienced demographic change at such a fast pace.

One consequence of sustaining human fertility at or below the replacement rate for a generation or more is that the population of the world's largest nation has aged rapidly. Equal to 22 years in 1980, median age had climbed to 32 years in 2005. Twenty years from now, it will exceed 40 years and there will be three senior citizens for each minor child (Eberstadt, 2009b, p. 138).

China will find it difficult to maintain high savings rates once there are more retirees, many intent on gradually depleting their bank accounts. Also, older workers tend to be less productive than their younger colleagues, so GDP growth can be expected to decelerate (Eberstadt, 2009b, p. 140). There undoubtedly will be strains on the state pension system, which to say the least is not a model of efficiency.

Frequently identified as a reason for low human fertility, Chinese population policies—in particular, limiting urban households to just one child—have had other effects, most notably an imbalance between the sexes. In other parts of the world, baby boys outnumber baby girls by 3 to 5 percent. In contrast, there are 120 male newborns for every 100 female newborns, mainly because female fetuses are aborted by couples desiring a male heir (Eberstadt, 2009b, p. 148).

But while today's couples are succeeding in reproducing a son, many of them will find themselves without any grandchildren at all, as the number of men with no prospects whatsoever of traditional matrimony increases with time.

11.3 Agricultural Development

Since agriculture and civilization have such a long history in China, India, Java (Indonesia's most populous island), and other Asian settings, population density is elevated in many parts of the region. True, there are Caribbean Islands as well as East African highlands where the number of rural inhabitants per square kilometer of farmland is about as high. Nevertheless, overall population density is much lower in Latin America and the Caribbean and Sub-Saharan Africa than in Asia. The same holds for the Middle East and North Africa, even though few agricultural landscapes are as crowded as Egypt's Nile River Valley.

Going along with elevated population densities are limited opportunities to increase the geographic domain of agriculture. Even where such opportunities exist, on some Indonesian islands for example, there are adverse environmental impacts of agricultural extensification (Chapter 5). Between these impacts and the direct expenses farmers must incur to clear away trees and other vegetation, no major increase in agricultural land use is anticipated in the region (Chapter 3).

Intensified Production

Scarce land and abundant labor in Asia have induced the kind of agricultural intensification characteristic of densely populated settings described in Chapters 3 and 9. Rural wages being low in China, India, and many other countries, incentives for agricultural mechanization are not strong. Despite an increase during the 1980s and 1990s, the number of tractors per square kilometer of arable land remains below levels in Latin America and the Caribbean and the Middle East and North Africa (Table 9.5). As indicated in the previous chapter, this indicator of mechanization is much higher in wealthy nations, where rural labor is scarce relative to other factors of production.

Where land rather than labor is the limiting factor, intensification in Asia has not been driven primarily by increased use of tractors and other equipment, which has little effect on production per hectare. Instead, yields have been raised thanks to fertilization, irrigation, and the planting of crop varieties that respond to increased applications of nutrients and water. As indicated in Table 9.5, the average fertilizer application rate in and around China exceeds average rates in every other part of the world and is more than ten times the application level in Sub-Saharan Africa. As reported in the same table, the irrigated portion of arable land in Asia even exceeds the irrigated portion in the Middle East and North Africa, where water is a limiting factor of crop production for obvious climatic reasons.

In a region as large and diverse as Asia, patterns of agricultural input use are bound to vary. These patterns are reported in Table 11.4, in which countries are ranked from richest to poorest—with Hong Kong and Singapore being omitted because these two small former colonies of Great Britain do not have much farmland. Average living standards and the rate of fertilizer application were positively correlated around 1980 (column 2). While still apparent, the correlation has since become less pronounced (column 3). In nations like the Philippines and Thailand, application rates around 1980 were modest by regional standards, but subsequently went up dramatically. Also, there are poor countries, such as Vietnam, where high population densities in rural areas (column 1) have a lot to do with intensive nutrient applications. Mongolia is an

Table 11.4 Agricultural inputs in Asia, 1979 to 2005

Country (ranked by GDP per capita in 2007)	Rural population density (rural inhabitants per sq. km of arable land) 2005 (1)	Fertilizer use (kg per ha of arable land)		Irrigated land (% of cropland)		Agricultural machinery (tractors per sq. km of arable land)	
		1979–1981 (2)	2003–2005 (3)	1979–1981 (4)	2001–2003 (5)	1979–1981 (6)	2003–2005 (7)
South Korea	569	392	501	59.6	47.1	0.1	13.4
Malaysia	462	462	820	6.7	—	0.8	2.4
Thailand	300	18	141	16.4	26.6	0.1	2.6
China	542	149	315	45.1	35.5	0.8	2.2
Sri Lanka	1,823	180	287	28.3	34.4	1.4	1.1
Indonesia	498	64	155	16.2	12.7	0.0	0.0
Philippines	553	64	145	12.8	14.5	0.2	1.1
Mongolia	95	8	4	6.7	7.0	0.8	0.4
India	489	35	114	22.8	32.7	0.2	1.6
Vietnam	927	30	331	25.6	33.9	0.4	2.5
Pakistan	477	52	160	72.7	81.1	0.5	1.9
Laos	411	3	—	13.3	17.2	0.1	0.1
Cambodia	303	5	3	5.8	7.0	0.1	0.1
Bangladesh	1,432	—	172	17.1	54.3	0.0	0.0
Nepal	968	10	26	22.5	47.1	0.1	0.2
Myanmar	331	11	1	—	17.9	—	0.1

Source: World Bank (2009b).

outlier, resembling the Former Soviet Union in that applications have declined since the fall of communism.

Columns 4 and 5 of Table 11.4 underscore the importance of hydrologic resources in Asia. Due to abundant precipitation, few Malaysian farmers irrigate. Rain-fed agriculture also predominates in Mongolia as well as Cambodia, which lacks the financial resources needed for a major buildup of canals and other infrastructure. In addition, the irrigated portion of arable land in China has declined since 1980, in part because of agricultural extensification in rain-fed areas but also because environmental limits on irrigated agriculture are being reached. Elsewhere in the region, an expansion has occurred. The irrigated portion of arable land has gone up in Bangladesh. A similar change has happened in Vietnam, in part because war-related damage to infrastructure has been repaired. Expansion even has occurred in Pakistan, where nearly three out of every four arable hectares were irrigated around 1980.

Except for Mongolia and Sri Lanka as well as the five poorest countries listed at the bottom of Table 11.4, substantial mechanization has occurred during the past three decades (columns 6 and 7). The number of tractors per square kilometer of arable land has proceeded farther in relatively wealthy nations. In South Korea (and Taiwan), a small portion of the workforce is employed in agriculture and the opportunity cost of farm labor reflects the high wages that can be earned in the industrial and service sectors. As in Japan and other countries belonging to the Organization for Economic Cooperation and Development (OECD), tractors (generally of small size) and other equipment are substituted for labor wherever possible. Similarly, agriculture is mechanized in Malaysia, which is more prosperous than its neighbors. Also, better living standards have led to a sharp increase in the ratio of tractors to agricultural land in Thailand.

Intensification, Extensification, and Output Growth

With fertilizer being applied at higher rates and a considerable investment being made in irrigation infrastructure, agriculture has intensified significantly in China and other parts of East Asia as well as in India and its South Asian neighbors. For all major crops other than fruits and vegetables, annual increases in planted area between 1970 and 2000 (Table 11.5, column 2) were considerably less than annual yield growth (column 3). Obviously, intensification was the primary driver of output growth, which except for minor grains was rapid during the same three decades (column 4).

As fertilization and irrigation have increased, production per hectare has improved. Of all the nations listed in Table 11.6, Malaysia is the only

Table 11.5 Trends in crop area, yield, and output in Asia, 1970 to 2000

	2000 production (million tons) (1)	Average annual change, 1970 to 2000		
		Area (2)	Yield (3)	Production (4)
East Asia				
A. Crops				
Rice	344	0.4	1.8	2.2
Vegetables	313	4.4	1.5	6.0
Roots & Tubers	239	0.1	1.2	1.3
Maize	127	1.0	2.7	3.8
Fruits	105	4.7	1.2	5.9
Wheat	100	0.1	4.0	4.1
Oil Crops	41	2.2	3.7	5.8
Other Cereals	15	−3.5	1.4	−2.2
B. Animal Products				
Total Meat	74	—	—	6.9
Total Eggs	26	—	—	7.7
Total Milk	16	—	—	6.1
Total Wool	0.3	—	—	2.8
South Asia				
A. Crops				
Rice	184	0.5	2.0	2.5
Wheat	98	1.4	2.8	4.3
Vegetables	71	1.7	1.2	3.0
Fruits	40	3.0	1.2	4.3
Pulses	15	0.3	0.2	0.5
Maize	14	0.4	1.0	1.6
Millet	10	−1.7	0.7	−1.0
Oil Crops	10	1.3	1.4	2.6
Sorghum	10	−1.6	0.7	0.5
B. Animal Products				
Total Milk	105	—	—	4.2
Total Meat	8	—	—	3.2
Total Eggs	2	—	—	6.2
Total Wool	1	—	—	0.9

Source: Dixon et al. (2001), pp. 182, 183, 228, and 230.

Table 11.6 Trends in cereal yields, farmed area, food production, value added per agricultural worker, and agricultural protection

Country (ranked by GDP per capita in 2007)	Cereal yield (kg per ha)		Arable and permanent cropland (1,000 hectares)		Food production index (1999–2001 = 100)		Value added per agricultural worker (2000 $)		Nominal rate of assistance*	
	1979–81 (1)	2005–07 (2)	1980 (3)	1999 (4)	1979–81 (5)	2003–05 (6)	1979–81 (7)	2003–05 (8)	1980–84 (9)	2000–04 (10)
South Korea	4,986	6,291	2,196	1,899	77.5	96.9	3,765	11,286	89.4	137.3
Malaysia	2,828	3,336	4,310	7,605	55.6	120.0	3,939	525	-4.6	1.2
Thailand	1,911	2,907	17,970	18,000	79.7	104.6	616	624	-2.0	-0.2
China	3,027	5,322	99,200	135,361	60.8	116.5	161	407	-45.2	5.9
Sri Lanka	2,462	3,636	2,147	1,900	98.1	103.3	642	702	-13.5	9.5
Indonesia	2,837	4,374	19,500	30,987	63.1	120.2	604	583	9.2	12.0
Philippines	1,611	3,164	9,920	10,050	86.1	114.3	1,381	1,075	-1.0	22.0
Mongolia	573	839	1,182	1,322	88.1	72.3	994	907	—	—
India	1,324	2,465	169,130	169,700	68.2	105.3	269	392	1.9	15.8
Vietnam	2,049	4,726	—	—	—	122.0	—	305	—	21.2
Pakistan	1,608	2,639	20,320	21,880	66.3	109.5	416	696	-6.4	1.2
Laos	1,402	3,517	880	955	70.3	115.2	—	459	—	—
Cambodia	1,006	2,482	3,046	3,807	48.9	111.1	—	314	—	—
Bangladesh	1,938	3,769	9,145	8,440	79.3	107.3	232	338	-3.3	2.7
Nepal	1,615	2,271	2,330	2,968	65.4	112.7	156	207	—	—
Myanmar	2,521	3,629	10,023	9,961	88.2	121.7	—	—	—	—

Sources: World Bank (2009b); FAO (1982), pp. 45–56; FAO (2002a), pp. 3–13; Anderson and Martin (2009), pp. 17–27.
*NRA = relative effect (%) of government policies on gross returns to crop and livestock production.

one with modest growth in cereal yields (columns 1 and 2) during the past three decades. This exception arises because the country's agricultural sector lacks comparative advantage in rice and other grains and instead produces tree crops like palm oil and rubber efficiently. Measures to raise cereal yields have not been rewarding and, as in South America, the capital intensity of agriculture has facilitated agriculture's geographic expansion (columns 3 and 4).

Besides Malaysia, with its three-quarters increase in farmed area, just a few countries in the region have experienced agricultural extensification of any consequence since 1980. The two most important are China and Indonesia, with growth of 36 percent and 59 percent, respectively. Agricultural land use also increased by 25 percent in Cambodia and 27 percent in Nepal. In Indonesia, extensification exceeded growth in per hectare output of cereals. In Nepal, the two were comparable. In contrast, yield increases accounted for most of the growth in grain production in China and Cambodia.

Thanks mainly to yield-driven increases, the food production index (Table 11.6, columns 5 and 6) has more than doubled since 1980 in Cambodia, China, Laos, Malaysia, Pakistan, and Vietnam. In countries such as these, output growth has outstripped demographic expansion, which means of course that per capita food supplies have increased. Value added per agricultural worker remains low where population density is high and GDP per capita is modest, although large relative increases have been registered by China, India, and other nations (Table 11.6, columns 7 and 8). In South Korea, where well-paid jobs in nonagricultural sectors have driven up the opportunity cost of farm labor (see above), this productivity measure has reached a high level. In addition, value added per agricultural worker is inflated in that country for exactly the same reason that it is in Japan. To be specific, an obvious lack of comparative advantage for South Korean and Japanese agriculture is masked by agricultural protectionism, which keeps domestic commodity prices well above international values.

Something else that South Korea has in common with the prosperous island-nation just to its east is an elevated **nominal rate of assistance (NRA)** for agriculture (Table 11.6, columns 9 and 10), which is a measure of the subsidies and protection from imports that governments provide farmers. The more general trend in the region has been from implicit taxation (indicated by a negative NRA) in the early 1980s to mild protection by the early twenty-first century.

11.4 Dietary Change, Consumption Trends, and Food Security

As demographic expansion decelerates, food demand trends in Asia are being driven less by simple increases in the number of mouths to be fed and more by other forces. Among the latter are changes in the age-composition of the population. For example, children, who need to eat less than young and middle-aged adults, are becoming less numerous. At the same time, there are more elderly people, whose nutritional requirements are not as great as those of younger adults. Something else that has a substantial effect on food demand in Asia is income growth. Income elasticities of demand for livestock products are greater than income elasticities for most other foods (Chapter 2), so the former becomes a more important part of the human diet as living standards improve.

Dietary change in Asia can be identified by using Rask's (1991) approach of converting all food and beverage consumption into cereal equivalents (Chapter 10). As indicated in Table 11.7, consumption of plant-derived products (column 1) did not change dramatically during the 1980s and 1990s. Noticeable increases occurred in China, Indonesia, India, Myanmar, and Nepal. Elsewhere, however, there were small increases, no change at all, or (in Hong Kong, Sri Lanka, and South Korea) modest declines. Growth in per capita consumption, then, was almost entirely the result of the rising importance of livestock products

Table 11.7 Food consumption trends in Asia, 1979 through 1999

Country (ranked by income per capita in 1999)	Years	Plant product consumption per capita (in cereal equiv. tons) (1)	Animal product consumption per capita (in cereal equiv. tons) (2)	Total food consumption per capita (in cereal equiv. tons) (3)	Total food self-sufficiency* (4)
Hong Kong	1979–1981	0.26	1.14	1.40	0.48
	1989–1991	0.27	1.33	1.60	0.44
	1997–1999	0.24	1.55	1.79	0.25
South Korea	1979–1981	0.32	0.28	0.60	0.79
	1989–1991	0.31	0.48	0.80	0.69
	1997–1999	0.31	0.74	1.04	0.73
Malaysia	1979–1981	0.27	0.45	0.72	1.17
	1989–1991	0.26	0.60	0.86	1.35
	1997–1999	0.27	0.74	1.01	0.98
Thailand	1979–1981	0.24	0.33	0.57	1.33
	1989–1991	0.22	0.37	0.59	1.28
	1997–1999	0.25	0.42	0.66	1.23

(continued)

Table 11.7 (*Continued*)

Country (ranked by income per capita in 1999)	Years	Plant product consumption per capita (in cereal equiv. tons) (1)	Animal product consumption per capita (in cereal equiv. tons) (2)	Total food consumption per capita (in cereal equiv. tons) (3)	Total food self-sufficiency* (4)
Philippines	1979–1981	0.23	0.33	0.56	0.81
	1989–1991	0.24	0.36	0.60	0.74
	1997–1999	0.23	0.42	0.65	0.68
China	1979–1981	0.25	0.16	0.41	0.97
	1989–1991	0.27	0.32	0.59	0.98
	1997–1999	0.29	0.61	0.91	0.97
Sri Lanka	1979–1981	0.26	0.15	0.41	0.79
	1989–1991	0.24	0.15	0.39	0.70
	1997–1999	0.25	0.18	0.44	0.65
Indonesia	1979–1981	0.25	0.12	0.37	0.95
	1989–1991	0.29	0.16	0.46	1.02
	1997–1999	0.32	0.19	0.51	0.97
India	1979–1981	0.22	0.15	0.38	0.97
	1989–1991	0.26	0.19	0.45	1.00
	1997–1999	0.26	0.22	0.48	1.01
Vietnam	1979–1981	0.23	0.16	0.39	0.93
	1989–1991	0.24	0.23	0.46	1.04
	1997–1999	0.26	0.30	0.56	1.07
Pakistan	1979–1981	0.22	0.35	0.57	0.70
	1989–1991	0.23	0.44	0.67	0.73
	1997–1999	0.24	0.51	0.74	0.76
Bangladesh	1979–1981	0.22	0.10	0.32	0.93
	1989–1991	0.23	0.10	0.33	0.91
	1997–1999	0.24	0.11	0.35	0.89
Laos	1979–1981	0.23	0.15	0.38	0.97
	1989–1991	0.23	0.18	0.41	0.99
	1997–1999	0.23	0.25	0.48	1.02
Cambodia	1979–1981	0.19	0.09	0.28	0.82
	1989–1991	0.20	0.20	0.40	0.97
	1997–1999	0.21	0.22	0.43	0.96
Nepal	1979–1981	0.20	0.26	0.46	1.00
	1989–1991	0.27	0.29	0.55	0.99
	1997–1999	0.25	0.29	0.53	0.99
Myanmar	1979–1981	0.26	0.17	0.43	1.08
	1989–1991	0.29	0.16	0.46	0.96
	1997–1999	0.31	0.18	0.49	1.00

*Total food self sufficiency equals total domestic food production divided by total domestic food comsumption.

(column 2). In terms of cereal equivalents, goods like meat and eggs account for large shares of total food consumption in Hong Kong, Malaysia, and South Korea. In all but a few nations where GDP per capita is under $3,000, most calories consumed by humans continue to be plant derived, even though relative increases in livestock consumption during the last two decades of the twentieth century exceeded relative growth in plant-derived calories. Only in Bangladesh, which has the lowest per capita consumption of food in the region, have the livestock and plant-derived portions of the human diet remained steady.

Particularly where people were eating more livestock products in the late 1990s than they did 20 years earlier, food self-sufficiency—defined simply in Table 11.7 (column 4) as domestic production (expressed in cereal equivalents) divided by domestic consumption (column 3)—has diminished. In South Korea, the ratio of production to consumption slid from 0.79 in 1979 to 1981 to 0.73 in 1997 to 1999. During the same period, there were declines in Malaysia (1.17 to 0.98), Thailand (1.33 to 1.23), and the Philippines (0.81 to 0.68). The ratio went up markedly in Cambodia and Vietnam, which were still recovering from war-related devastation in the late 1970s. Pakistan is a special case in that per capita livestock consumption is significantly above levels in nearby countries with similar living standards. The same nation also has become a little more (not less) food self-sufficient even though a lot more animal products are being eaten. Another special case is Sri Lanka, which has experienced much less dietary change although its ratio of production to consumption has deteriorated significantly. Elsewhere, the ratio changed little during the 1980s and 1990s.

Where accelerated GDP growth is happening, especially in non-agricultural sectors, a self-sufficiency ratio below 1.00 is not a great concern. There is no need to worry about food supplies in Hong Kong and South Korea, for example, because these countries can easily afford the imports needed to close gaps between national production and domestic consumption. By the same token, feed grains can be purchased from other nations so that the growing demand for animal products can be satisfied. The situation is different, though, in poorer countries. If agricultural imports are not affordable and if food aid is not available in sufficient quantities, then food insecurity is a concern if agricultural production consistently lags behind food consumption. This seems to be the situation in Pakistan, for instance.

In most countries where the ratio of domestic production to consumption went up or down a little or did not change at all, per capita livestock consumption has not grown very much, remains modest, or both. However, the situation is different in China, where the ratio stood at 0.97 as the HRS was being adopted and remained at exactly the same level two decades later, on the eve of the twenty-first century. No one argues that failure to achieve a higher degree of self-sufficiency heralds impending famine in Asia's largest nation. Fifteen years ago, environmentalist Lester Brown contended that rapid economic growth in China could cause the country's agricultural imports to balloon, thereby driving up global commodity prices. The end result, he added, would be aggravated food insecurity in poorer parts of the world, most notably Sub-Saharan Africa (Brown, 1994). Other analysts subsequently discounted this possibility. In a statistical analysis, Huang, Rozelle, and Rosegrant (2000) found no reason to expect a major increase in Chinese imports of feed grains and other commodities, since consumption growth is not expected to exceed increases in production by a wide margin. Thus, the troubling developments that Brown (1994) foresaw remain unlikely.

Finally, we note that frequently used measures of food security generally correlate in Asia with improved diets. These measures are provided in Table 11.8, in which countries are ranked according to food consumption per capita (column 1). In places with the highest average consumption, infant mortality (column 2), the prevalence of childhood malnutrition (columns 3 and 4), and the incidence of poverty (column 5) are all negligible. This is true even where the Gini index of income concentration (column 6) is elevated, as it is in Hong Kong. As always, there are interesting exceptions to general tendencies. One is Sri Lanka, which is justly famous for its public health and family planning services—as reflected in infant mortality and poverty rates that are unexpectedly low in light of modest per capita food consumption. A less positive outlier is Pakistan. There, food consumption per capita exceeds levels in the Philippines and Thailand, where average incomes are higher. Nevertheless, infant mortality and the prevalence of child malnutrition are both inordinately high. These disturbing indicators are related in part to the sluggish GDP growth that India's western neighbor registered during the 1990s.

Something else can be discerned from Table 11.8, which is that low economic inequality—as expressed by the Gini coefficient—does not correlate with limited poverty. In general, countries with coefficients below 40 have high poverty rates, just as the incidence of poverty is low where the inequality measure is above 40. Clearly, economic equality is not a good thing if it is only a consequence of widespread economic deprivation. By the same token, inequality is not discouraging if this is the result of a dynamic economy in which the earnings of some people are being pulled up faster than those of others. The latter experience happens to have been shared by a number of Asian nations in recent years.

Table 11.8 Food insecurity indicators in Asia

Country (ranked by column 1)	Food consumption per capita in 1997–1999* (in cereal equivalent tons) (1)	Infant mortality rate in 2007 (per 1,000 live births) (2)	Malnutrition Prevalence			Gini coefficient in late 1990s or early 2000s (6)
			Percent of abnormally short 5-year-olds in late 1990s or early 2000s (3)	Percent of abnormally underweight 5-year-olds in late 1990s or early 2000s (4)	Percent of population living on $1.25/day or less in late 1990s or early 2000s (5)	
Hong Kong	1.79	—	—	—	—	43
South Korea	1.04	4	—	—	2.0	32
Malaysia	1.01	10	—	—	2.0	38
China	0.91	19	22	7	15.9	42
Pakistan	0.74	73	42	31	22.6	31
Thailand	0.66	6	16	7	2.0	42
Philippines	0.65	23	34	21	22.6	45
Vietnam	0.56	12	36	20	21.4	38
Nepal	0.53	43	49	39	55.1	47
Indonesia	0.51	25	29	24	16.7	39
Myanmar	0.49	74	41	30	—	—
India	0.48	54	48	44	41.6	37
Laos	0.48	56	48	36	44.0	33
Sri Lanka	0.44	16	18	23	14.0	41
Cambodia	0.43	70	44	28	40.2	41
Bangladesh	0.35	47	48	39	49.6	31
Mongolia	—	35	28	5	22.4	33

Source: World Bank (2009b).
*From Table 11.7, column (3).

11.5 Summary

By the turn of the twenty-first century, Asia was moving to the center of the world economic stage, thanks to substantial increases in GDP never matched in any other part of the world. Since diminished natural increase has coincided with economic expansion, there has been a corresponding improvement in average living standards. It is undeniable that hundreds of millions of Asians remain impoverished. Also, economic performance in some countries compares unfavorably to regional trends, just as growth has been interrupted at times in nations where GDP per capita is much higher today than what it was a generation ago. Nevertheless, the region's economic accomplishments are impressive—impressive enough that issues that formerly aroused debate and concern only in the OECD, such as food safety and environmental protection, are becoming progressively more important in South Korea, Taiwan, and other nations.

Agriculture, which is the focus of much of the environmental debate, has had an important role in Asia's development. The Green Revolution allowed famine to be averted in places like the Indian subcontinent, even when average incomes were not rising quickly. During the last two decades of the twentieth century, China's economic transformation began with the defiance of doctrinaire communism in the countryside. To be sure, challenges remain. One is to meet the increased demand for livestock products resulting from higher living standards. Another is to take full advantage of all technologies for agricultural development, including genetic modification (Box 11.3).

Asia can be counted on to address challenges such as these. As it does so, the continent will grow more prosperous and food secure. The entire world will benefit as well, as improved living standards cause hundreds

---

Box 11.3 GM foods in India

Opposed strenuously by environmental activists at home and abroad, agricultural biotechnology could be harnessed immediately in the world's second most populous nation.

As described in Chapter 3, farmers' claims on water are substantial in India and neighboring nations. Aside from trimming irrigation subsidies, which cause hydrologic resources to be squandered (Chapter 5), new technological approaches are needed for the sake of environmentally sustainable agricultural development, particularly since food demands are certain to continue growing for many years.

> India's agricultural scientists are rising to the challenge—for example, by developing a genetically modified variety of the purple eggplant, which is a key ingredient of curry dishes. But Luddites still have the ear of the Environment Minister. In February 2010, he banned the new variety, which is pest resistant because it contains *Bacillus thuringiensis* (Bt) bacteria (Chapter 5).
>
> This decision has been denounced—not only by scientists but also by farmers aware that the adoption of Bt cotton has allowed India to become the world's second-leading producer of that commodity as it has reduced pesticide applications. But opponents of agricultural biotechnology are probably fighting a losing battle. Sooner or later, India will join the ranks of nations where genetically-modified foods are produced and consumed (Chandran, 2010).

of millions of Asians to be better customers of goods and services produced in other regions.

Key Words and Terms

nominal rate of assistance (NRA)
population policy

state-owned enterprise (SOE)

Study Questions

1. What obstacles to sustained economic growth must still be overcome in China? In India?
2. What Asian countries have experienced the greatest relative declines in human fertility in recent decades?
3. Why does a policy of one child per family no longer make economic sense for China?
4. Why is fertilizer used more intensively in Asia than in other parts of the world?
5. In what ways was the Green Revolution particularly well suited to conditions in Asia?
6. Analyze the declines in food self-sufficiency—defined as domestic consumption (in cereal equivalents) divided by domestic production—that a number of Asian countries experienced in the late twentieth century.

12

Latin America and the Caribbean

Settled right after the last Ice Age, when low sea levels allowed the New World's first inhabitants to travel from Siberia to Alaska, the Western Hemisphere experienced agricultural revolutions unconnected with those of Eurasia and Africa. Domestication of crops such as corn and potatoes gave rise to a series of civilizations in Mesoamerica and the Andes. During the 1520s and 1530s, the Spanish conquest of the Aztec and Incan empires ushered in three centuries of colonial rule, which came to an end on the American mainland early in the nineteenth century.

Sharp ethnic and class divisions dating back to the conquest remained largely intact long after independence was achieved. Also, Latin America and the Caribbean have undergone prolonged economic stagnation at times and prospered in other periods. For example, per capita gross domestic product (GDP) declined by one-third in Mexico between 1800 and 1850 (Engerman and Sokoloff, 1997), which set the stage for the loss of the northern third of the country to the United States during the 1840s as well as a French invasion in the 1860s. However, there was an upswing in commerce during the last third of the nineteenth century and the early part of the twentieth, as the Industrial Revolution created new demands for primary commodities produced throughout the Western Hemisphere and as transoceanic shipping costs were driven down by the introduction of steam power, metal hulls, refrigeration, and other advances. Globalization came to a halt during the 1930s, when protectionism in the United States and other leading nations caused international trade to collapse.

Shaken by the Great Depression, Argentina, Brazil, Mexico, and other countries held back from full engagement of the global economy after the Second World War, and instead tried to develop by sheltering domestic industry from foreign competitors. This approach, called import-substituting industrialization (ISI), eventually led to large fiscal and trade deficits (Chapter 7) and the **Debt Crisis** of the 1980s (Fraga, 2004). Many years were required for the region to come to terms with that crisis and the protectionist approach to development that precipitated it. Also, Latin America and the Caribbean are still dealing with the pronounced socioeconomic inequality that is in part a legacy of Iberian colonialism.

12.1 Trends in GDP per Capita

Although Asia has experienced greater relative improvements during the last three or four decades, living standards south of the U.S. border still compare favorably with those of other developing regions. GDP per capita in the region was $9,700 in 2007, as opposed to $5,000 in East and Southeast Asia, $2,530 in South Asia, $7,400 in the Middle East and North Africa, and $1,870 in Sub-Saharan Africa (Table 9.1).

Latin America and the Caribbean will cease being the most prosperous part of the developing world if it undergoes another period like the 1980s, when GDP expansion failed to match population increases in most countries (Table 12.1, column 1). Prior to the **Lost Decade**, the region had sizable trade and fiscal deficits, because of attempts to stimulate industrialization by manipulating market forces. Deficits were covered thanks to heavy borrowing from international banks, which were eager to lend **petrodollars** deposited by oil exporters that had benefited from energy price increases. Export growth during the 1980s (column 2) continued to be anemic. Moreover, foreign banks refused to lend anywhere in the region after Mexico stopped paying interest on its foreign debt in 1982. As a result, a number of governments slashed spending on infrastructure, education, and public health in addition to printing money to cover the difference between remaining expenditures and tax revenues. Printing money in turn ignited inflation (column 3), which helped bring capital accumulation to a halt in all but a few nations (column 4).

Aside from Colombia, the only positive outlier of importance during these years was Chile. Major steps toward free trade having been taken during the 1970s, annual export growth in the country was nearly 7 percent during the 1980s. Though high by Asian standards, Chilean inflation compared favorably to what neighboring lands were suffering. Thus, conditions for investment were relatively encouraging. Thanks to an

Table 12.1 Economic trends in Latin America and the Caribbean from 1980 to 2007

Country (ranked by column 1)	A. Average annual rate of growth, 1980–90 (%)			Country (ranked by column 5)	B. Average annual rate of growth, 1990–99 (%)				Country (ranked by column 9)	C. Average annual rate of growth, 2000–07 (%)			Remittances ÷ exports (%), 2004–2007 (12)	Freedom status in 2008** (13)
	GDP per capita (1)	Exports (2)	Inflation (3)	Gross invest. (4)		GDP per capita (5)	Exports (6)	Inflation (7)	Gross invest. (8)		GDP per capita (9)	Exports* (10)	Inflation* (11)		
Chile	2.6	6.9	20.7	4.3	Chile	5.7	9.7	8.6	11.4	Peru	3.6	8.2	3.6	6.9	F
Colombia	1.5	7.5	24.8	0.6	Dom. Rep.	3.8	7.5	9.8	7.4	Panama	3.5	4.9	1.4	1.1	F
Dom. Rep.	0.9	4.5	21.6	4.3	Peru	3.7	9.0	28.7	9.0	Ecuador	3.1	5.7	4.9	20.6	PF
Jamaica	0.8	5.4	18.6	4.1	Argentina	3.6	8.7	6.2	9.1	Dom. Rep.	3.1	4.2	14.0	24.0	F
Brazil	0.7	7.5	284.0	0.2	Uruguay	3.0	7.0	36.0	8.9	Costa Rica	2.9	7.9	9.9	4.6	F
Costa Rica	0.3	6.1	23.6	5.2	El Salvador	2.8	11.7	8.1	7.2	Colombia	2.9	4.6	7.5	12.8	PF
Uruguay	−0.2	4.3	61.3	−8.2	Panama	2.4	0.0	2.0	12.1	Chile	2.8	6.3	6.0	–	F
Honduras	−0.4	1.1	5.7	−0.7	Costa Rica	2.1	9.7	16.8	3.4	Honduras	2.7	4.6	9.6	40.7	PF
Ecuador	−0.5	5.4	36.4	−2.9	Bolivia	1.8	4.9	9.4	10.1	Argentina	2.5	5.3	9.4	0.8	F
Paraguay	−0.5	12.2	24.4	−1.4	Guatemala	1.6	6.4	10.7	5.0	Venezuela	2.3	−0.1	25.1	0.2	PF
El Salvador	−0.9	−3.4	16.3	2.2	Brazil	1.5	4.9	264.3	3.1	Uruguay	2.3	6.0	8.2	1.4	F
Mexico	−1.0	7.0	71.5	−3.4	Colombia	1.4	5.2	20.5	7.5	Nicaragua	1.7	9.3	7.8	38.9	PF
Venezuela	−1.5	2.8	19.3	−5.4	Nicaragua	1.4	10.3	38.5	12.6	Brazil	1.7	8.9	8.0	1.5	F
Panama	−1.6	−0.9	1.9	−12.8	Mexico	0.9	14.3	19.3	3.9	Bolivia	1.4	7.8	6.0	16.1	PF
Guatemala	−1.7	−1.8	14.6	−2.1	Honduras	0.3	2.0	19.8	6.0	Paraguay	1.2	4.6	9.9	5.5	PF
Haiti	−2.1	1.2	7.5	−3.4	Ecuador	0.1	4.4	33.7	1.1	Mexico	1.2	7.1	7.7	8.3	F
Argentina	−2.2	3.8	391.1	−8.3	Paraguay	−0.3	5.1	13.7	1.5	El Salvador	1.2	6.2	3.1	69.4	F
Bolivia	−2.2	1.0	327.2	−10.7	Venezuela	−0.5	5.6	47.6	2.9	Guatemala	1.0	1.5	4.6	48.7	PF
Peru	−2.5	−1.6	231.3	−5.0	Jamaica	−0.8	0.1	25.8	3.9	Jamaica	0.0	–	9.3	38.6	F
Nicaragua	−4.6	−3.9	422.3	−4.5	Haiti	−0.8	2.4	23.3	1.7	Haiti	−1.1	–	14.5	165.3	PF

Sources: World Bank (2001), pp. 278–279 and 294–295; World Bank (2009b); and Freedom House (2008).

*Trends in exports and inflation are for 2000–2004.

**F = free; PF = partly free; and NF = not free.

economic revival after 1984, yearly increases in GDP per capita averaged 2.6 percent for the entire decade.

In contrast, triple-digit inflation plagued some countries and threatened elsewhere. There has never been an inflation-ridden setting in which all prices march up in unison. Consequently, businesses that face inflation must continuously review the terms under which output is sold and inputs are purchased. At the same time, households must be on constant alert lest inflation erode the real value of their savings. Toward the end of the Lost Decade, the costs of this vigilance had risen high enough that the political will required for macroeconomic reform was mustered in places like Argentina, Brazil, and Peru. Fiscal deficits were trimmed, which reduced money-creation and eased inflation. This was a key element of structural adjustment (Chapter 8), which the International Monetary Fund (IMF) was prescribing for indebted countries. Other prescriptions for recovery included trade liberalization and deregulation of domestic markets, freer capital flows, the privatization of state-owned enterprises (SOEs), and stronger property rights (Fraga, 2004).

Once implemented, structural adjustment facilitated economic growth. Diminished inflation (Table 12.1, column 7) stimulated investment (column 8). Trade revived as well, thanks in no small part to major reductions in tariff and nontariff barriers.[1] Annual export growth during the last decade of the twentieth century (column 6) exceeded what it had been during the 1980s in most countries. Between 1990 and 1997, GDP per capita for Latin America and the Caribbean as a whole rose by 2.0 percent per annum (Ocampo, 2004). Of the nations listed in Table 12.1, Haiti, Jamaica, Paraguay, and Venezuela were the only four where human numbers rose faster than GDP between 1990 and 2000 (column 5).

Living standards stagnated for about five years beginning in 1997. Feeling vindicated as they have reflected on what they often called the lost half-decade, critics of market-friendly policies complain that GDP expansion after structural adjustment turned out to be modest and uneven (Ocampo, 2004). Fraga (2004) disputes this, contending that nations with sustained macroeconomic stability because of sound fiscal and monetary policies also have enjoyed consistent economic expansion. As already indicated, Chile founded this club. Colombia, which is the

---

[1] Lora (2001, cited by Fraga, 2004) reports that the average tax on imports for the 12 largest Latin American nations fell from 49 percent in 1985 to less than 20 percent in 1994. Whereas nontariff barriers applied to 38 percent of imports in the middle 1980s, this share was a little over 6 percent 10 years later.

third most populous nation south of the U.S. border, also has had sound economic policies for a long time. With the victories achieved over left- and right-wing rebels and terrorists under President Alvaro Uribe, whose father was a victim of political murder and who won office in 2002, the Colombian economy has grown at a steady clip since the turn of the twenty-first century (Table 12.1, column 9).

In Latin America's two leading nations, sound macroeconomic management and market-friendly policies are now sufficiently entrenched to have survived two or more changes in government. Mexico abandoned protectionism when it joined the **North American Free Trade Agreement (NAFTA)** on January 1, 1994. Suffering from some of the worst inflation in the world two decades ago, Brazil now has a currency that is at least as stable as the U.S. dollar and has experienced economic growth by embracing globalization. One consequence of this global economic engagement is that the country now exports more agricultural commodities than any other tropical or subtropical nation (*The Economist*, 2010a). For instance, soybeans from Brazil have helped China to satisfy its burgeoning demand for livestock products.

Another place to transcend its inflationary and protectionist past is Peru. Alan Garcia, the country's current leader, presided over skyrocketing prices and economic decline during an earlier term in office, in the 1980s. Elected again in 2006, he has embraced the macroeconomic stability and openness to global markets that his two immediate predecessors practiced. As a result, the Peruvian economy has expanded at a fast pace since the early 1990s (Table 12.1, columns 5 and 9).

Brazil, Chile, Colombia, Mexico, and Peru are five of the seven Latin American and Caribbean nations, which as a group produce about 90 percent of the region's GDP. The other two are Argentina and Venezuela. Much wealthier than its neighbors from the late 1800s through the middle of the twentieth century (Engerman and Sokoloff, 1997), Argentina maintained an inflexible exchange rate all through the 1990s, which created currency overvaluation and hurt the balance of trade. Meanwhile, public-sector indebtedness mounted, which set the stage for a severe financial crisis in late 2001. GDP per capita grew after the turn of the century, although less than during the previous decade (Table 12.1, columns 5 and 9). Even though few countries have the resources that Argentina possesses, its economic fortunes have been eclipsed by those of Brazil and Chile—a pair of nations with inferior living standards not so long ago.

Like Russia and various Middle East nations, Venezuela has benefited from rising prices of oil (its principal export) since 2002. However, the country's economic performance has been undermined due to socialist experimentation by its autocratic leader, Hugo Chávez, who won office in 1998 largely because of popular frustration over many years of stagnating living standards. Chávez also has clamped down on independent media and has curtailed individual liberties, so Venezuela can at best be regarded as partly free (Table 12.1, column 13). The same is true of other nations in the region governed by his allies, such as Bolivia, Ecuador, and Nicaragua. By and large, these nations have small economies and populations and many of them, like other small countries in the region, receive substantial foreign exchange in the form of remittances (column 12).

Heavy-handed state control of the economy and other features of Chávez's **Bolivarian Revolution** closely resemble the policies responsible for prolonged recession throughout the region during the 1980s. To repeat, Brazil, Mexico, and other nations have experienced sustained economic improvement since jettisoning these policies. This improvement is measured not just in terms of growth over time in GDP per capita, but also in terms of greater resilience in the face of periodic downturns in the global business cycle (Box 12.1). GDP per capita might have improved since 2000 in places like Venezuela and Ecuador, thanks mainly to higher energy prices. However, recent history is not reassuring about the long-term returns of the approach they have adopted. Indeed, a single event, such as sliding oil prices, is enough to cause these economies to stall, thereby providing yet another lesson about the consequences of not adhering to the standard model (Chapter 8).

Box 12.1 Prosperous economies sneeze, but Latin America no longer catches pneumonia

Downturns in the United States and other leading economies used to devastate developing nations in the Western Hemisphere. The Great Depression of the 1930s was a case in point. After several decades of expanded trade and rising living standards, Latin America suffered a collapse of export prices and volumes as well as economic contraction due to a worldwide slump, brought about by the spread of protectionism and other misguided policies at the beginning of the decade (Chapter 6). Likewise, the recession of the early 1980s marked the beginning of a Lost Decade for the region.

(continued)

> (*continued*)
>
> As reported in Chapter 10, events have unfolded differently in recent years. Prosperous economies have recovered slowly from the downswing in the global business cycle that began in 2007. In contrast, Brazil, Chile, Colombia, Peru, and other nations south of the United States largely avoided prolonged declines in GDP and resumed growth well before the end of 2009.
>
> What explains the region's newfound resilience? The short answer has to do with the standard model. Perhaps because they learned painful lessons about the consequences of macroeconomic mismanagement just two decades ago, in the late 1980s and early 1990s, national authorities in Brasilia, Lima, and other capitals have been careful to avoid pitfalls such as chronic fiscal deficits and unrealistic exchange rates (Porzecanski, 2009).
>
> Experts used to be dispatched from Washington to those same capitals to implement structural adjustment. Perhaps Latin American economists should now be given the opportunity to return the favor, for the sake of ending macroeconomic mismanagement in the United States.

12.2 Population Dynamics

As in Asia and the Middle East and North Africa, noneconomic measures of human well-being have improved at least as much as average incomes in Latin America and the Caribbean in recent decades. Fraga (2004) points to gains in school enrollment and adult literacy as well as infant survival and life expectancy in the late twentieth century. Even in places like Venezuela, these noneconomic measures have improved even during times when GDP per capita was falling. In general, noneconomic indicators of well-being have converged toward indicators in prosperous nations in spite of the absence of economic convergence, conventionally defined.

Reduced Human Fertility

Mortal threats to life are contained at least as well in Latin America and the Caribbean as in the rest of the Global South. Life expectancy at birth in the region is 73 years—versus 70 to 72 years in East and Southeast Asia, the Middle East and North Africa, and Eastern Europe and the

Former Soviet Union, 65 years in South Asia, and 51 years south of the Sahara (Table 9.2). As is to be expected, progress against mortal illness has been followed by diminished human fertility.

Though less dramatic than the fertility decline that Asia experienced in the late twentieth century (Table 11.2), total fertility rates (TFRs) have fallen markedly since 1980 throughout the Western Hemisphere (Table 12.2, columns 2 and 3). Distinct from demographic conditions in Asia, where human fertility has reached the replacement level in nearly every country with a GDP per capita of $3,000 or more, no similar threshold exists in Latin America and the Caribbean, at least not yet. Families are larger in Panama, Venezuela, and other countries with living standards (column 1) well above Asian norms. The same is true of poorer nations, including Bolivia and Honduras. Only five countries have TFRs of 2.2 births per woman or less, although it is significant that the two most populous nations—Brazil and Mexico—are among those five.

Fertility differences cannot really be ascribed to the historical predominance of Roman Catholicism in the region. TFRs will soon reach 2.1 births per woman in Argentina and Colombia, each of which has a constitution that establishes Catholicism as the state religion. In these countries and all their hemispheric neighbors, religious affiliation is not preventing women and their partners from acting on the desire to have fewer children. A stronger sociocultural influence may be ethnicity. In particular, large families are the rule among indigenous peoples, which have been marginalized since the Spanish conquest. This is reflected in high TFRs for Bolivia, Guatemala, and Paraguay, where these groups remain impoverished and comprise a majority of the population. But nearly everywhere else, women generally are bearing fewer than three children each.

Outside of Haiti and a few other countries, average incomes in Latin America and the Caribbean are the envy of most of the developing world. The region is also quite urbanized (Table 12.2, column 4), much more so than Asia for example. In addition, levels of educational attainment do not differ greatly between the sexes. True, male illiteracy (column 5) is appreciably lower than female illiteracy (column 6) in a handful of places: El Salvador and Guatemala in Central America as well as Bolivia and Peru in the Andes. Elsewhere, illiteracy rates are comparable. In spite of an increase since 1980, female participation in the labor force (columns 7 and 8) is a little below that of Asia. Nevertheless, contraceptive use (column 9) is high. Indeed, choices about birth control made by women south of the U.S. border are pretty much the same as choices made by females in affluent parts of the world.

Table 12.2 Human fertility and its determinants in Latin America and the Caribbean from 1980 to 2007

Country (ranked by column 1)	2007 GDP per capita, PPP (constant 2005 international $)	Total fertility rate		Urban % of total population in 2007	Adult illiteracy rate in early 2000s (%)		Female % of labor force		Contraceptive prevalence rate in 2000–2007 (% women 15–49)
		1980	2007		Males	Females	1980	2006	
	(1)	(2)	(3)	(4)	(5)	(6)	(7)	(8)	(9)
Mexico	13,307	4.6	2.1	77	6	9	27.3	35.6	71
Chile	13,108	2.7	1.9	88	3	4	29.3	36.0	58
Argentina	12,502	3.3	2.3	92	2	2	27.4	41.1	—
Venezuela	11,480	4.2	2.6	93	5	5	27.4	38.9	77
Panama	10,757	3.7	2.6	72	6	7	30.2	37.1	—
Uruguay	10,592	2.7	2.0	92	3	2	30.8	43.5	84
Costa Rica	10,510	3.6	2.1	63	4	4	22.9	34.5	96
Brazil	9,034	4.0	2.2	85	10	10	31.1	43.3	77
Colombia	8,109	4.0	2.4	74	8	7	25.2	46.1	78
Peru	7,400	4.9	2.5	71	5	15	26.4	43.9	46
Ecuador	7,035	5.0	2.6	65	13	17	21.1	40.0	73
Dominican Republic	6,333	4.3	2.4	68	11	10	25.1	43.8	61
Jamaica	5,741	3.7	2.4	53	19	9	46.5	44.0	69
El Salvador	5,481	4.9	2.7	60	15	20	33.2	38.8	67
Guatemala	4,308	6.1	4.2	48	21	32	24.4	37.1	43
Paraguay	4,186	5.2	3.1	60	4	7	30.7	45.2	73
Bolivia	3,972	5.5	3.5	65	4	14	32.8	45.0	58
Honduras	3,585	6.2	3.3	47	16	17	27.6	31.7	65
Nicaragua	2,427	6.1	2.8	56	34	33	31.0	31.0	69
Haiti	1,090	6.1	3.8	45	48	52	44.4	33.4	32

Source: World Bank (2009b).

Natural Increase

Diminished mortality long ago ceased being the main driver of natural increase in Latin America and the Caribbean. Haiti, where public sanitation and health care are appalling, is about the only place where lower death rates (Table 12.3, column 2) could be achieved, through investment in drinking water systems, hospitals and clinics, and so forth. The same sort of investment also could drive down the crude death rate (CDR) by one or two per thousand in Bolivia, Guatemala, and other poor nations. Elsewhere, that rate cannot fall anymore and, if anything, will rise this century as populations age. Uruguay is an unusual case. More like a

Table 12.3 Population trends in Latin America and the Caribbean

Country (ranked by column 1)	2007 GDP per capita, PPP (constant 2005 international $) (1)	Crude death rate in 2007 (per 1,000 population) (2)	Crude birth rate in 2007 (per 1,000 population) (3)	Average annual population growth in 2006–2007 (%) (4)
Mexico	13,307	5	19	0.5
Chile	13,108	5	15	1.5
Argentina	12,502	8	18	1.3
Venezuela	11,480	5	22	2.3
Panama	10,757	5	21	2.0
Uruguay	10,592	9	15	0.5
Costa Rica	10,239	4	18	2.4
Brazil	9,034	6	19	1.7
Colombia	8,109	6	19	1.8
Peru	7,400	6	21	1.8
Ecuador	7,035	5	21	1.9
Dom. Rep.	6,333	6	23	1.9
Jamaica	5,741	6	17	0.9
El Salvador	5,481	6	23	1.5
Guatemala	4,308	6	33	2.4
Paraguay	4,186	6	25	2.5
Bolivia	3,972	8	27	2.1
Honduras	3,585	6	28	2.5
Nicaragua	2,427	5	25	2.0
Haiti	1,090	9	28	2.0

Source: World Bank (2009b).

member of the Organization for Economic Cooperation and Development (OECD) in demographic terms than a developing country, it has the region's highest death rate, equaled only by Haiti's. In addition, Uruguay's CDR is bound to rise as its population continues to mature. The same thing is happening in Argentina.

With CDRs at or near their respective nadirs and with TFRs equal to or approaching the replacement level in all large countries as well as a number of smaller states, natural increase in Latin America and the Caribbean is mainly a consequence of demographic momentum, which is yet another demographic parallel with Asia. A gap between the birth rate (Table 12.3, column 3) and the CDR persists even in Argentina, Uruguay, and other places where the demographic transition is far advanced. In other countries, adolescents and young adults comprise a large segment of the population, so crude birth rates (CBRs) exceed CDRs and natural increase will persist farther into the future.

Perhaps the most important demographic difference between Asia and the developing nations of the Western Hemisphere has to do with international migration. The attraction of remunerative employment in the United States, Spain, and other prosperous countries has caused millions of people to relocate from the Caribbean, Central America, Mexico, and even the Andean Region. This immigration, which diminishes overall population growth (Table 12.3, column 4), is unlikely to cease as long as international differences in GDP per capita remain sizable.

12.3 Agricultural Development

In contrast to the general demographic congruence between Asia and Latin America and the Caribbean, agricultural differences between the two regions are considerable. These differences derive largely from varying resource endowments. The scope for increasing farmed area is much greater in the New World—in South America, to be specific—than in China, India, and neighboring lands (Chapter 3). But just as natural resources are abundant in much of the region, labor is relatively scarce. The ratio of rural population to arable land is elevated in two Andean nations (Colombia and Ecuador), four countries in Central America (Costa Rica, El Salvador, Guatemala, and Honduras), as well as the Dominican Republic, Haiti, and Jamaica in the Caribbean. Elsewhere, ratios under 200 people per square kilometer are the norm. Few agricultural landscapes in Asia are as lightly populated.

Within Latin America and the Caribbean, agriculture has developed in diverse ways. By no means do modes of production in the temperate, southern reaches of the American mainland seem exotic to farmers visiting from the United States. The crops are the same: corn, soybeans, and wheat in Argentina and Uruguay and soybeans in southern and central Brazil. Machinery is widely used, which is expected since labor is scarce. Rural areas in Argentina and Uruguay, for example, are more sparsely populated than the U.S. countryside, where there are 33 people per square kilometer of arable land (Table 10.5). Also, commercial farming is complemented by agribusinesses that supply a wide range of inputs and marketing services efficiently. Chilean agriculture has expanded rapidly in large part because local service providers are adept at adding value to the horticultural products in which the country holds a comparative advantage.

Considerable differences exist between agriculture in Latin America's Southern Cone and farming closer to the equator. In part, the differences have to do with commodities produced in the tropics and subtropics: sugar and bananas on lowland plantations and coffee at higher altitudes. But in other places, people raise crops that their distant ancestors domesticated, but which now are produced more efficiently far away in temperate latitudes. One indigenous food is corn, which was first cultivated thousands of years ago in Mesoamerica. Another is the potato, which is from the Andes. By and large, production of these traditional crops in their places of origin involves limited use of machinery, commercial fertilizer, and other modern inputs. Yields are minimal.

Rural hinterlands in the Andes and Mesoamerica are among the poorest parts of the hemisphere. Even after most Native Americans died from diseases brought over to the New World by Columbus and the Europeans who followed him, the two areas offered large subjugated populations that Spaniards could put to work in mines and on large estates (Engerman and Sokoloff, 1997). In spite of land reforms during the twentieth century, most rural Native Americans in places like Southern Mexico (e.g., the state of Chiapas, where there was an uprising in the middle 1990s), Guatemala, Peru, and Bolivia possess tiny holdings, called *minifundios*. Given the choice, many of these people would work outside agriculture. However, their lack of educational credentials as well as the economic isolation associated with a deficient road network relegate them to subsisting on their own limited harvests.

People from outside the region who visit the Andean or Mesoamerican countryside are often struck by the juxtaposition of commercial farms, which can be quite profitable, and impoverished *minifundios*. The same juxtaposition occurs in the Caribbean (although most of the region's workforce is now employed in light manufacturing, tourism, and other nonagricultural sectors), northeastern Brazil, and other places. The mix of modern, commercial farming, often on large estates, and subsistence or near-subsistence production on small parcels must be kept in mind when reviewing national-level data on the use of agricultural inputs. Similarly, interpretation of regional data requires cognizance of the marked differences between one part of Latin America and another.

Factor Use

With respect to some categories of agricultural input use, it is hard to distinguish between the northern- and southern-most reaches of the Western Hemisphere. In Canada, which has a comparative advantage in grain production, fertilizer applications averaged 58 kilograms per hectare between 2003 and 2005 (Table 10.5). This is between contemporaneous rates (Table 12.4, column 3) for Argentina (48 kilograms per hectare) and Uruguay (132 kilograms per hectare), which also compete successfully in international grain markets. Argentina and Australia resemble each other as well. For example, the irrigated portion of arable land (column 5) is identical in Argentina (5.0 percent) and the antipodal continent (4.9 percent).

For Chile, which does not produce grain as efficiently as do its neighbors east of the Andes, a better benchmark from outside the region is New Zealand. Although dairy products and sheep are mainstays of agriculture in the latter nation, both specialize in horticultural products. Rates of fertilizer application are high in both countries. Differences in the irrigated portion of farmland—over four-fifths in Chile versus 4 percent in New Zealand—relate to bone-dry conditions along much of South America's Pacific Coast. These conditions oblige Chilean farmers, not to mention their counterparts in coastal Peru and Ecuador, to divert water onto their fields from rivers and streams flowing out of nearby mountains. In contrast, rainfall is generous in New Zealand—in fact, excessive enough to preempt all farming in some places.

Differences between factor use in the Southern Cone and factor use in Australia, Canada, and New Zealand relate mainly to the high opportunity cost of labor in the latter nations. For example, ratios of tractors to land (Table 12.4, columns 6 and 7) are much higher among OECD members. However, farming is considerably more mechanized in

Table 12.4 Agricultural inputs in Latin America and the Caribbean, 1979 to 2005

Country (ranked by GDP per capita 2007)	Rural population density (rural inhabitants per sq. km of arable land) 2005 (1)	Fertilizer use (kg per ha of arable land)		Irrigated land (% of cropland)		Agricultural machinery (tractors per sq. km of arable land)	
		1979–81 (2)	2003–05 (3)	1979–81 (4)	2000–05 (5)	1979–81 (6)	2000–03 (7)
Mexico	98	57	73	20.2	22.8	0.5	1.3
Chile	104	34	291	31.1	81.3	0.9	2.7
Argentina	12	4	48	5.7	5.0	0.7	0.9
Venezuela	77	70	175	10.1	16.9	1.3	1.9
Panama	172	69	42	5.0	6.2	1.2	1.5
Uruguay	19	56	132	5.4	14.1	2.4	2.7
Costa Rica	737	265	853	12.1	20.5	2.1	3.1
Brazil	50	1	154	3.0	4.4	1.2	1.3
Colombia	592	81	294	7.7	22.4	0.8	1.0
Peru	213	38	85	32.3	27.9	0.4	0.4
Ecuador	353	47	173	24.8	27.7	0.4	1.1
Dominican Republic	383	57	—	11.7	20.9	0.2	0.2
Jamaica	722	123	43	10.1	6.6	2.1	1.8
El Salvador	406	138	90	4.6	5.0	0.6	0.5
Guatemala	466	73	129	5.0	6.5	0.3	0.3
Paraguay	58	4	58	3.4	1.9	0.5	0.4
Bolivia	108	2	6	6.6	4.1	0.2	0.2
Honduras	342	17	54	3.8	5.6	0.2	0.5
Nicaragua	125	42	32	4.7	2.8	0.2	0.2
Haiti	683	4	—	6.4	8.4	0.0	0.0

Source: World Bank (2009b).

Argentina, Chile, and Uruguay than in other parts of the hemisphere, where GDP per capita is lower.

Aside from the three southern-most states in the region, all Latin American and Caribbean countries are tropical or subtropical. Some of the latter are similar to Asian nations in the same latitudes. One of these is Colombia, where a high ratio of agricultural workers to arable land reflects a comparative advantage in the labor-intensive production of cut flowers. In this northern Andean nation, fertilizer use averages nearly 300 kilograms per hectare—an application rate comparable to Chile's. Another country where agriculture has Asian characteristics is Costa Rica, which has a higher ratio of rural population to arable land than any other nation listed in Table 12.4. Exceeding 850 kilograms per hectare, fertilizer applications in Costa Rica are elevated. By Asian standards, development of water resources for agriculture is limited. However, precipitation is abundant in this small state, which nevertheless irrigates a larger portion of its arable land than all but six other countries in the Western Hemisphere. One of the six is Mexico, in part because of irrigated production of fruits and vegetables for the North American market.

Elsewhere in the American tropics and subtropics, there are few resemblances with Asia. Even though Brazil has small-holder farming, along with the poverty that normally accompanies this activity, agriculture in Latin America's leading nation is mainly commercial and directly reflects the relative scarcity of land and labor. No more than one in every 25 hectares of arable land is irrigated, and crop production is mechanized. Widespread use of machinery and the development by national scientists of crop varieties that thrive in local conditions also have allowed Brazilians to extend soybean farming into the central part of the country, thereby increasing the competition faced by exporters in the United States and elsewhere.[2]

Other countries in the low latitudes have relative endowments of land and people closer to those of Asia, but have experienced less agricultural intensification. One reason for this is the prevalence of *minifundios*, which use fertilizer and irrigation sparingly. Limited intensification is also a result of structural transformation (Chapter 7). For example, improved employment prospects outside of agriculture, including for rural

---

[2] As more and more grasslands along the southeastern fringes of the Amazon Basin are converted to soybean fields, new routes to foreign markets are being developed. One option is to pave a road that runs north to Santarem, a port on the Amazon River located a few hundred kilometers upstream from the Atlantic Ocean, although this could accelerate deforestation in areas that are still hard to reach as long as the road remains unpaved.

households, are a reason for declining fertilizer applications in El Salvador, Panama, and Jamaica. Even though rural population densities are high in each of these three places, the option of nonagricultural work discourages spending for the sake of higher land productivity.

Intensification, Extensification, and Output Growth

Although yield growth has had a greater impact than expanded agricultural land use on production, the difference between the two in Latin America and the Caribbean has not been as great as in Asia (Chapter 11). From 1970 to 2000, yield increases (Table 12.5, column 3) greatly exceeded the pace of agricultural extensification (column 2) for corn, rice, and wheat. In contrast, growth in planted area was the main reason for increased output (column 4) of oil crops, mainly because of dramatic expansion (from a small initial base) in the area planted to soybeans (including genetically-modified varieties) in Brazil and other nations. As in

Table 12.5 Trends in crop area, yield, and output in Latin America, 1970 to 2000

	2000 Production (million tons) (1)	Average annual change, 1970–2000 (%)		
		Area (2)	Yield (3)	Production (4)
A. Crops				
Fruits	99	2.8	0.1	2.8
Maize	76	0.3	2.1	2.3
Roots & Tubers	53	−0.1	0.4	0.2
Vegetables	32	1.3	1.8	3.3
Wheat	24	0.4	2.1	2.5
Rice	23	−0.1	2.3	2.2
Oil Crops	16	3.1	2.4	5.7
Fibers	2	−3.8	2.8	−1.1
B. Animal Products				
Total Milk	60	—	—	2.9
Total Meat	31	—	—	3.5
Total Eggs	5	—	—	4.3
Total Wool	0.2	—	—	−2.0

Source: Dixon et al. (2001), pp. 269–270.

other regions, extensification has been a major driver of increased fruit and vegetable production.

Where impoverished peasants with small holdings and using few purchased inputs do a large share of the farming, trends in production per hectare have been disappointing. Cereal yields have actually fallen in Guatemala, Haiti, and Jamaica (Table 12.6, columns 1 and 2). The only three nations in Mesoamerica and the Caribbean to score more than a 33 percent gain between 1979 and 2007 are the Dominican Republic, El Salvador, and Mexico.

In contrast, yields rose by 50 percent or more during the same period where commercial producers predominate and where they have found it in their interest to use more fertilizer and other purchased inputs. This is obviously the case in the Southern Cone and Brazil as well as in Venezuela and Colombia. This last country does not produce grain (other than rice) as efficiently as its trading partners in temperate latitudes. Instead, Colombian agriculture specializes in coffee and, these days, cut flowers. The global dominance achieved by producers of the latter good is indicated both by the greenhouses that extend for kilometers out from the airports of Bogotá and other major cities and by the tags reporting country of origin in florists' shops throughout North America and Europe.

In a statistical analysis, Southgate (1994) has shown that rapid increases in crop yields tend to coincide with slow geographic expansion of agriculture in the Western Hemisphere. By the same token, sluggish growth in production per hectare often goes hand in hand with the rapid conversion of forests and other habitats to farmland. This finding is generally consistent with land-use trends reported in columns 3 and 4 of Table 12.6. To be sure, there are some nations, including Bolivia as well as smaller countries in Mesoamerica and the Caribbean, where virtually all soils that lend themselves to crop production were occupied by farmers long ago. In these nations, agricultural extensification during the 1980s and 1990s was modest, even if production per hectare was not going up much. Conversely, extensification can easily occur in spite of appreciable yield growth if land that is well suited to farming is covered with trees and other natural vegetation. Illustrative in this regard is expansion of the area planted to soybeans in Brazil (Chapter 5).

Although further increases in the geographic domain of agriculture are possible in Brazil and a few of its neighbors, intensification has been the primary response of Latin American and Caribbean farmers to demand growth. Where yields have risen appreciably, in Brazil, the Southern Cone, and other places, production (Table 12.6, columns 5 and 6) has

Table 12.6 Trends in cereal yields, farmed area, food production, value added per agricultural worker, and agricultural protection

Country (ranked by GDP per capita 2007)	Cereal yield (kg per ha)		Arable and permanent cropland (1,000 hectares)		Food production index (1999–2001 = 100)		Value added per agricultural worker (2000 $)		Nominal rate of assistance*	
	1979–81	2005–07	1980	2005	1979–81	2003–05	1979–81	2003–05	1980–84	2000–04
	(1)	(2)	(3)	(4)	(5)	(6)	(7)	(8)	(9)	(10)
Mexico	2,164	3,145	23,000	25,000	66.6	109.0	2,109	2,793	3.8	11.6
Chile	2,124	6,073	3,836	1,950	53.6	113.9	2,330	5,309	7.2	5.8
Argentina	2,184	4,077	26,000	28,500	66.3	107.2	6,334	10,072	−19.3	−14.9
Venezuela	1,904	3,364	2,957	2,650	59.9	94.7	3,997	6,331	—	—
Panama	1,524	1,938	435	548	84.8	103.1	2,243	3,904	—	—
Uruguay	1,644	4,223	1,403	1,370	77.1	111.9	—	7,973	—	—
Costa Rica	2,498	3,110	283	225	46.3	104.2	2,293	4,506	—	—
Brazil	1,496	3,205	45,000	59,000	49.5	122.8	1,019	3,120	−25.7	4.1
Colombia	2,452	3,792	3,712	2,004	64.8	112.3	2,456	2,749	5.0	25.9
Peru	1,946	3,604	3,220	3,700	48.7	112.0	976	1,481	—	—
Ecuador	1,633	2,751	1,542	1,348	54.0	109.7	1,539	1,676	5.9	10.1
Dominican Republic	3,024	4,426	1,070	820	93.1	108.7	2,071	4,586	−30.7	2.5
Jamaica	1,667	1,211	135	174	73.5	98.9	1,429	1,889	—	—
El Salvador	1,702	2,785	558	660	67.8	103.7	1,853	1,638	—	—
Guatemala	1,578	1,502	1,270	1,440	52.9	106.7	2,239	2,623	—	—
Paraguay	1,535	2,201	1,620	4,200	48.6	116.4	1,268	2,623	—	—
Bolivia	1,183	1,938	1,943	3,050	52.0	115.3	668	773	—	—
Honduras	1,170	1,519	1,484	1,068	78.9	158.3	895	1,483	—	—
Nicaragua	1,475	1,880	1,070	1,925	61.7	123.7	—	2,071	—	−4.2
Haiti	1,009	922	780	780	104.8	101.7	—	—	—	—

Sources: World Bank (2009b); FAO (1982), pp. 45–56; FAO (2002a), pp. 3–13; Anderson and Valdés (2008), pp. 16–22.

*NRA = relative effect (%) of government policies on gross returns to crop and livestock production.

increased at least as much as human numbers. In contrast, output growth has been disappointing in poorer parts of the region, where human numbers have risen even as yields have stagnated.

A similar pattern holds for trends in agriculture's total factor productivity (TFP). As in other parts of the world, value added per agricultural laborer (Table 12.6, columns 7 and 8) correlates with GDP per capita as well as agriculture's mechanization and commercial orientation. Needless to say, this indicator of productivity is higher in the Southern Cone, Venezuela, and Costa Rica, not to mention the more prosperous parts of Brazil, Colombia, and Mexico. In contrast, value added per agricultural worker is low in impoverished settings in the Andes, Mesoamerica, and the Caribbean.

When ISI was being pursued in Latin America and the Caribbean, incentives for crop production were correspondingly weak. This penalization continued into the early 1980s in Argentina, Brazil, and other countries (Table 12.6, column 9). Agriculture's nominal rate of assistance (NRA) is still negative in Argentina (column 10), which helps to explain its economic underperformance (Box 12.2). In contrast, across-the-board discrimination against the sector no longer occurs in Brazil and other nations.

Box 12.2 Dutch Disease tango

The standard case of Dutch Disease features energy production or mining as the source of domestic currency appreciation and agriculture among the sectors adversely affected by this development (Chapter 9). But in Argentina, agriculture plays a different role.

Although the country has been an important exporter of crops and beef since the 1800s, farming and ranching employ less than 1 percent of the national workforce (World Bank, 2009b). Also, much of the machinery used to achieve high levels of labor productivity in the sector is not produced domestically. In short, Argentine agriculture closely resembles the enclave industries that produce the exports responsible for Dutch Disease in countries with typical cases of the syndrome.

With 92 percent of Argentines living in Buenos Aires and other urban areas (Table 12.2, column 4), most of the population finds production and employment opportunities in manufacturing and

> other tradable sectors reduced as exports of beef, wheat, soybeans, and other commodities drive up the value of the peso. Moreover, urban dwellers rarely if ever respond to a boom in agriculture by migrating to the countryside to farm or ranch.
> Instead, they often support measures that reduce agricultural earnings and diminish incentives for crop and livestock production. Typical of these measures are the export restrictions that authorities in Buenos Aires implemented a few years ago as commodity prices spiked in global markets (Chapter 4).

12.4 Dietary Change, Consumption Trends, and Food Security

As is the case in Asia, population growth is having a smaller impact on food demand in Latin America and the Caribbean with each passing decade. The influence of improved living standards on per capita consumption is not as great, for the simple reason that GDP per capita has not been increasing as rapidly. However, the retail side of the food economy has been greatly affected by the proliferation of supermarkets and other large stores.

As Reardon and Berdegué (2002) have documented, large retail outlets, which accounted for 10 to 20 percent of all food sales in 1990, had reached three-fifths of the total a decade later. This growth has far-reaching implications for small growers, since large retailers prefer to do business with suppliers that can provide steady and sizable flows of high-quality produce. Moreover, there is little chance of a return to small shops and traditional markets. Employing managerial techniques nearly identical to those in the OECD, supermarket chains keep track of inventories and manage deliveries with considerable efficiency. Also, they are better at controlling quality. This is an important selling point for wealthy, middle-class, and even lower-middle-class households, which comprise the majority of consumers in Latin America and the Caribbean.

Just as people south of the U.S. border buy food in stores much like those in an affluent country, dietary composition, which we measure in terms of cereal equivalents of different foods and beverages (Chapter 10), is changing in some places much as it is in affluent settings. For example, per capita consumption of animal products (Table 12.7, column 2) has

Table 12.7 Food consumption trends in Latin America and the Caribbean, 1979 through 1999

Country (ranked by GDP per capita in 1999)	Years	Plant product consumption per capita (in cereal equivalent tons) (1)	Animal product consumption per capita (in cereal equivalent tons) (2)	Total food consumption per capita (in cereal equivalent tons) (3)	Total food self-sufficiency* (4)
Argentina	1979–1981	0.25	2.3	2.55	1.26
	1989–1991	0.24	1.84	2.08	1.3
	1997–1999	0.25	1.81	2.07	1.41
Chile	1979–1981	0.26	0.69	0.95	0.69
	1989–1991	0.24	0.77	1.01	0.77
	1997–1999	0.26	0.99	1.25	0.99
Mexico	1979–1981	0.3	0.71	1.02	0.87
	1989–1991	0.3	0.74	1.05	0.79
	1997–1999	0.3	0.85	1.15	0.75
Costa Rica	1979–1981	0.25	0.81	1.06	1.19
	1989–1991	0.27	0.81	1.08	1.09
	1997–1999	0.27	0.76	1.02	1.12
Uruguay	1979–1981	0.21	1.98	2.2	1.98
	1989–1991	0.19	1.74	1.93	1.74
	1997–1999	0.21	2.06	2.27	2.06
Brazil	1979–1981	0.27	0.74	1.01	1.05
	1989–1991	0.27	0.91	1.18	1.02
	1997–1999	0.27	1.12	1.4	1.07
Dominican Republic	1979–1981	0.23	0.44	0.67	0.97
	1989–1991	0.23	0.49	0.72	0.88
	1997–1999	0.23	0.57	0.79	0.79
Colombia	1979–1981	0.23	0.63	0.86	0.96
	1989–1991	0.23	0.67	0.91	0.97
	1997–1999	0.25	0.69	0.94	0.91
Panama	1979–1981	0.21	0.83	1.05	0.78
	1989–1991	0.21	0.81	1.02	0.84
	1997–1999	0.22	0.81	1.03	0.84
Venezuela	1979–1981	0.27	0.9	1.16	0.78
	1989–1991	0.24	0.7	0.94	0.87
	1997–1999	0.22	0.71	0.93	0.86
Peru	1979–1981	0.21	0.39	0.61	1.18
	1989–1991	0.2	0.35	0.55	1.36
	1997–1999	0.25	0.4	0.66	2.04

(*continued*)

Table 12.7 (*Continued*)

Country (ranked by GDP per capita in 1999)	Years	Plant product consumption per capita (in cereal equivalent tons) (1)	Animal product consumption per capita (in cereal equivalent tons) (2)	Total food consumption per capita (in cereal equivalent tons) (3)	Total food self-sufficiency* (4)
El Salvador	1979–1981	0.23	0.33	0.56	0.87
	1989–1991	0.25	0.29	0.54	0.85
	1997–1999	0.25	0.34	0.6	0.85
Paraguay	1979–1981	0.24	1.13	1.37	0.95
	1989–1991	0.23	1.04	1.27	1.24
	1997–1999	0.23	1.28	1.51	1.05
Guatemala	1979–1981	0.25	0.22	0.46	1.17
	1989–1991	0.26	0.24	0.51	1.11
	1997–1999	0.24	0.30	0.53	1.03
Jamaica	1979–1981	0.26	0.54	0.8	0.61
	1989–1991	0.25	0.61	0.86	0.58
	1997–1999	0.26	0.64	0.9	0.56
Ecuador	1979–1981	0.23	0.5	0.73	0.92
	1989–1991	0.25	0.49	0.74	0.97
	1997–1999	0.26	0.63	0.9	0.99
Honduras	1979–1981	0.22	0.38	0.6	1.22
	1989–1991	0.24	0.34	0.57	1.03
	1997–1999	0.23	0.37	0.6	0.83
Nicaragua	1979–1981	0.23	0.51	0.73	1.01
	1989–1991	0.23	0.31	0.54	0.95
	1997–1999	0.24	0.26	0.5	0.88
Bolivia	1979–1981	0.2	0.74	0.94	0.9
	1989–1991	0.21	0.71	0.92	0.99
	1997–1999	0.21	0.74	0.95	1.08
Haiti	1979–1981	0.22	0.21	0.43	0.84
	1989–1991	0.2	0.18	0.37	0.76
	1997–1999	0.21	0.19	0.39	0.67

*Total food self sufficiency equals total domestic food production divided by total domestic food consumption.

been on a downward path in Costa Rica. The same thing is happening in Argentina, where consumption of meat, dairy goods, and other livestock products 25 years ago was elevated even by rich countries' standards. Elsewhere in the region, there is a positive correlation between how many animal products people eat and their earnings, exactly as in Asia and other parts of the developing world. During the 1980s and 1990s, rising consumption of these goods accounted for the entire increase in per capita food intake (column 3) in Brazil, Chile, Mexico, and a number of other countries. In Haiti and Venezuela, GDP per capita has been contracting, so diminished per capita food intake mainly reflects reduced consumption of livestock products.

In light of the agricultural potential of Latin America and the Caribbean, which is not close to being fully exploited, increases in per capita food intake do not always result in reduced food self-sufficiency (Table 12.7, column 4), defined here as in other regionally focused chapters as cereal-equivalent production divided by cereal-equivalent consumption. Mexico exhibits a pattern familiar in Asia, where growth in per capita consumption mainly has to do with livestock products and usually coincides with declines in food self-sufficiency. Of course, Mexico deals with the latter decline exactly as various countries on the other side of the Pacific do, by trading its nonagricultural exports (e.g., petroleum and tourism services) for imported food. A different approach makes sense in Brazil, Chile, and other nations, where food self-sufficiency has gone up as diets have improved. In the latter countries, a comparative advantage in agriculture is exploited to the benefit of living standards, generally, and how well people eat, specifically.

Diets are meager and food self-sufficiency is low or declining in most, though not all, of the poorest countries in the Western Hemisphere. In Central America, El Salvador, Guatemala, and Honduras cope with this problem in part by exporting nonagricultural goods, including clothing and other products of assembly plants—called *maquilas* in Spanish—in which output is manufactured from imported materials by local workers using imported machinery. But these countries also export people, who send money home so that relatives can buy food and other necessities, build houses, and start small businesses. Similarly, hundreds of thousands of Nicaraguans now live in Costa Rica, performing agricultural and other work disdained by people accustomed to higher standards of living. Migration and remittances (Table 12.1, column 12) are common in the Caribbean as well, as indicated by the risks that Haitians run to reach the shores of Florida and by their willingness to cut sugarcane and perform other menial work across the border in the Dominican Republic.

No less than in other parts of the world, frequently used measures of food security correlate with per capita food consumption in Latin America and the Caribbean. There are some positive outliers in the region. The most celebrated case is Costa Rica, which has experienced less economic inequality than its neighbors in times past and has invested heavily in education and health care since disbanding its armed forces in 1949. The country's infant mortality rate (Table 12.8, column 2) actually compares favorably to those of OECD members, including the United States. Receiving much less international attention, though, is Chile. Its GDP per capita (Table 12.2, column 3) is a little above Costa Rica's, as is its Gini coefficient (Table 12.8, column 6). Nevertheless, the incidence of extreme poverty (column 5) is identical in the two countries, and the infant mortality rate is actually lower in Chile.

There are no Sri Lankas among the nations listed in Table 12.8—that is, places where GDP per capita is low and yet economic inequality and hunger are limited (Chapter 11). In contrast, a number of Latin American and Caribbean nations have indicators of food insecurity that are elevated in light of average incomes. One example is Guatemala, which has a GDP per capita above $4,000. Its infant mortality rate is 29 per thousand live births. Also, nearly one in every five Guatemalan children is severely underweight (column 4) and more than one-third of the population is extremely poor. Obviously, substandard food consumption as well as abundant evidence of food insecurity, such as elevated infant mortality, go hand in hand with the hemisphere's lowest living standards in Haiti.

12.5 Summary

Long more prosperous than the rest of the developing world, Latin America and the Caribbean is now overshadowed by Asia, which is far more populous and where living standards are rising at a much faster clip. In spite of differences in economic trajectories, the two regions resemble each other demographically. Death rates have fallen to low levels and family size has plummeted. Human numbers will peak in a generation or two.

Thanks to a generous endowment of natural resources relative to the region's population, agricultural development in Latin America and the Caribbean has differed substantially from what Asia has undergone. In wealthy parts of the region with low rural population densities, such as the temperate Southern Cone, crop production is mechanized and farm commodities are exported. Some settings in the tropics and subtropics have a comparative advantage in traditional crops, such as coffee, bananas, and sugar, as well as new products, such as cut flowers. But

Table 12.8 Food insecurity indicators in Latin America and the Caribbean

Country (ranked by column 1)	Total food consumption per capita 1997–1999* (in cereal equivalent tons) (1)	Infant mortality rate in 2007 (per 1,000 live births) (2)	Malnutrition prevalence		% of population living on $1.25/day or less in late 1990s or early 2000s (5)	Gini coefficient in late 1990s or early 2000s (6)
			Abnormally short % of 5-year-olds in late 1990s or early 2000s (3)	Abnormally under-weight % of 5-year-olds in late 1990s or early 2000s (4)		
Mexico	1.15	29	16	3	2.0	48
Chile	1.25	8	2	1	2.0	52
Argentina	2.07	15	8	2	4.5	50
Venezuela	0.93	17	—	—	3.5	43
Panama	1.03	18	22	6	9.5	55
Uruguay	2.27	12	14	6	2.0	46
Costa Rica	1.02	10	—	—	2.4	47
Brazil	1.40	20	7	2	5.2	55
Colombia	0.94	17	16	5	16.0	58
Peru	0.66	17	31	5	7.9	50

(continued)

Table 12.8 (*Continued*)

Country (ranked by column 1)	Total food consumption per capita 1997–1999* (in cereal equivalent tons) (1)	Infant mortality rate in 2007 (per 1,000 live births) (2)	Malnutrition prevalence			% of population living on $1.25/day or less in late 1990s or early 2000s (5)	Gini coefficient in late 1990s or early 2000s (6)
			Abnormally short % of 5-year-olds in late 1990s or early 2000s (3)	Abnormally under-weight % of 5-year-olds in late 1990s or early 2000s (4)		
Ecuador	0.90	20	29	6	4.7	54
Dominican Republic	0.79	31	12	4	5.0	50
Jamaica	0.90	26	4	3	2.0	46
El Salvador	0.60	21	25	6	11.0	50
Guatemala	0.53	29	54	18	11.7	54
Paraguay	1.51	24	—	—	6.4	53
Bolivia	0.95	48	32	6	19.6	58
Honduras	0.60	20	30	9	18.2	55
Nicaragua	0.50	28	25	8	15.8	52
Haiti	0.39	57	30	19	54.9	60

Source: World Bank (2009b).
*From Table 12.7, column (3).

throughout the Andes, northeastern Brazil, Mesoamerica, and some parts of the Caribbean, rural poverty is severe. Ironically, few people are more deprived than indigenous producers of corn and potatoes, who are the descendants of people who domesticated these same crops.

Food insecurity south of the U.S. border, which is not as widespread as undernourishment in other developing regions, has little to do with agricultural output. Aside from Haiti, every Latin American and Caribbean nation can feed itself from domestic production and imports, paid for with nonagricultural exports and international migration and remittances. As in other parts of the world, reducing hunger depends primarily on the alleviation of poverty.

Key Words and Terms

Bolivarian Revolution
Debt Crisis
Lost Decade
maquila

minifundio
North American Free Trade Agreement (NAFTA)
petrodollars

Study Questions

1. What did import-substituting industrialization in Latin America and the Caribbean have to do with the Lost Decade of the 1980s?
2. Is Hugo Chávez's Bolivarian Revolution a promising approach for Latin America?
3. Can a case be made that economic growth in Latin America and the Caribbean has had little positive effect on life expectancies and other noneconomic measures of human well-being?
4. How do human fertility and population growth in the region compare with TFRs and demographic expansion in Asia?
5. Why are patterns of agricultural development and factor use so heterogeneous in Latin America and the Caribbean?
6. Compare and contrast the supply side of the food economy, particularly beyond the farm gate, in Latin America and the Caribbean with the same set of activities in OECD nations.
7. Are indicators of food insecurity south of the U.S. border higher or lower than one would expect, based solely on GDP per capita in the region?

13

The Middle East and North Africa

The river valleys of the Middle East and North Africa having been cradles of civilization, the region also was the source of the world's three great monotheistic religions: Judaism, Christianity, and Islam. The youngest of these, Islam was expansionary from the time of its founding in the seventh century and spread quickly through Arab lands, Persia (now Iran), and present-day Turkey. The religion eventually won adherents in southeastern Europe, across central and southern Asia, and south of the Sahara.

During the Middle Ages, the Islamic world eclipsed Christendom in astronomy, mathematics, philosophy, and knowledge of agriculture. In part, this was because of greater tolerance of religious minorities among those embracing the faith of Mohammed than among the followers of Jesus. Islamic dominance did not stop immediately once Europe underwent its Renaissance and began exploring and colonizing the wider world. Arab occupation of the Iberian Peninsula, which lasted for centuries, did not come to a complete end until 1492, as Columbus was embarking on the first of his four voyages across the Atlantic Ocean. The Ottoman Turks, who had conquered the Arabs after converting to their religion, besieged Vienna in the late 1600s.

The ascendancy of Europe was clear a little more than a century later, when the Ottomans opened permanent embassies in London, Vienna, Berlin, Saint Petersburg, and Paris—the first any Islamic polity had seen fit to establish in any occidental capital (Lewis, 2002, p. 40). Overshadowed by Europe and its offshoots for more than 200 years now, the Middle East and North Africa continue to struggle with modernity. Representative government, separation of religion and state, women's rights, and tolerance of dissent are all exceptions to the rule in the region. So are the institutional underpinnings of free markets (Kuran, 2004).

These deficiencies have not been fully compensated by a bounteous endowment of oil and natural gas. The Middle East and North Africa have gained hundreds of billions of dollars by developing fossil fuels. Even nations with meager resources have benefited as their citizens have found employment in members of the **Organization of Petroleum Exporting Countries (OPEC)**. Yet overall economic accomplishments have been disappointing, in part because of Dutch Disease (Chapter 9). As documented in the **Arab Human Development Report (AHDR)**, the combined gross domestic product (GDP) of all 22 Arab countries,[1] from Morocco (across the Strait of Gibraltar from Spain) to Oman (across the Straits of Hormuz from Iran), which have a combined population of nearly 300 million, was $531 billion in 1999. The same year, Spain, which has fewer than 50 million inhabitants and is by no means the richest place in Western Europe, had a GDP of $596 billion (UNDP, 2002a, p. 85).

13.1 Political Realities and Economic Trends

This lack of success would have been surprising from the perspective of the middle and late 1970s. In October 1973, Israel repulsed a surprise attack by Egypt and Syria, which had received abundant military assistance from the Soviet Union, during the Yom Kippur War. Arabian oil exporters retaliated against the United States and other Western nations that had provided arms to the Jewish state (the only affluent democracy in the region) by cutting petroleum exports. This caused oil prices to triple in late 1973 and early 1974. With higher prices sustained after the export embargo came to an end, Arab economies boomed. Between 1975 and 1980, GDP per capita rose from $1,845 to $2,300 thanks to annual growth of 5.6 percent (UNDP, 2002a, p. 88).

The oil bonanza was renewed in early 1979, when the overthrow of the Shah of Iran precipitated another price spike. However, prosperity based exclusively on selling fossil fuels internationally at elevated values did not last. Price surges led to exploration and development in the North Sea, Alaska's North Slope, the Gulf of Mexico, and other non-OPEC settings. As new sources of petroleum came online, prices fell and then stayed low until after the turn of the twenty-first century. This would not have caused growth to cease in oil-exporting countries had windfalls from the 1970s and early 1980s been harnessed effectively for economic expansion and diversification. However, a large portion of these

---

[1] The governments of three of the 22 countries—Djibouti, Mauritania, and Sudan—identify themselves as Arab, even though non-Arab populations are sizable in each place. In this book, we follow the convention of various international agencies, which is to include the three in Sub-Saharan Africa.

windfalls was swallowed up by hegemonic and bureaucratic states, which tended to favor consumption by favored elites over investments favoring the general population. Growth consequently stalled or even reversed during the last two decades of the twentieth century.

To be sure, the expansion of government and the corresponding stunting of private economic activity in nations with abundant resources have happened outside the Middle East and North Africa, in Nigeria (Chapter 15) and Venezuela (Chapter 12) for example. Nevertheless, governmental hegemony has been especially pronounced in the Arab world. This probably relates to the weak loyalty that many Arabs feel for their respective nation-states, many of which were carved out of the long-decaying Ottoman Empire after the First World War by European negotiators who drew international boundaries in the region arbitrarily. Lacking the sort of legitimacy that governments have where people enjoy a strong sense of national identification and citizenship, Arab states compensate by overreaching. In the economic sphere, key sectors, such as the oil industry and banking, are state run and glaringly inefficient as a result. Also, private enterprise, relegated in many countries to traditional commerce, tends not to be internationally competitive since it has grown dependent on the protection from foreign competition that governments provide (Yousef, 2004). In addition, corruption abounds as protection from imports is sought and won by privileged firms and individuals.

The most egregious overreaching is in the realm of security. As a portion of GDP, expenditures on the armed forces in the Middle East and North Africa are triple or quadruple the levels recorded in other parts of the world, including the United States and old Soviet Union. Frequently rationalized in terms of threats from Israel, these expenditures are more directly a response to the menace posed by co-religious neighbors, as exemplified by the Iran-Iraq War of the 1980s, Saddam Hussein's invasion of Kuwait, and Syria's long occupation of Lebanon. Military spending is also motivated by concerns about internal security. This security is threatened when an appreciable segment of the population harbors a loyalty to pan-Arabism or, more frequently these days, Islamic fundamentalism, either of which is antithetical to the nation-state. Regardless of the source of threats, most governments operate extensive networks of secret police and informers, muzzle the press, and curb freedoms of assembly and association. Of course, the repression of civil liberties breeds public frustration and alienation, which can in turn add to rulers' discomfort and lead to even more repression and military spending.

Now that the Soviet Union no longer exists, the Middle East and North Africa comprise the most tyrannized part of the world. Fundamental freedoms are lacking in 11 of the region's 19 countries (Table 13.1, column 13). To say the least, repression after the fraudulent election of 2009 casts doubts on Iran's partly-free ranking. In contrast, electoral politics have revived in Iraq since the U.S. invasion of 2003 drove Saddam Hussein from power; if violence continues to diminish, as it has done since 2007, then the country will deserve to be counted with Algeria, Jordan, and Tunisia as partly free. Turkey, which aspires to membership in the European Union (EU), is similarly categorized. The only free nation in the region is Israel, where everyone—including people who are not Jewish—enjoys democracy and individual rights.

With oil prices falling off the peaks attained shortly after the Iranian Revolution and then plummeting after 1985 (Box 13.1), living standards eroded in the Middle East and North Africa. No more than six of the 19 countries listed in Table 13.1 reported a positive change in GDP per capita during the 1980s (column 1); of those six, Oman was the sole exporter of fossil fuels. The other 13 nations either reported a decline or failed to furnish macroeconomic data (presumably because living standards were deteriorating). Within this group, Jordan and Syria (where average incomes fell), Lebanon (where heavy fighting precluded documentation of economic trends), and Yemen (which is poorer than

Box 13.1 Using the oil weapon during the 1980s

During the first half of 1985, President Ronald Reagan, Defense Secretary Caspar Weinberger, and CIA Director William Casey lobbied the rulers of Saudi Arabia, where oil was being produced for about $1.50/barrel and sold for twenty times that amount, to increase output, so that prices would decline. U.S. officials were motivated primarily by their country's economic stake in cheaper fuel. However, lower energy prices were also part of a broader strategy of the Reagan administration to undermine the Soviet Union, which earned a large share of its foreign currency from sales of oil and natural gas.

Interested in good relations with the United States and fiercely anticommunist, the Saudis ramped up output in 1985, from 2 million barrels/day to 9 million by the end of the year. As a result, the price of a barrel of petroleum sank from $30 to $12, thereby hastening the Soviet Union's demise (Schweizer, 1994, pp. 216–220, 233, 242–243).

Table 13.1 Economic trends in the Middle East and North Africa from 1980 to 2007

Country (ranked by column 1)	A. Average annual rate of growth, 1980–90 (%)				Country (ranked by column 5)	B. Average annual rate of growth, 1990–99 (%)				Country (ranked by column 9)	C. Average annual rate of growth, 2000–07 (%)			Remittances ÷ exports (%), 2004–2007	Freedom status in 2008**
	GDP per capita (1)	Exports (2)	Inflation (3)	Gross Invest. (4)		GDP per capita (5)	Exports (6)	Inflation (7)	Gross Invest. (8)		GDP per capita (9)	Exports* (10)	Inflation* (11)	(12)	(13)
Oman	4.4	—	−3.6	—	Lebanon	5.9	15.6	24.0	18.4	Kuwait	4.3	3.1	10.2	—	PF
Turkey	3.1	16.9	45.3	5.3	Tunisia	3.0	5.1	4.7	3.4	Iran	3.8	8.3	17.8	1.2	NF
Egypt	2.9	5.2	11.7	2.7	Syria	2.9	4.7	8.7	7.9	Oman	3.5	7.8	7.2	0.2	NF
Morocco	2.0	6.8	7.2	2.5	Turkey	2.6	11.9	77.9	4.6	Tunisia	3.4	4.1	2.8	7.6	NF
Israel	1.7	5.5	101.5	2.2	Egypt	2.5	3.1	9.1	6.7	Morocco	3.4	6.9	0.9	25.0	PF
Tunisia	0.8	5.6	7.4	−1.8	Israel	2.1	9.1	10.6	5.5	Bahrain	3.2	—	10.0	—	PF
Algeria	−0.2	4.1	8.0	−2.3	Iran	1.8	0.2	26.7	1.4	Turkey	3.1	6.1	25.0	0.8	PF
Iran	−0.9	—	—	—	Oman	1.0	—	—	—	Jordan	3.1	6.6	2.6	32.7	PF
Jordan	−1.2	5.9	7.0	7.3	Morocco	0.5	3.0	3.2	1.5	Algeria	2.3	4.5	10.1	3.3	NF
Syria	−1.8	3.6	15.3	−7.0	Jordan	0.4	7.4	3.2	3.4	Egypt	2.3	11.5	6.5	19.4	NF
Kuwait	−3.1	−2.3	−2.4	−4.5	Algeria	−0.6	2.2	19.0	0.2	Qatar	2.2	—	14.6	—	NF
Saudi Ar.	−5.2	—	−3.7	—	Yemen	−1.0	10.2	26.0	7.7	U. Ar. Em.	2.1	7.4	7.6	—	NF
Libya	−6.3	—	—	—	U. Ar. Em.	−1.5	—	—	—	Lebanon	1.7	11.0	1.4	81.5	PF
U. Ar. Em.	−7.7	0.0	0.7	−8.7	Saudi Ar.	−1.8	—	1.4	—	Syria	1.7	2.6	6.3	5.8	NF
Lebanon*	—	—	—	—	Kuwait*	—	—	—	—	Libya	1.7	—	19.1	0.0	NF
Yemen*	—	—	—	—	Bahrain*	—	—	—	—	Saudi Ar.	1.2	1.9	7.5	—	NF
Bahrain*	—	—	—	—	Qatar*	—	—	—	—	Israel	1.0	7.6	1.3	1.5	F
Qatar*	—	—	—	—	Libya*	—	—	—	—	Yemen	0.8	13.7	14.1	30.4	PF
Libya*	—	—	—	—	Iraq*	—	—	—	—	Iraq	—	—	12.3	—	NF
Iraq*	—	—	—	—											

Sources: World Bank (2001), pp. 278–279 and 294–295; World Bank (2009b); and Freedom House (2008).

*Trends in exports and inflation are for 2000–2004.

**F = free; PF = partly free; and NF = not free.

any of its regional neighbors) do not belong to OPEC. The other eight all have sizable energy resources. In Kuwait, Libya, Saudi Arabia, and the United Arab Emirates, GDP per capita was lower in 1990 than it had been in 1980. The same was probably true of Iraq and Qatar. Clearly, the strategy of channeling oil export revenues to loss-leading enterprises owned and (mis)managed by the state (Bennett, 2003) yielded negative dividends once fossil fuels grew cheaper (Yousef, 2004).

The economic slide was arrested for OPEC members during the last decade of the twentieth century. After contracting during the 1980s, GDP per capita for the oil exporters held steady in the ensuing decade. Meanwhile, growth among the nonexporters, which had been positive though anemic during the 1980s, picked up in the 1990s. None of the half-dozen countries with the fastest increases in GDP per capita in the latter period (Table 13.1, column 5) belong to OPEC. Instead, economic expansion was achieved as nations without oil traded and invested more.

Rising energy prices since the turn of the twenty-first century have benefited the Middle East and North Africa. It is significant, though, that no economy in the region has experienced as much improvement in living standards as China and a number of other Asian nations that import oil. The only place in the region where GDP per capita went up faster than 4 percent per annum from 2000 to 2007 was Kuwait (Table 13.1, column 9), where many years were needed to recover from the invasion that Saddam Hussein ordered in 1990 and the Iraqis' forced removal in 1991 during the Persian Gulf War. The second-fastest growth was registered in Iran, which has ignored production quotas agreed to by OPEC in order to shore up nonoil sectors and to finance its nuclear ambitions.

As in earlier years, some of the countries experiencing the fastest growth in GDP per capita since the turn of the twenty-first century do not belong to OPEC. Exports of textiles and garments have increased from Egypt, Jordan, Morocco, and Tunisia, thanks partly to trade agreements signed with the European Union and the United States. In addition, these same nations along with others that lack abundant oil have benefited from the money sent home by emigrants working in Europe, in Morocco's case, or OPEC members in the Persian Gulf (Table 13.1, column 12).

13.2 Population Dynamics

Few countries illustrate the arrested growth in GDP per capita that oil exporters experience better than Saudi Arabia. However, the country also has benefited from energy resource development. Petrodollars have been used to

Human Fertility

Many of the factors that normally coincide with lower total fertility rates (TFRs) are at work in the Middle East and North Africa. One of these is urbanization (Table 13.2, column 4), which outside of Egypt and Yemen has proceeded quite far. Also, average incomes (column 1) are by no means the lowest in the world, and not just in countries with abundant energy resources. Indeed, more than a few places where the majority of the population lives in cities and has little fear of extreme poverty bear more than a passing demographic resemblance to, say, the leading nations of Latin America. Especially in the Maghreb (i.e., Morocco, Algeria, and Tunisia), Iran, and Turkey, half or more of all females aged 15 to 49 years use contraceptives (column 9). Also, female labor force participation has increased in 13 of the region's 19 countries.

But just as there are parts of Latin America and the Caribbean where TFRs are still well above 2.1 births per woman for cultural and other reasons (Chapter 12), human fertility remains elevated in some places notwithstanding widespread urbanization and relatively comfortable living standards. Clearly, Saudi Arabia is a case in point. That country's form of Islam is conservative. This militates against female empowerment, as is indicated by pronounced gender discrepancies in literacy rates (Table 13.2, columns 5 and 6) and limited participation by women in the labor force (columns 7 and 8).

Saudi Arabia is an outlier, not just for the world as a whole but also for the Middle East and North Africa. TFRs have fallen to the replacement level in two of the region's three largest countries: Iran, where there were 6.6 births per woman in 1980, and Turkey, which by regional standards had a low TFR, 4.3 births per woman, 30 years ago (Table 13.2, columns 2 and 3). Precipitous declines in fertility have happened in all but a few Arab nations—including Kuwait, Lebanon, and Tunisia, where the level required for generational replacement has been reached. Only in Yemen, which is the only impoverished country in the region and where there were nearly nine births per woman in 1980, does the TFR still exceed 5.0 births per woman.

Natural Increase

Other than in Yemen, death rates (Table 13.3, column 2) have reached minimal levels in the Middle East and North Africa. These rates might

Table 13.2 Human fertility and its determinants in the Middle East and North Africa from 1980 to 2007

Country (ranked by column 1)	2007 GDP per capita, PPP (constant 2005 international $) (1)	Total fertility rate 1980 (2)	Total fertility rate 2007 (3)	Urban % of total population in 2007 (4)	Adult illiteracy rate 2007 (%) Males (5)	Adult illiteracy rate 2007 (%) Females (6)	Female % of labor force 1980 (7)	Female % of labor force 2007 (8)	Contraceptive prevalence rate in 2000–2007 (% women 15–49) (9)
Qatar	70,716	5.7	2.7	96	6	10	8.7	15.5	43
U. Ar. Em.	51,586	5.4	2.3	78	11	9	5.0	14.5	28
Kuwait	45,152	5.3	2.2	98	5	7	12.7	24.0	50
Bahrain	28,069	4.9	2.3	88	10	14	10.7	21.5	62
Israel	24,824	3.2	2.9	92	—	—	36.7	46.3	—
Saudi Arabia	21,659	7.1	3.4	83	11	21	8.0	14.9	21
Oman	21,546	7.2	3.0	72	11	23	14.7	19.6	32
Libya	13,565	7.3	2.7	77	6	22	15.1	23.5	45
Turkey	11,825	4.3	2.2	68	4	19	35.2	26.8	71
Iran	10,346	6.6	2.0	68	13	23	19.5	29.2	79
Lebanon	9,546	4.1	2.2	87	7	14	28.1	25.6	58
Algeria	7,310	6.7	2.4	65	16	34	20.4	31.9	61
Tunisia	7,102	5.2	2.0	66	14	31	18.8	26.5	63
Egypt	5,052	5.4	2.9	43	25	42	19.1	25.3	59
Jordan	4,628	7.0	3.6	78	5	13	17.9	16.9	57
Syria	4,260	7.3	3.1	54	10	24	23.0	20.8	58
Morocco	3,880	5.6	2.4	56	31	57	21.4	24.7	63
Yemen	2,205	8.7	5.5	30	23	60	28.7	24.4	28
Iraq	—	6.5	4.3	67	16	36	16.5	—	50

Source: World Bank (2009b).

Table 13.3 Population trends in the Middle East and North Africa

Country (ranked by column 1)	2007 GDP per capita, PPP (constant 2005 international $) (1)	Crude death rate in 2007 (per 1,000 population) (2)	Crude birth rate in 2007 (per 1,000 population) (3)	Average annual population growth in 2006–2007 (%) (4)	International migrants ÷ total population in 2005 (%) (5)
Qatar	70,716	2	16	0.9	90
U. Ar. Em.	51,586	1	16	2.8	70
Kuwait	45,152	2	18	1.8	74
Bahrain	28,069	3	17	2.8	38
Israel	24,824	6	21	1.7	38
Saudi Arabia	21,659	4	25	0.7	27
Oman	21,546	3	22	2.7	27
Libya	13,565	4	23	2.4	10
Turkey	11,825	7	19	1.7	2
Iran	10,346	6	18	3.2	3
Lebanon	9,546	7	18	0.5	18
Algeria	7,310	5	21	0.8	1
Tunisia	7,102	6	17	2.6	0
Egypt	5,052	6	24	2.6	0
Jordan	4,628	4	29	1.3	43
Syria	4,260	3	27	2.0	7
Morocco	3,880	6	21	0.8	0
Yemen	2,205	7	38	0.3	2
Iraq	—	—	—	—	0

Source: World Bank (2009b).

even be going up in Turkey and a few other places where the demographic transition has advanced far enough for average ages to be rising.

With death rates stable or nearly so, natural increase is being driven by changes in birth rates (Table 13.3, column 3). Typical of settings where TFRs greatly exceeded the replacement level in the recent past, demographic momentum helps cause births to outpace deaths in the Middle East and North Africa. Subtracting death rates from birth rates, natural increase is close to 10 per thousand in Iran and Turkey as well as a pair of Arab nations: Lebanon and Tunisia. In several countries, including Jordan, Saudi Arabia, and Syria, the rate exceeds 20 per thousand. Few nations anywhere in the world are experiencing more natural increase than Yemen, where the annual rate is 31 per thousand.

As mentioned earlier in this chapter, remittances are large relative to export earnings in a number of countries. By the same token, the international migration that makes remittances possible also affects demographic trends. Where emigration has occurred (e.g., Jordan, Morocco, and Yemen), population growth (Table 13.3, column 4) is smaller than the difference between birth and death rates. Conversely, an opposite relationship between population growth and natural increase holds in several states in the Persian Gulf, such as Kuwait and the United Arab Emirates, where immigrants comprise large segments of the population (column 5).

13.3 Agricultural Development

Agricultural revolutions in the Western Hemisphere gave humankind corn, potatoes, and other widely used crops (Chapter 12). But for Europe and its overseas offshoots, the emergence of farming thousands of years ago in the valleys of the Nile, Euphrates, and Tigris Rivers was also of fundamental and lasting importance. Various crops, including wheat, were domesticated, as were sheep and goats. The agricultural significance of the Middle East and North Africa also relates to the region's having served as a conduit for the westward diffusion of rice, sugar, and other species originally cultivated in Asia (Dixon and Gulliver with Gibbon, 2001, pp. 83–87).

Farming in Egypt and present-day Iraq began alongside rivers because of productive land in this setting—to be specific, land made fertile because of the deposition of soil eroded from the upper reaches of drainage basins (Ethiopian highlands, in the case of the Nile). Riparian agriculture also makes sense in the Middle East and North Africa because of sparse precipitation in much of the region. In Egypt, around the Persian Gulf, and in many other places, there is no rain-fed crop production. Instead, the water supplied to farm fields must be diverted from rivers or extracted from underground aquifers.

In recent times, agricultural development has been greatly influenced in various nations by the presence of energy resources. Exploitation of these resources has created Dutch Disease (Chapter 9), as exports of oil and natural gas have strengthened national currencies and thereby discouraged the production of tradable goods other than fossil fuels. Among these goods are farm commodities. A few petroleum-rich nations in the Middle East and North Africa have done nothing to alleviate the agricultural disincentives resulting from Dutch Disease. But others have compensated farmers with subsidies, such as underpriced irrigation water or low prices for energy used to run pumps. Where this has been done, abundance of one subsurface resource (i.e., petroleum) has contributed to mounting scarcity of subsurface deposits of water, which actually is the most prized and vital resource in the region (Box 13.2).

> ### Box 13.2 Extracting oil and water in the Arabian Peninsula
>
> The government of Saudi Arabia charges farmers almost nothing for the water they use to raise wheat and other crops. But while irrigation subsidies are not at all unique (Chapter 5), the possible consequences for the country and its neighbors could be dire.
>
> The Arabian Peninsula has no rivers of consequence and all its freshwater resources are in deep aquifers, which recharge slowly. Once groundwater is depleted, hydrologic resources for households and other consumers will have to be extracted from the Red Sea and Persian Gulf, at a high cost.
>
> The future expenses of desalinization will probably be defrayed by extracting and exporting fossil fuels, which today are enabling governments such as Saudi Arabia's to subsidize the irrigation that is in large measure responsible for groundwater depletion.

Factor Use

Another reason for irrigation's spread in the Middle East and North Africa is that no part of the world outside of Asia has been affected more by the Green Revolution. The ratio of rural population to arable land in the region is comparable to ratios in South, Southeast, and East Asia (Chapter 9). In the Nile River Valley and a few other parts of Egypt with agriculture, there are well over 1,000 rural people per cultivated hectare (Table 13.4, column 1). Where the countryside is this densely populated, abundant irrigation and fertilization, which raise yields, are critical for agricultural development. In some countries, rates of fertilizer application (columns 2 and 3) have increased since the early 1980s.[2] But elsewhere, these rates have declined. Algeria and Libya, for example, are two examples of fossil-fuel producers that make little attempt to counteract Dutch Disease. Diminished fertilization in the former nation and only a marginal increase in the latter should be interpreted in terms of weak

---

[2] The 18-fold increase in the fertilizer application rate that happened in Jordan between 1980 and 2003–2005 coincided with a sizable decline in farmed area during the same period (see below).

Table 13.4 Agricultural inputs in the Middle East and North Africa, 1979 to 2005

Country (ranked by GDP per capita in 2007)	Rural population density (rural inha-bitants per sq. km of arable land) 1979–2005	Fertilizer use (kg per ha of arable land)		Irrigated land (% of cropland)		Agricultural machinery (tractors per sq. km of arable land)	
		1979–1981	2003–2005	1979–1981	2001–2005	1979–1981	2003–2005
	(1)	(2)	(3)	(4)	(5)	(6)	(7)
Qatar	203	223	1,050	75.6	—	1.9	0.4
U. Ar. Em.	1,430	225	553	100.0	29.9	1.1	0.6
Kuwait	287	450	1,063	—	72.2	2.2	0.7
Bahrain	4,204	82	368	17.8	66.7	0.4	0.7
Israel	184	238	2,001	49.3	—	8.1	7.6
Saudi Ar.	123	23	106	28.9	42.7	0.1	0.3
Oman	1,152	47	342	74.5	90.0	0.4	0.4
Libya	78	36	51	10.7	21.9	1.3	2.2
Turkey	99	53	84	9.5	19.7	1.7	4.3
Iran	138	28	93	35.5	47.2	0.6	1.6
Lebanon	289	166	162	28.3	31.3	1.4	4.5
Algeria	162	28	17	3.4	6.9	0.7	1.3
Tunisia	128	21	46	4.8	7.2	0.8	1.4
Egypt	1,394	286	671	100.0	100.0	1.6	3.2
Jordan	638	41	729	11.1	27.5	1.5	3.3
Syria	181	25	86	9.6	24.3	0.5	2.2
Morocco	160	27	57	15.0	15.4	0.3	0.6
Yemen	990	9	5	19.9	33.0	0.3	0.4
Iraq	—	17	—	32.1	58.6	0.4	1.3

Source: World Bank (2009b).

incentives for agriculture in the face of national currencies made strong by exports of oil and natural gas.

Likewise, the irrigated portion of cropland (Table 13.4, columns 4 and 5) in Algeria and Libya is modest relative to what other nations in the region have achieved. Aside from reflecting poor incentives for farming, limited irrigation is explained in part by the absence of major rivers in those two countries. Also, Algeria has upland areas with enough precipitation to allow for rain-fed farming complemented by limited, small-scale irrigation. Similar environments exist in Morocco, Tunisia, the Levant (Israel, Jordan, Lebanon, and Syria), and Turkey, as well as northern Iraq and Iran. Approximately 27 million upland farmers produce cereals and legumes and harvest fruit, olives, and other tree crops (Dixon and Gulliver with Gibbon, 2001, p. 88). Elsewhere in the Middle East and North Africa, irrigation has played a crucial role in developing the agricultural sector, generally, and in raising yields, specifically.

As in other parts of the world, agricultural mechanization depends on the opportunity cost of rural labor, as reflected in GDP per capita. Israel, where living standards are comparable to those of many members of the Organization for Economic Cooperation and Development (OECD), has the highest ratio of tractors to arable land in the region (Table 13.4, columns 6 and 7). Several countries have experienced a considerable increase in machinery use since 1980, including Egypt, Iran, Syria, and Turkey. As a result, the tractor-to-land ratio in the Middle East and North Africa has risen above those of Latin America and the Caribbean, Asia, and Sub-Saharan Africa (Chapter 9).

Intensification, Extensification, and Trends in Output

By and large, the Middle East and North Africa have a limited comparative advantage in agriculture. Fruit and vegetables, raised in many countries primarily for local markets, are the main crops and, as in other parts of the world, increases in this category of farm output (Table 13.5, column 4) have resulted more from agricultural extensification (column 2) than from higher yields (column 3). The opposite is true of cereals—including wheat, which has been the staff of life in the region for millennia.

Egyptian agriculture is noteworthy for its intensity, with crop yields extremely high by global as well as regional standards. As already mentioned, all the country's limited farmland is irrigated. In addition, the natural fertility of soils in the Nile River Valley has been enhanced by the application of fertilizer at high rates. Fertilization rose from less than 300 kilograms three decades ago nearly to 700 kilograms per hectare in recent years (Table 13.4, columns 2 and 3), which explains much of the 88 percent increase in cereal yields during the same period (Table 13.6, columns 1 and 2). This intensification greatly exceeded relative growth in cropland between 1980 and 1999 (columns 3 and 4) and is the main reason why annual food production has nearly tripled since 1980 (columns 5 and 6).

Natural conditions for agriculture are considerably different in Turkey and Iran, which along with Egypt are the three most populous countries in the region. Compared to its neighbors to the south, Turkey has a favorable environment for rain-fed farming, as indicated by limited irrigation

Table 13.5 Trends in crop area, yield, and output in the Middle East and North Africa, 1970 to 2000

	2000 Production (million tons) (1)	Average annual change, 1970 to 2000		
		Area (2)	Yield (3)	Production (4)
A. Crops				
Vegetables	44	2.3	1.7	4.0
Fruits	30	3.0	1.0	4.1
Wheat	23	0.4	1.8	2.2
Rice	9	1.3	1.3	2.6
Maize	8	0.3	3.3	3.6
Barley	4	0.0	−0.7	−0.7
Pulses	2	0.8	0.0	0.8
Oil Crops	1	0.8	1.8	2.7
B. Animal Products				
Milk	17	—	—	3.4
Meat	6	—	—	4.5
Eggs	2	—	—	5.4

Source: Dixon et al. (2001), pp. 94–95.

in the country.[3] Also, fertilizer application rates are below 100 kilograms per hectare. Nevertheless, crop yields are moderately high, after having risen since 1980. Even though agricultural land use has diminished, food production has increased by more than one-half. Iran, which exports petroleum, has made a substantial investment in irrigation. This has helped cause cereal yields to approach 2,500 kilograms per hectare in recent years, after barely exceeding 1,000 kilograms 30 years ago. With farmed area going up by 27 percent, the threefold increase in food production that has occurred during the past three decades is mainly a consequence of agricultural intensification.

There has been less agricultural development in the four North African nations west of Egypt. Although rainfall is sparse in Libya, that country has

---

[3] Irrigation will expand substantially in Turkey as the national government proceeds with its project to dam the upper courses of the Tigris and Euphrates Rivers. This project greatly concerns Iraq and Syria, through which these two waterways flow on their way to the Persian Gulf. Their worry is that water availability downstream from Turkey will diminish. This impact will occur while the new reservoirs are being filled. Water availability in Syria and Iraq will also be reduced later on because of evaporation from the same reservoirs and because of increased diversions of irrigation water to Turkish farmland (Beaumont, 1998).

Table 13.6 Trends in cereal yields, farmed area, food production, and value added per agricultural worker

Country (ranked by GDP per capita in 2007)	Cereal yield (kg per ha)		Arable and permanent cropland (1,000 hectare)		Food production index (1999–2001=100)		Value added per agricultural worker (2000 $)	
	1979–1981 (1)	2005–2007 (2)	1980 (3)	2005 (4)	1979–1981 (5)	2003–2005 (6)	1979–1981 (7)	2001–2006 (8)
Qatar	2,623	3,609	4	18	25.4	105.0	—	—
U. Ar. Em.	2,224	2,000	16	64	9.6	64.7	10,048	25,841
Kuwait	3,124	2,621	1	15	39.6	115.0	—	13,521
Bahrain	—	—	2	2	130.9	123.0	—	—
Israel	1,840	3,153	325	317	68.9	115.0	—	—
Saudi Arabia	820	4,464	1,890	3,500	24.4	111.7	2,151	15,780
Oman	982	3,215	23	62	36.2	93.0	—	1,302
Libya	430	617	1,753	1,750	60.4	99.0	—	—
Turkey	1,869	2,529	25,354	23,830	68.1	105.1	—	1,846
Iran	108	2,480	12,981	16,533	40.0	114.0	1,313	2,561
Lebanon	1,307	2,666	210	186	55.1	103.3	—	29,950
Algeria	656	1,465	6,875	7,450	53.3	123.3	1,317	2,225
Tunisia	828	1,320	3,191	2,729	55.7	115.0	1,414	2,700
Egypt	4,053	7,624	2,286	3,000	44.2	113.3	1,033	2,072
Jordan	521	1,263	299	184	42.3	119.0	984	1,360
Syria	1,156	1,749	5,230	4,873	61.5	120.3	—	3,261
Morocco	811	986	7,530	8,480	56.1	126.3	919	1,746
Yemen	1,038	886	1,366	1,515	52.8	105.3	—	328
Iraq	832	794	5,250	5,750	—	118.7	—	2,102

Source: World Bank (2009b); FAO (1982), pp. 45–56; FAO (2002a), pp. 3–13.

not made a large investment in irrigation. In addition, fertilizer use is modest, so cereal yields are low. Starting from a modest base in 1980, food production has increased, although Libya does the economically sensible thing of exporting fossil fuels and using a portion of the proceeds to import farm goods. Algeria is in much the same position, although it has more land suited to rain-fed farming. Neither of the other two Maghreb nations has oil or natural gas. Food production has doubled in Morocco and Tunisia during the past three decades due to a combination of intensification and extensification.

Along with other sufferings the country has undergone, Iraqi agriculture deteriorated significantly under the despotism of Saddam Hussein.

Although an increase was reported in the irrigated portion of arable land, canals and other infrastructure in fact fell into disrepair and output per hectare consequently registered a decline. The food production index for 2003 through 2005 (Table 13.6, column 6) indicates a sharp increase in agricultural output between the turn of the twenty-first century and the years immediately following the U.S. invasion of the country. This figure is believable if yields, which have been modest in recent years (column 2), were even lower in the late 1990s, when the tyrant still held power.

More encouraging developments have occurred in two of the region's smallest states. With less than a half million hectares of farmland apiece, Israel and Lebanon both have good irrigation systems. They also register high levels of fertilizer use and small-scale mechanization. Having come up with **drip irrigation**, a technique that saves water, Israel has been able to make its arid lands bloom, thereby allowing it to export a wide range of horticultural products to Europe. Lebanon has entered this market as well, which is the main reason why value added per agricultural worker has risen far above levels in those neighboring countries for which data are available (Table 13.6, columns 7 and 8). The two larger Levantine nations, Syria and Jordan, have moderate cereal yields, yet a good record of food production in recent decades.

In the Arabian Peninsula, Yemen stands out. As already indicated, the country lacks energy resources and has the lowest living standard in the region. Irrigation, fertilizer use, and mechanization are all limited and cereal yields, which were slightly above 1,000 kilograms per hectare 30 years ago, have stagnated. In contrast, every other peninsular nation is richly endowed with oil and natural gas. Farming in Kuwait, Oman, Qatar, and the United Arab Emirates consists mainly of niche operations that supply fresh fruit and vegetables to local consumers. The largest country on the peninsula, Saudi Arabia attempted a major expansion of farming after oil prices rose in the early 1970s. A portion of the earnings generated by petroleum exports was used to underwrite generous subsidies for fertilizer, agricultural machinery, and above all else irrigation. Cereal yields multiplied and, as can be seen in Table 13.6, food production in the early twenty-first century was nearly five times what it had been 20 years earlier.

Not least among the demerits associated with subsidizing inefficient irrigation are its adverse environmental impacts. If water is applied to farm fields without proper drainage, land may become water-logged. Another problem is soil salinization, which occurs as irrigation water evaporates (as it does quickly in arid places close to the equator) and leaves the minerals it contains to collect on the ground. Furthermore, providing underpriced groundwater to farmers and other users accelerates the depletion of aquifers, with extraction far outpacing **recharge**. This problem is arising not only in Saudi Arabia, but in Jordan, Libya, and Yemen as well (Dixon and Gulliver with Gibbon, 2001, p. 91). The long-term costs of aquifer depletion in places as dry as these are sure to dwarf whatever advantages are derived in the short term from irrigated agriculture that is both uneconomical and unsustainable.

13.4 Dietary Change, Consumption Trends, and Food Security

The Middle East and North Africa possess abundant energy resources and have living standards superior to those of many lands to the south and east. Due mainly to the scarcity of water, the region's comparative advantage in agriculture is limited and food imports are substantial. Outside of a few pockets of poverty, such as Yemen, food insecurity is not a great concern.

Applying Rask's (1991) methodology for converting food and beverage consumption into cereal equivalents (Chapter 10), we find the usual linkage between average income and per capita food consumption. For one thing, animal products comprise a large part of the diet (in terms of cereal equivalents) where GDP per capita is high (Table 13.7, columns 2 and 3). Such is the case in Israel as well as Kuwait and the United Arab Emirates, in the Persian Gulf. Conversely, consumption of meat, eggs, and dairy goods is limited in poorer nations, such as Morocco, where consumption of plant products (column 1) is comparable to consumption of animal products.

Since average incomes have grown slowly in many countries and contracted in a number of others, per capita food intake has increased dramatically in recent years. One of the exceptions is Egypt, where average food consumption in the country climbed by more than one-fifth during the last two decades of the twentieth century, with the intake of animal products accounting for most of the increase. In contrast, the human diet, generally, and consumption of meat, eggs, and dairy products, specifically, have deteriorated noticeably in some oil-based economies. Libya is one of these countries. Iraq is another.

Egyptian experience is distinctive in that the ratio of domestic food production to food consumption (Table 13.7, column 4) has gone up even as the denominator of this indicator of self-sufficiency has been increasing. Similar change, on a more modest scale, has occurred in Algeria. In five other countries, the ratio of production to consumption has risen: Iran, Jordan, Lebanon, Saudi Arabia, and the United Arab Emirates. But in each of these, the increase in the ratio has been modest, per capita food

Table 13.7 Food consumption trends in the Middle East and North Africa, 1979 through 1999

Country (ranked by GDP per capita 1999)	Years	Plant product consumption per capita (in cereal equivalent tons) (1)	Animal product consumption per capita (in cereal equivalent tons) (2)	Total food consumption per capita (in cereal equivalent tons) (3)	Total food self-sufficiency* (4)
Israel	1979–1981	0.29	1.02	1.31	0.68
	1989–1991	0.32	1.02	1.33	0.73
	1997–1999	0.33	1.10	1.44	0.66
Bahrain	1979–1981	—	—	—	—
	1989–1991	—	—	—	—
	1997–1999	—	—	—	—
Kuwait	1979–1981	0.26	1.20	1.46	0.44
	1989–1991	0.22	0.96	1.18	0.34
	1997–1999	0.28	1.24	1.51	0.43
Oman	1979–1981	—	—	—	—
	1989–1991	—	—	—	—
	1997–1999	—	—	—	—
Saudi Arabia	1979–1981	0.28	0.67	0.95	0.30
	1989–1991	0.28	0.65	0.94	0.41
	1997–1999	0.29	0.65	0.94	0.43
Tunisia	1979–1981	0.30	0.38	0.68	0.75
	1989–1991	0.34	0.46	0.79	0.75
	1997–1999	0.35	0.50	0.85	0.60
Iran	1979–1981	0.28	0.50	0.79	0.72
	1989–1991	0.30	0.43	0.74	0.74
	1997–1999	0.31	0.45	0.77	0.82
Turkey	1979–1981	0.33	0.55	0.88	1.02
	1989–1991	0.37	0.57	0.94	0.99
	1997–1999	0.36	0.53	0.89	0.99
Algeria	1979–1981	0.28	0.34	0.62	0.51
	1989–1991	0.30	0.45	0.75	0.54
	1997–1999	0.31	0.42	0.73	0.55
Lebanon	1979–1981	0.26	0.67	0.93	0.44
	1989–1991	0.32	0.53	0.84	0.48
	1997–1999	0.32	0.55	0.88	0.55
Jordan	1979–1981	0.27	0.46	0.73	0.33
	1989–1991	0.29	0.49	0.78	0.37
	1997–1999	0.29	0.45	0.74	0.43
Egypt	1979–1981	0.31	0.34	0.65	0.66
	1989–1991	0.34	0.38	0.72	0.67
	1997–1999	0.36	0.46	0.82	0.75
Morocco	1979–1981	0.30	0.34	0.64	0.77
	1989–1991	0.33	0.41	0.73	0.88
	1997–1999	0.33	0.37	0.70	0.76
Syria	1979–1981	0.29	0.58	0.87	0.85
	1989–1991	0.32	0.49	0.82	0.80
	1997–1999	0.34	0.51	0.84	0.86

(*continued*)

Table 13.7 (*Continued*)

Country (ranked by GDP per capita 1999)	Years	Plant product consumption per capita (in cereal equivalent tons) (1)	Animal product consumption per capita (in cereal equivalent tons) (2)	Total food consumption per capita (in cereal equivalent tons) (3)	Total food self-sufficiency* (4)
Yemen	1979–1981	—	—	—	—
	1989–1991	—	—	—	—
	1997–1999	—	—	—	—
Iraq	1979–1981	0.30	0.39	0.69	0.52
	1989–1991	0.33	0.30	0.65	0.49
	1997–1999	0.27	0.13	0.40	0.52
Libya	1979–1981	0.33	1.05	1.39	0.72
	1989–1991	0.33	0.65	0.98	0.56
	1997–1999	0.34	0.59	0.92	0.67
Qatar	1979–1981	—	—	—	—
	1989–1991	—	—	—	—
	1997–1999	—	—	—	—
U. Ar. Em.	1979–1981	0.30	1.42	1.72	0.26
	1989–1991	0.26	1.26	1.52	0.35
	1997–1999	0.27	1.34	1.60	0.40

*Total food self sufficiency equals total domestic food production divided by total domestic food consumption.

intake has not gone up (or has actually diminished), or both. Turkey is the only nation in the region where self-sufficiency is consistently achieved, although distortions of agricultural trade there are not great (Chapter 10) and edible imports and exports are sizable. Elsewhere in the Middle East and North Africa, importing one-third to two-thirds of the domestic food supply is not at all unusual.

The tie between economic well-being and food security that holds for other parts of the world also applies to the Middle East and North Africa. As in other places, per capita food consumption (Table 13.8, column 1) is inversely related to standard indicators of food insecurity, such as infant mortality (column 2) and the percentage of children aged five years of less who are so underweight or short as to be categorized as malnourished (columns 3 and 4).

Information is limited about how many people in the Middle East and North Africa are living below the international threshold of extreme poverty. But data provided by a number of countries indicate that the incidence of poverty (Table 13.8, column 5) is not high in the region. This and other indicators of food insecurity compare favorably with what one finds in other parts of the developing world for various reasons. GDP per capita is not especially low. Also, part of the wealth created by fossil-fuel development has been used to pay for hospitals and other public services and to lift people out of extreme poverty. Moreover, as indicated in the AHDR, the Muslim tradition of charity has beneficial effects (UNDP, 2002a, p. 91).

Table 13.8 Food insecurity indicators in the Middle East and North Africa

Country (ranked by column 1)	Total food consumption per capita 1997–99* (in cereal equivalent tons) (1)	Infant mortality rate in 2007 (per 1,000 live births) (2)	Malnutrition prevalence		% of population living on $1.25/day or less in late 1990s or early 2000s (5)	Gini coefficient in late 1990s or early 2000s (6)
			Percent of abnormally short 5-year-olds in late 1990s or early 2000s (3)	Percent of abnormally underweight 5-year-olds in late 1990s or early 2000s (4)		
U. Ar. Em.	1.60	7	—	—	—	—
Kuwait	1.51	9	—	—	—	—
Israel	1.44	4	—	—	—	39
Saudi Arabia	0.94	20	—	—	—	—
Libya	0.92	16	21	4	—	—
Turkey	0.89	21	16	4	2.7	43
Lebanon	0.88	26	15	3	—	—
Tunisia	0.85	18	—	—	2.6	41
Syria	0.84	15	28	9	—	—
Egypt	0.82	30	24	5	2.0	32
Iran	0.77	29	—	—	2.0	38
Jordan	0.74	21	12	4	2.0	38
Algeria	0.73	33	22	10	6.8	35
Morocco	0.70	32	23	10	2.5	41
Iraq	0.40	36	28	7	—	—
Qatar	—	12	—	—	—	—
Bahrain	—	9	—	—	—	—
Oman	—	11	13	11	—	—
Yemen	—	55	59	13	17.5	38

Source: World Bank (2009b).
*From Table 13.7, column (3).

One might be tempted to offer similar explanations for the last measure reported in Table 13.8, which is the Gini coefficient of income inequality (column 6). Among the countries for which estimates of this index are available, inequality is intermediate to low, definitely comparing favorably with what one finds in Sub-Saharan Africa and Latin America and the Caribbean (Chapter 9). In Yemen's case, modest inequality helps to explain why less than one-sixth of the population is living on less than $1.25/day, even though GDP per capita is little more than $2,000. However, Gini estimates are unavailable for much of the Middle East and North Africa—presumably because accurate estimates would reveal that income inequality was pronounced in Saddam-era Iraq (where infant mortality was appalling) and continues to be so in Kuwait, Libya, Saudi Arabia, and other nonreporting nations.

13.5 Summary

Like the rest of the developing world, the Middle East and North Africa are wrestling with the establishment of the rule of law and representative government as well as a market economy. This task has been made more difficult by a generous endowment of energy resources. Revenues from international sales of oil and natural gas have provided various governments with the means to interfere with private enterprise and to suppress the rights of individual citizens. The results of this interference and suppression include poor economic performance for years after the petroleum boom of the 1970s came to an end.

Demographic change, in contrast, has been dramatic. Outside of a handful of countries where large families are still the rule, fertility rates have fallen dramatically—to the replacement level in Iran, Turkey, and a couple other countries. Since these rates were very high a generation ago, demographic momentum is strong. As long as this is the case, human numbers will continue to increase. Since GDP and employment are expanding slowly in the region, some of this growth will spill over into Europe, where natural increase is all but over (Chapter 10).

Much of the Middle East and North Africa is arid or nearly so. Accordingly, agricultural comparative advantage in many places does not extend far beyond raising fruits and vegetables for local markets. Moreover, crop and livestock production has been discouraged in some nations because of Dutch Disease, and offering subsidies to ameliorate the effects of currencies strengthened by fossil-fuel exports generally has been unwise. The worst consequence of subsidizing agriculture has been to accelerate the mining of water—undoubtedly the most precious resource in the region—from underground aquifers (Box 13.2).

If the presence of energy resources has lowered incentives for domestic agricultural production, these also provide the financial wherewithal for food imports. Except in Turkey, agricultural imports exceed agricultural exports throughout the region. The absence of food self-sufficiency does not create hunger. To the contrary, none of the standard indicators of food insecurity, such as infant mortality and the poverty rate, are elevated. If the Middle East and North Africa can define and follow a path toward freer markets and freer societies, hunger will continue diminishing.

Compared to what one finds in Asia, Latin America and the Caribbean, and Sub-Saharan Africa, poverty is not widespread in the Middle East and North Africa. Thus, the terrorism originating in the region has other causes. It is to be hoped that, if political and economic freedoms are respected and living standards improve, advocates of violence will win fewer recruits.

Key Words and Terms

Arab Human Development Report (AHDR)
drip irrigation

Organization of Petroleum Exporting Countries (OPEC)
recharge

Study Questions

1. Has oil wealth created a lasting economic boom in the Middle East and North Africa?
2. How has the lack of political freedom in the Middle East and North Africa correlated with the region's economic performance?
3. Is it accurate to characterize the Middle East and North Africa as a region where human fertility is uniformly high?
4. Describe the impacts of Dutch Disease on Middle Eastern and North African agriculture.
5. How has the presence of one subterranean resource—oil and natural gas—contributed to depletion of another resource, namely water?
6. Should Middle Eastern and North African nations strive for food self-sufficiency?
7. Are indicators of food security in the region above or below what one would expect, in light of average incomes?

14

Eastern Europe and the Former Soviet Union

Considerable though the challenge of modernization has been in the Middle East and North Africa, no part of the world has gone through a more profound transformation in recent years than Eastern Europe and the Former Soviet Union. Entirely dominated by steel, mining, and other traditional industries, which were rigidly locked into production technologies from the 1930s that made heavy use of conventional energy (Kotkin, 2001, p. 17), the region's economies were strictly controlled by bureaucrats for decades and proved incapable of harnessing advances in electronic and information technology that created unprecedented prosperity elsewhere in the world in the late twentieth century. Stagnation was setting in during the 1970s and living standards deteriorated noticeably in a number of countries, including the USSR, for several years before the Berlin Wall fell in 1989 (Aslund, 2002, p. 41).

The events of 1989 kindled great expectations behind the **Iron Curtain** (as Winston Churchill described the boundary between free and communist Europe at the outset of the Cold War). So did the defeat in 1991 of an attempted *coup d'etat* by communists in Russia. But rather than being marked by spreading affluence, the last decade of the twentieth century was a time of mounting hardship, with gross domestic product (GDP) plummeting in fledgling nations carved out of the USSR as well as in many former Soviet satellites.

In some ways, the transition from central planning to a market economy in Eastern Europe and the Former Soviet Union was more difficult than the recovery of West Germany (which was spared occupation by the Red Army) and Japan after the Second World War. True, factories and infrastructure in the defeated axis powers were in a shambles in 1945. However, a *tabula rasa* had been created by the deracination of totalitarian institutions, not to mention top fascist officials. Under U.S.

supervision, the rule of law and other elements of an institutional framework favorable to economic progress were put firmly in place. At the same time, development was bolstered by substantial foreign aid from the United States, through the Marshall Plan.

The situation was vastly different 45 years later in the Former Soviet Union and its erstwhile satellites, which suffered neither defeat in war nor occupation by a conquering army. Especially in Russia and other parts of the old USSR, the communist party elite, the *nomenklatura*, did not go away, but instead drew on old connections to secure privileged positions in the new regime and to grab natural resource industries and other valuable properties for themselves. Often resorting to bribery and strong-arm tactics, the same group was unenthusiastic about establishing the rule of law, which helps to explain the slow development of private property rights and judicial mechanisms for contract enforcement. Moreover, the old Soviet bloc did not receive significant funding from overseas descendants of emigrants, as China benefited from investment by overseas Chinese after Mao's demise (Chapter 11). Other capital inflows, including foreign aid, were similarly limited.

Even if an institutional *tabula rasa* had existed in Russia and neighboring lands in the early 1990s, the postcommunist transition was never going to be easy. A way had to be found to root out deeply ingrained cultural habits required for survival in a self-styled workers' paradise with pervasive central planning—habits such as avoiding even the appearance of entrepreneurial initiative. Furthermore, economic waste of epic dimensions presented an enormous challenge. Somewhat tentatively, McKinnon (1991) observed as the USSR was expiring that "industries producing finished goods might well exhibit negative value added at world market prices" (p. 165). Subsequent analysis (Reed, 2004, pp. 224–225) confirmed that, indeed, the Soviet economy was fundamentally an engine of value-destruction (as opposed to value-creation), in the straightforward sense that GDP was less than the combined value of resources and inputs harnessed by the economy. Thoroughly unsustainable, value-destruction had to end. But just as inevitably, dismantlement of an entire economic system, even a woefully inefficient one, involved considerable dislocation and belt-tightening.

To summarize, the damage wrought under communism was enormous, in economic terms and every other respect. Many countries and peoples subjected to this system have yet to make a full recovery.

14.1 Economic Trends since the Fall of Communism

As long as development was gauged in terms of the physical output of traditional industries, the economic shortcomings of communism were not immediately obvious. Through the 1970s, Soviet leaders could brag about steady increases in the production of iron and steel, tractors, and so forth—increases that at times matched the growth of conventional manufactured output in the Organization for Economic Cooperation and Development (OECD)—even though a great deal of communist production lay unused and rusting away at factories or in "customers'" parking lots. However, a service-dominated economy, one that developed and adopted technology at a rapid pace, proved incompatible with central planning. The ground lost during the waning years of communism to the capitalist West is difficult to determine, for the simple reason that communist officials never measured or reported GDP per capita accurately. Nevertheless, the gap in living standards was widening due to economic stagnation or worse east of the Iron Curtain.

Far less ambiguous was the erosion of living standards after inefficient, centrally planned systems collapsed. During the 1990s, GDP fell by 15 to 20 percent in Eastern Europe. In the old Soviet heartland, the contraction was more severe, generally in the range of 30 to 50 percent (Aslund, 2002, p. 118). There were exceptions, however. One of the positive standouts was Poland, where opposition to communism was galvanized in 1979 by the triumphal return of John Paul II less than a year after he became Pope. **Solidarity,** a free labor union, was soon organizing openly. Subsequently, during the 1980s, organized dissent grew in Czechoslovakia (which divided amicably into the Czech Republic and Slovakia in 1993) and Hungary. Not coincidentally, reform-minded cadres were ready in each of these countries to undertake the institutional transformation needed to put market economies on a sound footing soon after the Berlin Wall came down. In case these cadres needed encouragement, the clear signal that reform would be rewarded with membership in the European Union (EU) and the **North Atlantic Treaty Organization (NATO)** added to the impetus for change.

As indicated in Table 14.1 (column 1), the Czech Republic, Hungary, Poland, and Slovakia are among the minority of nations in Eastern Europe and the Former Soviet Union where GDP increased faster than human numbers during the last decade of the twentieth century. These four nations were among the first in the region to join the EU, having implemented pro-market reforms by the middle 1990s. Living standards improved as well in Albania, where a prior Maoist regime had created the worst economic deprivation in

Table 14.1 Economic trends in Eastern Europe and the Former Soviet Union from 1990 to 2007

Country (ranked by column 1)	A. Average annual rate of growth, 1990–1999 (%)				Country (ranked by column 5)	B. Average annual rate of growth, 2000–2007 (%)				
	GDP per capita (1)	Exports (2)	Inflation (3)	Gross investment (4)		GDP per capita (5)	Exports (6)	Inflation (7)	Remittances ÷ exports (%) in 2007 (8)	Freedom status in 2008** (9)
Poland	4.5	10.8	24.5	11.9	Azerbaijan	14.6	18.9	9.1	6	NF
Slovenia	2.5	−0.5	23.5	10.2	Turkmenistan*	13.6	13.6	16.4	—	NF
Albania	2.0	13.6	51.5	22.4	Armenia	11.3	16.3	3.1	48	PF
Slovakia	1.7	12.0	10.8	4.6	Latvia	8.4	7.3	7.1	5	PF
Hungary	1.3	8.2	20.7	8.4	Kazakhstan	8.3	8.9	14.3	1	NF
Macedonia	1.2	1.2	13.5	6.7	Georgia	8.2	9.4	6.4	22	F
Czech Republic	1.0	9.0	13.7	6.3	Lithuania	7.4	8.9	17.3	7	PF
Croatia	0.4	—	131.2	—	Ukraine	7.4	5.6	5.6	7	F
Estonia	−0.4	10.2	62.7	−1.8	Belarus	7.3	8.7	80.2	1	NF
Romania	−0.8	6.1	105.5	−11.8	Estonia	7.2	11.1	2.5	3	F
Bulgaria	−2.0	0.3	111.8	−0.9	Tajikistan	6.5	6.6	16.7	220	F
Lithuania	−3.8	2.9	90.4	8.8	Moldova	6.5	13.1	25.1	75	F
Latvia	−3.8	0.7	58.8	−4.4	Russia	6.1	9.0	5.4	1	F

(continued)

Table 14.1 (Continued)

Country (ranked by column 1)	A. Average annual rate of growth, 1990–1999 (%)				Country (ranked by column 5)	B. Average annual rate of growth, 2000–2007 (%)				
	GDP per capita (1)	Exports (2)	Inflation (3)	Gross investment (4)		GDP per capita (5)	Exports (6)	Inflation (7)	Remittances ÷ exports (%) in 2007 (8)	Freedom status in 2008** (9)
Armenia	−3.9	−21.5	269.5	−29.5	Romania	5.8	11.9	25.0	16	NF
Uzbekistan	−4.0	—	356.7	—	Bulgaria	5.5	9.3	22.5	9	NF
Belarus	−4.2	−11.1	449.9	−10.0	Slovakia	5.4	11.9	4.8	2	F
Kazakhstan	−5.3	4.3	255.7	−11.7	Uzbekistan	4.3	8.0	3.5	—	PF
Russia	−6.0	2.3	189.6	−13.3	Albania	4.3	11.3	2.1	49	F
Turkmenistan	−6.4	—	622.8	—	Croatia	4.2	6.4	3.4	6	F
Kyrgyzstan	−8.2	6.7	157.3	12.6	Czech Republic	3.9	11.1	5.0	1	F
Azerbaijan	−10.2	12.6	249.5	14.7	Slovenia	3.7	8.8	32.4	1	NF
Georgia	−10.3	9.8	513.0	51.2	Poland	3.6	10.1	3.8	6	F
Ukraine	−10.4	−3.6	440.0	−24.8	Hungary	3.5	12.3	6.6	2	F
Moldova	−11.3	4.8	1142.5	−20.0	Kyrgyzstan	2.9	1.5	11.9	43	PF
Tajikistan	−11.6	—	300.0	—	Macedonia	1.8	6.0	3.5	8	PF

Sources: World Bank (2001), pp. 278–279 and 294–295; World Bank (2009b); and Freedom House (2008).

*Trends in exports and inflation are for 2000–2005.

**F = free; PF = partly free; and NF = not free.

all of Europe. In addition, GDP per capita went up in three former parts of Yugoslavia, which broke up violently during the 1990s: Croatia, Macedonia, and Slovenia.[1] In general, countries where GDP grew faster than the population after the end of communism had solid growth in exports (column 2) and, thanks to the avoidance of triple-digit inflation (column 3), relatively high rates of investment (column 4). Much of this investment was undertaken by Western firms, which were attracted by pro-market reforms. In more than a few instances, foreign corporations modernized inefficient state companies after acquiring a full or partial ownership stake.

Much more troubled were countries where communists' stranglehold on power was longer, tighter, or (more often than not) both. Throughout the old USSR, the dissident-reform tradition was weak in some places and absent in many more, which allowed former communist *apparatchiks* to seize control as old regimes crumbled. This group routinely ignored civil and political liberties and showed little interest in establishing open, competitive capitalist economies. In places like Belarus and Ukraine, *apparatchiks* campaigned against reform or even subverted it. This helps to explain why large-scale, communist-era enterprises were never privatized or, when this happened, ended up in the hands of well-connected people thanks to officially sanctioned theft. Corruption proliferated in tightly controlled regimes characterized by predatory regulation. Furthermore, organized crime expanded, sometimes to meet demands for the enforcement of individual property rights and private contracts—a service that weak public institutions failed to provide.

No part of the Former Soviet Union avoided significant reductions in GDP per capita during the 1990s—reductions that coincided with modest population growth or demographic contraction in the same period (see below). The smallest declines occurred in the Baltic republics, which at the beginning of the Second World War had been swallowed up by the USSR—which in 1939 had seized eastern Poland as Nazi Germany (an ally at the time) conquered the western part of the country. Economic trends in Estonia, Latvia, and Lithuania were comparable to those of Bulgaria and Romania, which did not make as swift a transition from communism as places like Poland and Slovenia (Table 14.1, column 1). Elsewhere, inflation raged (column 3), which often led to negative rates of capital formation (column 4). With few exceptions, exports grew slowly or shrank (column 2). Annual rates of decline in GDP per capita ranged from

---

[1] Bosnia-Herzegovina and the remnant of Yugoslavia (Serbia and Montenegro) suffered prolonged armed conflict during the 1990s, which took a severe economic toll and impeded data collection.

4 percent or so in Armenia, Belarus, and Uzbekistan to more than 10 percent in Azerbaijan, Georgia, Moldova, Tajikistan, and Ukraine. The average yearly decline in Russia was 6 percent (column 1).

With living standards falling precipitously in the Former Soviet Union and places in Eastern Europe that bore a close resemblance to the old communist hegemon and with the ex-*nomenklatura* benefiting from various forms of corruption and theft, income distribution grew more skewed (Aslund, 2002, p. 311). In contrast, income inequality remained modest as GDP per capita rose in Central Europe and those parts of the old Yugoslavia that moved decisively away from communism. This accomplishment is explained by economic progress and broad access to education, as well as high living standards and modern institutions in some of these countries prior to the Second World War.

Rankings according to growth in GDP per capita have reshuffled in recent years. Central European nations such as Poland are still on a positive trajectory. However, the rankings from 2000 to 2007 (Table 14.1, column 5) were dominated by former Soviet republics. After the 1990s, partial rebounds in the latter were perhaps inevitable. Also, it is important to keep in mind that cumulative expansion since 2000 in many countries, including Russia and Ukraine, has not fully compensated for declines during the previous decade, which means that today's living standards are still below the levels of 20 years ago.

Rising energy prices have caused exports to skyrocket in Russia, Azerbaijan, and other oil and gas producers since the turn of the twenty-first century (Table 14.1, column 6). Furthermore, inflation has been ratcheted down (column 7), thereby allowing investment and growth to accelerate. As in other parts of the world, small countries where jobs are limited receive sizable remittances from citizens who have emigrated to find employment. Among those countries are Albania in southeastern Europe, Armenia and Georgia in the Caucasus, and Tajikistan and the Kyrgyz Republic in Central Asia (column 8). Of course, autocratic regimes that control vast subsoil resources and keep inflation in check can benefit from the sort of commodity boom that happened during the first decade of the twenty-first century. This explains the rapid growth of per capita GDP from 2000 to 2007 in Azerbaijan, Turkmenistan, and Kazakhstan—all energy exporters that are not free (column 9).

Two decades after the fall of communism, many of the underpinnings of sustained development remain insecure in the Former Soviet Union and some of its erstwhile satellites. Excessive borrowing in Eastern Europe since 2000 has threatened a number of currencies and hampered growth. In

Russia's Troubled Resurgence

addition, economic trends could reverse in nations that have benefited from rising commodity prices yet have not solidified the rule of law and other elements of the standard model for broad-based development—exactly as has happened in the Middle East and North Africa since the 1970s (Chapter 13).

The core of the old USSR, Russia was and remains profoundly debilitated by decades of communism. The worst consequences have been demographic, the country having lost population due to Stalin's mass murders in the 1930s as well as enormous casualties during the Second World War. In addition, wanton despoliation of the environment occurred around countless Soviet-era industrial and military sites. Less obvious perhaps, though no less damaging, have been the lasting cultural impacts of communism—such as the suppression of personal initiative and responsibility pithily expressed by the Soviet-era motto, "We pretend to work and you pretend to pay" (Ellman and Kontorovich, 1998, p. xxiv).

The cultural habits of communism undermined reform initiatives in the late twentieth century. Gregory and Zhou (2010) point this out in their analysis of the failure by Mikhail Gorbachev, the last Soviet leader, to introduce the same elements of capitalist agriculture that had emerged in the Chinese countryside during the late 1970s and early 1980s (Chapters 7 and 11). This initiative foundered because of the indifference of a rural population that much preferred salaried employment on state collectives over the risks of private farming.

Gorbachev's reform efforts, which were undertaken because the communist economic system was breaking down, were followed in the early 1990s by an attempt to deregulate prices and privatize state-owned enterprises (SOEs) after the demise of the Soviet Union. This attempt likewise failed to accomplish the fundamental purpose of establishing a robust market economy. Instead, valuable state industries and properties ended up in the hands of **oligarchs**, as these former-communist-leaders-turned-businessmen were called. The banks they owned made loans to the government, which then defaulted and consequently handed over shares in SOEs that had been used as collateral (Guriev and Rachinsky, 2005). At the same time, weaknesses of the Russian state led to the proliferation of organized crime as well as racketeering by official police forces administered by public authorities (Aslund, 2002, pp. 358–359).

Russia turned to a new leader at the dawn of the twenty-first century. Previously an officer of the Soviet security and intelligence service, Vladimir Putin reined in the oligarchs and lightened financial regulation during his two terms as president, which coincided with steady increases in energy prices, exports of oil and natural gas, and GDP. More worrying have been the autocratic tendencies of Putin (who became prime minister in 2008 after his second presidential term) and his associates (Stelzer, 2005), who often bully the **Near Abroad** of Central Asia, Belarus and Ukraine (which depends entirely on natural gas from Russia), and the Caucasus—including Georgia, which the Russian Army invaded in 2008. Other targets of intimidation are domestic. Not limited to oligarchs who have run afoul of Moscow's new leadership, these targets include the independent press.

Autocratic impulses and behavior have cost Putin and his successor, Dimitri Medvedev, little support as the economy has expanded. However, Russia's forward momentum has dissipated somewhat in recent years. Since energy prices peaked, in June 2008, GDP growth has decelerated markedly. In addition, foreign reserves, which grew to more than $500 billion, now add up to about half that amount, due to the efforts of national authorities to stimulate the economy and bail out friendly oligarchs experiencing financial stress. Vast though its natural resources are, Russia is no less immune than other commodity exporters to the hangover that follows a price spike.

14.2 Demographic Trends

As noted in Chapter 9, demographic realities in Eastern Europe and the Former Soviet Union resemble those on the other side of the old Iron Curtain more than demographic realities in the developing world. This remains the case today in countries such as Hungary and Poland, where crude death rates (CDRs), crude birth rates (CBRs), and total fertility rates (TFRs) are indistinguishable from rates in, say, Italy and Spain.

Human fertility also has fallen well below the replacement rate in much (though not all) of the old USSR, certainly including Russia. However, that country, along with the Baltic republics, Belarus, and Ukraine, are distinctive in having suffered a deterioration in life expectancy at birth and corresponding increases in death rates, especially for men (Box 14.1).

Human Fertility

The number of births per woman in different parts of Eastern Europe and the Former Soviet Union are similar to TFRs in adjacent areas with similar ethnic, religious, or other characteristics.

In places like Bulgaria, Poland, Russia, and Ukraine, populations are much like those of Western Europe. For one thing, they are largely urban

> **Box 14.1 No country for old comrades**
>
> In line with its Slavic neighbors just to its west, Russia has experienced a decline in male life spans. After no change during the 1980s, when life spans were going up in other parts of the world, a male infant born in 1989 could expect to live 64.2 years. But by 1994, this measure of human well-being had fallen to 57.6 years. Expected life spans for female newborns diminished as well, by 3.3 years during the same half-decade (Brainerd and Cutler, 2005).
>
> Underlying trends such as these have been increases in fatal traffic and work-related accidents, homicides, and suicides. Also having an impact is severe overconsumption of alcohol, generally, and hazardous home-brewed spirits, specifically (Leon et al., 2007). Communicable diseases—such as tuberculosis and acquired immune deficiency syndrome (AIDS), which is spread through a combination of drug use and sexual activity—are taking a toll as well. Furthermore, there has been an alarming rise in the incidence of cardiovascular illness. Risks of dying due to these causes and others are significantly higher in places like Russia and Ukraine than in Western Europe (Eberstadt, 2009a). Deterioration of the communist-era health care system, which some offer as an explanation for diminished life expectancies, has not really had a significant impact, according to statistical analysis (Brainerd and Cutler, 2005).

(Table 14.2, column 4). Also, women face few barriers to education or employment, female illiteracy rates being low and indistinguishable from male rates (columns 5 and 6) and women participating about as much as men in the labor force (columns 7 and 8). Even though GDP per capita (column 1) is generally lower than average incomes in Western Europe, TFR trends are about the same. At or close to the replacement level 30 years ago (column 2), births per woman have fallen since then to or below 1.6 in every Slavic nation carved out of the old USSR as well as in Eastern Europe, aside from the minor exception of Albania (column 3).

As happens in other places with low TFRs, women in these settings use contraception regularly (Table 14.2, column 9). In addition, abortion is widely accepted. In Russia, for example, there were two abortions for every three births in 2009 (Faulconbridge, 2010). Among other things, this is a consequence of the suppression of religion, which was active during the communist era.

Table 14.2 Human fertility and its determinants in Eastern Europe and the Former Soviet Union from 1980 to 2007

Country (ranked by column 1)	2007 GDP per capita, PPP (constant 2005 international $) (1)	Total fertility rate		Urban % of total population in 2007 (4)	Adult illiteracy in early 2000s (%)		Female % of labor force		Contraceptive prevalence rate in 2000–2007 (% women 15–49) (9)
		1980 (2)	2007 (3)		Males (5)	Females (6)	1980 (7)	2007 (8)	
Slovenia	26,294	2.1	1.4	49	0	0	46.2	45.9	74
Czech Republic	22,953	2.1	1.4	74	—	—	47.1	44.2	72
Slovakia	19,342	2.3	1.2	56	—	—	45.2	44.6	74
Estonia	19,327	2.0	1.6	69	0	0	50.6	49.8	70
Hungary	17,894	1.9	1.3	67	1	1	43.4	45.2	77
Lithuania	16,659	2.0	1.4	67	0	0	49.7	49.5	47
Latvia	16,317	1.9	1.4	68	0	0	50.8	48.2	48
Poland	15,634	2.3	1.3	61	0	1	45.3	45.4	49
Croatia	14,729	1.9	1.4	57	1	2	40.2	44.5	69
Russia	13,873	1.9	1.4	73	0	1	49.5	49.5	73
Romania	10,750	2.4	1.3	54	2	3	45.5	45.0	70
Bulgaria	10,529	2.1	1.4	71	1	2	45.3	46.2	42
Kazakhstan	10,259	2.9	2.4	58	0	1	47.6	49.3	51
Belarus	10,238	2.0	1.3	73	0	0	49.8	48.8	73
Macedonia	8,350	2.5	1.4	66	1	5	36.0	39.3	14
Azerbaijan	7,414	3.2	2.0	52	0	1	47.0	48.1	51

(continued)

Table 14.2 (*Continued*)

Country (ranked by column 1)	2007 GDP per capita, PPP (constant 2005 international $) (1)	Total fertility rate		Urban % of total population in 2007 (4)	Adult illiteracy in early 2000s (%)		Female % of labor force		Contraceptive prevalence rate in 2000–2007 (% women 15–49) (9)
		1980 (2)	2007 (3)		Males (5)	Females (6)	1980 (7)	2007 (8)	
Albania	6,707	3.7	1.8	46	1	1	38.7	41.9	60
Ukraine	6,529	2.0	1.2	68	0	0	50.2	49.3	67
Armenia	5,377	2.4	1.7	64	0	1	47.9	49.9	53
Turkmenistan	4,677	5.0	2.5	48	0	1	46.7	46.9	48
Georgia	4,403	2.3	1.4	53	—	—	49.2	46.3	47
Moldova	2,409	2.5	1.7	42	0	1	50.2	51.3	68
Uzbekistan	2,290	4.8	2.4	37	—	—	45.7	46.0	65
Kyrgyzstan	1,894	4.1	2.7	36	0	1	47.6	42.8	48
Tajikistan	1,657	5.7	3.3	26	0	0	46.9	46.6	38

Source: World Bank (2009b).

Fertility has declined as well in the southern reaches of the old USSR, although TFRs are still above the replacement level in Azerbaijan (in the Caucasus) and Turkmenistan, Uzbekistan, and other Central Asian republics. As in other parts of the world, larger families coincide in this setting with low average incomes and limited urbanization. Such is the case with the Kyrgyz Republic and Tajikistan, for example.

Natural Increase

Since TFRs in much of the old USSR and all of its European satellites were close to the replacement level in 1980 and since human fertility has continued to decline, there is little demographic momentum in Eastern Europe and the Former Soviet Union. Death rates (Table 14.3, column 2) exceed birth rates (column 3) in Albania as well as Azerbaijan and Central Asia. Elsewhere, there is no natural increase or, where life expectancies have declined because of accidents, alcoholism, and illness, there are more deaths than births.

Negative natural increase is accompanied by net emigration in Georgia and Moldova, thereby hastening population decline (Table 14.3, column 4). In other countries, immigrants outnumber emigrants, which means that population growth (be it positive or negative) exceeds natural increase. One of these countries is Russia—to which many ethnic Russians are returning from neighboring places where they or their forebears had settled during the days of the Soviet Union. Indeed, the country registered its first increase in human numbers after 14 years of decline in 2009 thanks largely to immigration (Faulconbridge, 2010). This sort of relocation is also reflected in shrinkage of the immigrant segment of the population in the three Baltic republics and other countries between 1990 and 2005 (columns 5 and 6).

The return of people to the countries of their birth or ancestors is bound to dwindle, so natural increase will have to revive if demographic contraction is to be avoided. One reason to do so is that a shrinking population is also an aging population. In the OECD, people and their elected representatives are wrestling with the funding of pensions for the elderly, who are becoming more numerous. Eastern Europe and the Former Soviet Union face similar issues, although without the high living standards enjoyed by their neighbors to the west. In short, this region is growing old without having grown rich first.

14.3 The Agricultural Sector

Along with GDP contraction, diminished life spans, and rock-bottom fertility rates, agricultural difficulties arose in Eastern Europe and the Former Soviet Union during communism's decline and fall. As with

Table 14.3 Population trends in Eastern Europe and the Former Soviet Union

Country (ranked by column 1)	2007 GDP per capita, PPP (constant 2005 international $) (1)	Crude death rate in 2007 (per 1,000 population) (2)	Crude birth rate in 2007 (per 1,000 population) (3)	Average annual population growth in 2006–2007 (%) (4)	International migrants ÷ total population	
					1990 (6)	2005 (7)
Slovenia	26,294	9	10	0.3	9	8
Czech Republic	22,953	10	11	0.3	4	4
Slovakia	19,342	10	10	0.1	2	2
Estonia	19,327	13	12	−0.1	24	15
Hungary	17,894	13	10	−0.2	3	3
Lithuania	16,659	14	10	−0.3	9	5
Latvia	16,317	15	10	−0.3	24	17
Poland	15,634	10	10	0.0	3	2
Croatia	14,729	12	9	0.0	10	15
Russia	13,873	15	11	−0.1	8	8
Romania	10,750	12	10	−0.1	1	1
Bulgaria	10,529	15	10	−0.3	0	1
Kazakhstan	10,259	10	20	0.6	22	20
Belarus	10,238	14	10	−0.4	12	11
Macedonia	8,350	9	11	0.0	5	6
Azerbaijan	7,414	6	18	0.4	5	3
Albania	6,707	6	16	0.1	2	3
Ukraine	6,529	16	10	−0.3	13	11
Armenia	5,377	10	13	0.0	19	16
Turkmenistan	4,677	8	22	0.6	8	5
Georgia	4,403	12	11	−0.4	6	4
Moldova	2,409	12	11	−0.4	13	12
Uzbekistan	2,290	5	21	0.7	8	5
Kyrgyzstan	1,894	7	23	0.4	14	6
Tajikistan	1,657	6	27	0.8	8	5

Source: World Bank (2009b).

other aspects of the postcommunist transition, some countries have succeeded in substituting capitalist farming for state-run systems, which predominated in most of the region for decades. Other nations, however, are still struggling with reform.

Reform in formerly communist nations can be measured using a ten-point scale developed by the World Bank. Applying this scale to farm ownership, liberalization of agricultural markets, and related factors in Eastern Europe and the Former Soviet Union, the **European Bank for Reconstruction and Development (EBRD)** has identified three sorts of countries (Table 14.4, column 1). Considerable progress has occurred in the first group, made up almost entirely of Central European and Baltic nations that generally have made a successful transition from communism (see above). From Hungary to Estonia, property rights of private farmers who had continued to operate during the communist era were strengthened, holdings confiscated when communists seized power after the Second World War were returned to the original owners or their heirs, or both (column 2). Private ownership of rural land is now the rule in these countries (column 3) and market distortions, as measured by producer subsidy equivalents (PSEs), are diminishing in some places (columns 5 and 6).

Countries in the second group, where reform is at an intermediate stage according to the EBRD, are in southeastern Europe (aside from Bulgaria, which is in the advanced group) as well as some of the old USSR. Private farming had survived under communism in three former Yugoslav republics: Bosnia-Herzegovina, Croatia, and Macedonia. In contrast, privatization of agricultural land has been slowed in the Former Soviet Union because the state has made farmland ownership unappealing. In Ukraine, for example, the public sector controls a large share of the country's marketing infrastructure and has limited exports for the sake of keeping food cheap in the country (von Cramon-Taubadel et al., 2008). For these and other reasons, rural people have been reluctant to assume the risks of ownership, either as individual farmers or as shareholders in old state farms that have been converted into private enterprises. As a result, no more than one-fifth of the agricultural holdings in the old USSR were fully privatized in 2005 (USAID, 2006) and, where privatization had occurred, the new owners tended to be former officials and managers.

Similar patterns and problems have emerged in the third group of countries. With the exception of the remnants of Yugoslavia (i.e., Serbia and Montenegro), every member of this last group was part of the USSR, where reformers have found it difficult to deal with the debilitating

Table 14.4 Indicators of agricultural reform in Eastern Europe and the Former Soviet Union in 2000

Countries (ranked by column 1)	World Bank reform index, 2001 (1)	Privatization strategy (2)	% Rural land in individual use circa 2000 (3)	Net agricultural trade (US $ millions) 2000 (4)	Producer subsidy equivalent (%)	
					1993 (5)	2000 (6)
A. Advanced						
Hungary	9.2	Both	51	1,153	22.0	20.0
Czech Republic	9.2	Restitution	26	−676	27.0	16.0
Slovenia	9.2	Indiv. pre-1990	na	−361	28.0	43.0
Estonia	9.0	Restitution	61	−199	−32.0	10.0
Latvia	9.0	Restitution	91	−297	−40.0	18.0
Slovakia	8.2	Restitution	9	−382	30.0	3.0
Poland	8.0	Indiv. pre-1990	84	−533	12.0	7.0
Lithuania	8.0	Restitution	87	−106	−37.0	9.0
Bulgaria	8.0	Restitution	56	139	−4.0	2.0
B. Intermediate						
Albania	7.6	Distribution	94	−221	na	na
Armenia	7.4	Distribution	90	−184	na	na
Romania	7.4	Both	85	−593	16.0	11.0
Croatia	7.0	Indiv. pre-1990	66	−288	na	na
Azerbaijan	6.6	Distribution	5	−16	na	na

(*continued*)

Table 14.4 (Continued)

Countries (ranked by column 1)	World Bank reform index, 2001 (1)	Privatization strategy (2)	% Rural land in individual use circa 2000 (3)	Net agricultural trade (US $ millions) 2000 (4)	Producer subsidy equivalent (%)	
					1993 (5)	2000 (6)
Georgia	6.6	Distribution	44	−69	na	na
Macedonia	6.6	Indiv. pre-1990	80	na	na	na
Bosnia-Herzegovina	6.2	Indiv. pre-1990	94	na	na	na
Kyrgyzstan	6.2	Distribution	37	na	na	na
Moldova	6.0	Distribution	20	179	na	na
Ukraine	6.0	Distribution	17	442	na	na
C. Rudimentary						
Russia	5.8	Distribution	13	5,689	−24.0	3.0
Kazakhstan	5.8	Distribution	24	138	na	na
Tajikistan	4.8	Distribution	9	na	na	na
Former Yugoslavia	4.8	Indiv. pre-1990	na	na	na	na
Uzbekistan	3.4	Distribution	14	na	na	na
Turkmenistan	2.0	Distribution	8	na	na	na
Belarus	1.8	Distribution	14	−603	na	na

Source: EBRD (2002), pp. 78–84.

legacy of collectivization (Gregory and Zhou, 2010). As a result, the hand of the state is still heavy in the countryside, to the detriment of agricultural development and overall economic progress (Box 14.2).

Box 14.2 What has become of government-run farms in Russia?

Ever characterized by fits and starts, agricultural reform in Russia is still very much a work in progress.

Socialized agriculture remained largely intact for several years after the fall of communism. By the late 1990s, however, the conversion of collectives and state farms into allegedly private corporations began. By 2005, 25,800 of the government-owned entities registered in 1990 had been reorganized. Due to modest consolidation, their number declined, to 24,000 or so. In addition, their use of land and labor fell by one-third and one-half, respectively. Their share of national output dropped as well, from 75 percent in 1990 to only 40 percent in 2002 (Gardner, 2005). Most of the slack was, and continues to be, taken up by production from small private plots (which were tolerated even after the Stalin-era collectivization) and private farms established in recent years.

Even though they receive inputs at subsidized prices, collectives and state farms that have been reorganized as corporations have elevated costs of production (Gardner, 2005). One reason for inefficiency is the lack of clear and enforceable property rights in much of the Russian countryside (Allina-Pisano, 2008), which among other things limits access to credit markets. At the same time, bankruptcy laws are poorly developed, which makes it difficult to liquidate poorly run enterprises and put those enterprises' real estate and other assets in the hands of effective managers (Gardner, 2005).

Due to these and other failures of agricultural reform, Russia does not feed itself and export farm products, as it is entirely capable of doing. Instead, the country must import a portion of its food supply.

Factor Use

Recent trends in the utilization of farm inputs reflect the transition away from state-run agriculture, halting and imperfect though this transition has been. Also influential has been sluggish demand resulting from demographic stagnation and falling standards of living.

Aside from Slovenia, Tajikistan, and a few other places, Eastern Europe and the Former Soviet Union feature low ratios of rural population to arable land (Table 14.5, column 1). Under these circumstances, yield-enhancing measures, such as fertilization and widespread irrigation, are not a high priority. In the late communist era, however, inputs often were applied with little concern for efficiencies and opportunity costs. For example, fertilizer applications (column 2) were frequently excessive. While applications remain heavy in a number of small countries with limited endowments of arable land, fertilizer use fell during and after the 1990s, as subsidies for the use of this input were eliminated. In Russia, application rates have fallen to 11 kilograms per hectare (column 3).

Since precipitation is fairly generous in much of the region, many countries have invested little in irrigation. Again, the main exceptions are nations with limited land resources for crop production, particularly in the Caucasus and Central Asia (Table 14.5, columns 4 and 5). After the Second World War, Soviet authorities undertook a large expansion of irrigated agriculture in Central Asia, where rainfall is not abundant, mainly for the sake of producing cotton for international as well as domestic markets. With large volumes of water diverted from rivers in the area, deserts spread and the Aral Sea dried up; another lasting environmental consequence of Soviet-era irrigation has been farmland salinization (Dixon and Gulliver with Gibbon, 2001, p. 130). Irrigation investment has yielded better returns in Armenia, Azerbaijan, and Georgia—the warmest parts of the old USSR, which these days export sizable quantities of fruits and vegetables (grown mainly by small farmers) to Moscow and other cities to the north.

Just as the elimination of communist-era subsidies has resulted in lower levels of fertilization, diminished subsidization has driven down ratios of farm machinery to arable land (Chapter 9). Nevertheless, agriculture is more mechanized in Eastern Europe and the Former Soviet Union than in any other part of the world, save nations belonging to the OECD. This is understandable, in light of the hypothesis of induced innovation, given that rural population densities are low in most of the region.

Agricultural Yields and Land Use and Trends in Output

Diminished use of fertilizer, machinery, and other non-land inputs to crop production has had a dampening effect on output per hectare. During the last three decades of the twentieth century, yields of barley,

Table 14.5 Agricultural inputs in Eastern Europe and the Former Soviet Union, 1979 to 2005

Country (ranked by GDP per capita in 2007)	Rural population density (rural inhabitants per sq. km of arable land) 2005 (1)	Fertilizer use (kg per ha of arable land)		Irrigated land (% of cropland)		Agricultural machinery (tractors per sq. km of arable land)	
		1979–1981 (2)	2003–2005 (3)	1979–1981 (4)	2001–2005 (5)	1979–1981 (6)	2003–2005 (7)
Slovenia	574	—	352	—	1.2	—	—
Czech Republic	89	—	140	—	0.7	—	2.9
Slovakia	170	—	77	—	3.8	—	1.6
Estonia	70	—	72	—	0.7	—	6.5
Hungary	74	291	115	3.6	2.5	1.1	2.6
Lithuania	60	—	147	—	0.4	—	6.6
Latvia	67	—	44	—	2.1	—	5.5
Poland	121	239	130	0.7	0.6	4.3	11.2
Croatia	174	—	137	—	0.7	—	15.7
Russia	32	—	11	—	3.6	—	0.4
Romania	108	145	43	21.9	3.2	1.5	1.8
Bulgaria	73	221	161	28.3	16.6	1.6	1.0
Kazakhstan	29	—	7	—	15.7	—	0.2
Belarus	50	—	146	—	2.0	—	1.0
Macedonia	124	—	40	—	9.0	—	9.5
Azerbaijan	221	—	11	—	69.1	—	1.3
Albania	301	156	92	53.0	50.5	1.7	1.2
Ukraine	47	—	19	—	6.6	—	1.1
Armenia	219	—	23	—	51.5	—	2.9
Turkmenistan	111	—	—	—	79.5	—	2.2
Georgia	265	—	20	—	44.0	—	2.2
Moldova	120	—	12	—	10.7	—	2.2
Uzbekistan	352	—	—	—	84.9	—	3.6
Kyrgyzstan	257	—	24	—	73.1	—	1.6
Tajikistan	518	—	—	—	68.3	—	2.4

Source: World Bank (2009b).

Table 14.6 Trends in crop area, yield, and output in Eastern Europe and the Former Soviet Union, 1970 to 2000

	2000 Production (million tons) (1)	Average annual change, 1970–2000 (%)		
		Area (2)	Yield (3)	Production (4)
A. Crops				
Wheat	111	−1.2	0.9	−0.4
Roots & Tubers	101	−1.2	−0.3	−1.5
Vegetables	71	−0.3	1.2	1.5
Barley	40	−0.4	0.0	−0.3
Other Grains	31	−1.8	0.8	−0.9
Fruits	30	−0.1	0.0	0.1
Maize	27	−0.5	0.1	−0.5
Oil Crops	7	0.8	0.3	1.0
Pulses	5	−2.5	0.3	−2.2
B. Animal Products				
Milk	103	—	—	−0.5
Meat	18	—	—	−0.2
Eggs	5	—	—	1.1

Source: Dixon et al. (2001), pp. 138–139.

roots and tubers, and a number of other crops either rose slowly or fell (Table 14.6, column 3). Since cultivated areas have declined as well (column 2), output levels in 2000 were generally lower than what these had been 30 years earlier (columns 1 and 4). These reductions coincided with level or downward trends in human numbers and GDP per capita. Furthermore, food consumption has been affected by the removal of subsidies at the retail level.

As a rule, relative declines in food output have been modest in Eastern Europe. Part of the reason for this outcome has been the success of agricultural reform in countries such as Hungary and Poland. Larger declines have occurred in the Former Soviet Union, even including the three Baltic republics. Demand reduction is a part of the explanation. So is the lack of agricultural reform in places like Belarus, Kazakhstan, Russia, and Ukraine. Likewise, policy-induced distortions are responsible for cereal yields that are no more than half their potential level (von Cramon-Taubadel et al., 2008). In contrast, minimal or absent reform has not prevented output from going up in response to population growth in Uzbekistan and Turkmenistan, where most agricultural land is irrigated (Table 14.5, column 5).

Value added per agricultural worker (Table 14.7, columns 7 and 8) is generally low. The major exceptions are in those Eastern European nations (e.g., the Czech Republic and Slovakia as well as Hungary and Slovenia) where a successful transition from communism has occurred before and after these nations joined the EU. In addition, irrigated production of fruit and vegetables for foreign markets has enhanced value added per worker in the Armenian countryside. In contrast, this measure of productivity is modest in Russia and Ukraine, where agricultural output falls well short of potential production.

Finally, a shift has occurred between the early 1980s, when many countries penalized agriculture (Table 14.7, column 9), and the early years of the twenty-first century, when all nations other than Ukraine subsidized the sector (column 10). Subsidization in Eastern Europe has had much to do with harmonization with EU policies (Chapter 10). The swing from a negative nominal rate of assistance (NRA) to a positive rate in Russia has derived largely from higher energy prices, which has loosened the government's purse strings (Anderson, 2009, pp. 269–274). Penalization has declined in Ukraine, although the national government still uses export controls to keep food prices low (see above).

14.4 Dietary Change, Consumption Trends, and Food Security

Just as the old USSR and its satellites constantly strove to match Western output of steel, tractors, and other manufactured items, communist leaders used to aim for levels of nourishment in Eastern Europe and the Former Soviet Union comparable to those on the other side of the Iron Curtain. So that everyone could eat well, rural producers were subsidized and prices paid by consumers were kept artificially low. When domestic harvests fell short, as happened during the 1970s, foreign reserves were spent on grain from North America and other noncommunist producers (Chapter 4).

As communism declined and fell, insulating the food economy from market forces grew untenable (Rask and Rask, 2004). The resulting decline in per capita food intake is difficult to discern since data for the 1980s are not available for the Former Soviet Union and much of Eastern Europe. But if trends in Bulgaria, Hungary, Poland, and Romania (Table 14.8, column 4) are any guide, the contraction during communism's waning decade was appreciable. The negative trend having continued during the 1990s, current food consumption per capita, though superior to the

Table 14.7 Trends in cereal yields, farmed area, food production, value added per agricultural worker, and agricultural protection

Country (ranked by GDP per capita 2007)	Cereal yield (kg per ha)		Arable and permanent cropland (1,000 hectares)		Food production index (1999–2001 = 100)		Value added per agricultural worker (2000 $)		Nominal rate of assistance*	
	1979–81 (1)	2005–07 (2)	1980 (3)	2005 (4)	1979–81 (5)	2003–05 (6)	1979–81 (7)	2003–05 (8)	1980–84 (9)	2000–04 (10)
Slovenia	—	5,498	—	176	—	104.7	—	43,437	64.0	80.0
Czech Republic	—	4,480	—	3,047	—	94.7	—	5,521	20.0	27.0
Slovakia	—	4,081	—	1,391	—	101.3	—	5,026	28.0	30.0
Estonia	—	2,644	—	591	—	99.0	—	3,130	−14.0	20.0
Hungary	4,519	5,135	5,027	4,600	116	105.4	2,834	6,987	19.0	34.0
Lithuania	—	2,626	—	1,906	—	103.7	—	4,760	−19.0	32.0
Latvia	—	2,669	—	1,092	—	111.0	—	2,704	−15.0	36.0
Poland	2,345	3,029	—	12,141	104	97.6	—	2,182	10.0	7.0
Croatia	—	5,215	—	1,110	—	93.3	—	9,975	—	—
Russia	—	1,865	—	121,781	—	110.4	—	2,519	−8.0	13.0
Romania	2,854	2,680	9,834	9,288	123	114.0	1,138	4,646	24.0	55.0
Bulgaria	3,853	3,011	3,827	3,173	160	99.0	1,855	7,159	−19.0	0
Kazakhstan	—	1,158	—	22,364	—	112.6	—	1,557	—	—
Belarus	—	2,767	—	5,455	—	114.7	—	3,153	—	—
Macedonia	—	2,938	—	566	—	101.0	—	3,487	—	—

(continued)

Table 14.7 (*Continued*)

Country (ranked by GDP per capita 2007)	Cereal yield (kg per ha)		Arable and permanent cropland (1,000 hectares)		Food production index (1999–2001 = 100)		Value added per agricultural worker (2000 $)		Nominal rate of assistance*	
	1979–81 (1)	2005–07 (2)	1980 (3)	2005 (4)	1979–81 (5)	2003–05 (6)	1979–81 (7)	2003–05 (8)	1980–84 (9)	2000–04 (10)
Azerbaijan	—	2,637	—	18,432	—	128.1	—	1,143	—	—
Albania	2,500	3,580	585	578	72	109.3	816	1,449	—	—
Ukraine	—	2,368	—	32,452	—	111.7	—	1,702	−21.0	−11.0
Armenia	—	1,605	—	495	—	122.6	—	3,692	—	—
Turkmenistan	—	3,065	—	2,300	—	142.3	—	—	—	—
Georgia	—	1,650	—	802	—	104.3	—	1,790	—	—
Moldova	—	2,113	—	1,848	—	113.3	—	816	—	—
Uzbekistan	—	4,042	—	4,700	—	118.3	—	1,800	—	—
Kyrgyzstan	—	2,523	—	1,284	—	103.6	—	979	—	—
Tajikistan	—	2,275	—	930	—	150.0	—	409	—	—

Sources: World Bank (2009b); FAO (1982), pp. 45–56; FAO (2002a), pp. 3–13; Anderson and Swinnen (2008), pp. 10–21.
*NRA = relative effect (%) of government policies on gross returns to crop and livestock production.

Table 14.8 Food consumption trends in Eastern Europe and the Former Soviet Union, 1979 through 1999

Country (ranked by income per capita 1999)		Plant product consumption per capita (in cereal equivalent tons) (1)	Animal product consumption per capita (in cereal equivalent tons) (2)	Total food consumption per capita (in cereal equivalent tons) (3)	Total food self-sufficiency* (4)
Slovenia	1992–1993	0.24	1.18	1.42	0.99
	1998–1999	0.24	1.39	1.63	0.94
Czech Republic	1993–1994	0.25	1.43	1.67	1.09
	1998–1999	0.28	1.26	1.54	1.04
Hungary	1979–1981	0.26	1.30	1.56	1.19
	1989–1991	0.27	1.34	1.61	1.29
	1997–1999	0.27	1.04	1.32	1.16
Slovakia	1993–1994	0.25	1.18	1.42	0.99
	1998–1999	0.27	1.07	1.34	0.97
Estonia	1992–1993	0.18	1.51	1.69	1.23
	1998–1999	0.27	1.11	1.38	0.91
Poland	1979–1981	0.27	1.43	1.71	0.92
	1989–1991	0.26	1.39	1.66	0.94
	1997–1999	0.29	1.10	1.38	0.97
Lithuania	1992–1993	0.24	1.47	1.70	1.24
	1998–1999	0.27	1.03	1.30	1.10
Croatia	1992–1993	0.22	0.67	0.88	0.88
	1998–1999	0.25	0.57	0.81	0.92
Latvia	1992–1993	0.23	1.52	1.74	1.22
	1998–1999	0.25	0.81	1.06	0.94
Russia	1992–1993	0.26	1.26	1.52	0.81
	1998–1999	0.26	0.98	1.23	0.70
Bulgaria	1979–1981	0.33	1.11	1.44	1.09
	1989–1991	0.30	1.33	1.63	1.02
	1997–1999	0.25	0.99	1.24	0.94
Romania	1979–1981	0.28	1.00	1.28	0.94
	1989–1991	0.27	0.98	1.25	0.84
	1997–1999	0.29	0.85	1.14	0.81
Kazakhstan	1992–1993	0.28	1.35	1.62	1.07
	1998–1999	0.20	0.98	1.19	1.06
Belarus	1992–1993	0.26	1.52	1.77	1.10
	1998–1999	0.27	1.31	1.58	1.00
Ukraine	1992–1993	0.30	1.11	1.41	1.09
	1998–1999	0.26	0.77	1.03	1.05
Turkmenistan	1992–1993	0.25	0.86	1.10	0.72
	1998–1999	0.26	0.76	1.02	0.92
Azerbaijan	1992–1993	0.22	0.51	0.72	0.76
	1998–1999	0.21	0.46	0.68	0.77

(continued)

Table 14.8 (*Continued*)

Country (ranked by income per capita 1999)		Plant product consumption per capita (in cereal equivalent tons) (1)	Animal product consumption per capita (in cereal equivalent tons) (2)	Total food consumption per capita (in cereal equivalent tons) (3)	Total food self-sufficiency* (4)
Armenia	1992–1993	0.17	0.45	0.61	0.77
	1998–1999	0.22	0.50	0.71	0.60
Georgia	1992–1993	0.20	0.43	0.63	0.76
	1997–1999	0.24	0.48	0.72	0.77
Uzbekistan	1992–1993	0.26	0.68	0.93	0.84
	1998–1999	0.29	0.73	1.02	0.81
Kyrgyzstan	1992–1993	0.21	1.02	1.24	0.97
	1998–1999	0.27	0.94	1.20	0.99
Tajikistan	1992–1993	0.24	0.44	0.67	0.68
	1998–1999	0.21	0.24	0.45	0.57

*Total food self sufficiency equals total domestic food production divided by total domestic food consumption.

amounts eaten by most Africans, Asians, and Latin Americans, is well below levels in the OECD.

Consumption of plant products has not changed much (Table 14.8, column 1), aside from noticeable increases in Estonia, the Kyrgyz Republic, and a few other places, reflecting the substitution of these items for meat, eggs, and dairy goods. Most dietary adjustments have involved the latter commodities, demand for which is more income elastic. In several countries where per capita consumption of livestock products was still elevated when the Berlin Wall was pulled down, major reductions occurred during the next several years. Among these countries were the Czech Republic, Poland, Russia, and Kazakhstan (column 2). A noticeable decline also occurred in nations like Tajikistan and Turkmenistan, where consumption of livestock products was more limited at the end of the communist era. In contrast, meat, eggs, and dairy goods were eaten in greater quantities in Slovenia, which is the wealthiest nation in the entire region, as well as Armenia, Georgia, and Uzbekistan. The same four countries were the only exceptions to the general pattern of average food intake being lower in the late 1990s than it had been at the beginning of the decade (column 3).

Sizable though declines in per capita food consumption have been, these generally have been smaller than reductions in domestic output. Of course, a downward adjustment in the ratio of domestic production to consumption (Table 14.8, column 4) is not necessarily to be regretted, especially if GDP per capita is rising because of increased specialization and trade, as has occurred in Slovenia. On the other hand, seeing that the ratio has remained above 1.00 after a successful transition away from communism, including in the agricultural sector, convinces one that a nation has a comparative advantage in food production. This is the case with the Czech Republic, Hungary, and Lithuania, for example.

Agricultural comparative advantage clearly has been neutralized in some parts of the old USSR because of the halting reform (or worse) of state-run systems. One such setting is Ukraine, which was a major exporter when the international grain trade began in the middle 1800s (Morgan, 1979, pp. 27–28). Although domestic output still exceeds consumption, the gap between the two would be much greater if a more decisive break were made with communist agriculture. The same lesson holds for Russia, where the failure of reform in the countryside caused the self-sufficiency ratio to drop from 0.81 in the early 1990s to 0.70 at the end of the decade.

As living standards have deteriorated, food insecurity, which formerly was all but unknown in Eastern Europe and the Former Soviet Union, has become a problem in the region. While infant survival is excellent in a number of countries, the infant mortality rate, which is inversely related to per capita food intake, is higher elsewhere, particularly in the Caucasus and Central Asia (Table 14.9, column 2). Likewise, a sizable cohort of children is severely underweight in the southern reaches of the old USSR (columns 3 and 4). Extreme poverty is now detectable in the region as well. One out of every 50 Eastern Europeans and Russians is living on $1.25/day or less (column 5). The poverty rate is nearly 10 percent in Moldova, exceeds 20 percent in Tajikistan, and is barely under 50 percent in Uzbekistan.

Some of Eastern Europe and the Former Soviet Union has avoided major change in another indicator of food insecurity, which is income inequality. A sharp break with communism has not prevented nations such as the Czech Republic and Hungary from maintaining low Gini coefficients. The reason for this is that educational opportunities long have been widely available, thereby equipping much of the population for success in new market economies. Gini indices in the 20s and low 30s are also still observed in countries that have not pursued reform vigorously, including Belarus. The reason, as Aslund (2002) puts it, is that the state-controlled system has been able to "keep income differentiation at bay" (p. 312). In contrast, Gini coefficients have risen in Russia and much of the Caucasus and Central Asia, mainly because incomplete reforms have created economic distortions that breed inequality. Primarily because of changes in this last group of nations, the Gini coefficient for the region as a whole rose from 24 in the late 1980s to 33 (comparable to what one finds in much of Western Europe) in the middle 1990s (Aslund, 2002, p. 311).

Table 14.9 Food insecurity indicators in Eastern Europe and the Former Soviet Union

Country (ranked by column 1)	Total food consumption per capita 1997–99* (in cereal equivalent tons) (1)	Infant mortality rate in 2007 (per 1,000 live births) (2)	Malnutrition prevalence		% of population living on $1.25/day or less in late 1990s or early 2000s (5)	Gini coefficient in late 1990s or early 2000s (6)
			Percent of abnormally short 5-year-olds in late 1990s or early 2000s (3)	Percent of abnormally underweight 5-year-olds in late 1990s or early 2000s (4)		
Slovenia	1.63	3	—	—	2.0	31
Belarus	1.58	12	4	1	2.0	28
Czech Republic	1.54	3	3	2	—	25
Estonia	1.38	4	—	—	2.0	36
Poland	1.38	6	—	—	2.0	35
Slovakia	1.34	6	—	—	—	26
Hungary	1.32	6	—	—	2.0	30
Lithuania	1.30	7	—	—	2.0	36
Bulgaria	1.24	10	9	2	2.0	29
Russia	1.23	13	—	—	2.0	38
Kyrgyzstan	1.20	34	18	3	21.8	33
Kazakhstan	1.19	28	18	5	3.1	34
Romania	1.14	13	13	4	2.0	32
Latvia	1.06	7	—	—	2.0	36

(continued)

Table 14.9 (*Continued*)

Country (ranked by column 1)	Total food consumption per capita 1997–99* (in cereal equivalent tons) (1)	Infant mortality rate in 2007 (per 1,000 live births) (2)	Malnutrition prevalence			% of population living on $1.25/day or less in late 1990s or early 2000s (5)	Gini coefficient in late 1990s or early 2000s (6)
			Percent of abnormally short 5-year-olds in late 1990s or early 2000s (3)	Percent of abnormally underweight 5-year-olds in late 1990s or early 2000s (4)		
Ukraine	1.03	20	23	4	2.0	28
Turkmenistan	1.02	45	—	—	—	41
Uzbekistan	1.02	36	20	4	46.3	37
Croatia	0.81	5	—	—	2.0	29
Georgia	0.72	27	—	—	13.4	41
Armenia	0.71	22	18	4	10.6	34
Azerbaijan	0.68	34	24	14	2.0	36
Tajikistan	0.45	57	33	15	21.5	34
Macedonia	—	15	12	2	2.0	39
Albania	—	13	39	17	2.0	33
Moldova	—	16	11	3	8.1	36

Source: World Bank (2009b).
*From Table 14.8, column (3).

14.5 Summary

While it had accomplishments, including the elimination of hunger and the improvement of education, communism harmed Eastern Europe and the old USSR enormously, just as it has damaged every other place where it has been imposed. Aside from atrocious loss of life during the collectivization of Soviet agriculture in the early 1930s, economic output contracted severely once centrally planned systems, which were mechanisms for value-destruction instead of value-creation, wound down and finally collapsed. The costs of systematic failure were hidden as long as communist authorities, who were never keen on the publication of honest data, remained in charge. These costs only became apparent, in the form of sizable reductions in GDP per capita, after Marxist states imploded. During the 1990s, the only nations to avoid these reductions were those that made a decisive break with the socialist past.

Where the economy deteriorated, because the postcommunist transition was poorly handled, there was a demographic crisis, at least in nations with a Slavic heritage. Life spans diminished and human fertility fell to levels unknown anywhere else in the world, including Western Europe. Deaths exceed births in a number of countries and natural increase will soon swing negative in several more.

As with other parts of the economy, agriculture generally has prospered where a successful transition from communism has happened. Such is the case in Central Europe, where private farming is now reestablished. In contrast, the sector has languished where the communist past has not been dealt with effectively. Yields have grown slowly or not at all and farmed areas have diminished. Output of most commodities was lower at the turn of the twenty-first century than what it had been three decades earlier.

Consumption of food, especially livestock products, has fallen as well, both because retail-level subsidies are largely gone and because of declines in population and living standards. Practically unheard of between the Second World War and the waning days of the USSR, food insecurity is now a problem, as indicated by the presence of severely underweight children and people struggling to survive on no more than $1.25/day—especially in the Caucasus and Central Asia. By the same token, income inequality has risen in various places, most notably Russia.

As indicated by the nineteenth-century origins of the international cereals trade, Eastern Europe and the Former Soviet Union have a comparative advantage in agriculture. If the region can exploit its opportunities, it will feed itself easily, while also competing as an exporter in the global marketplace. Realizing this potential, however, will require seeing through the unfinished business of overcoming the malign legacies of communism.

Key Words and Terms

apparatchiks
European Bank for Reconstruction and Development (EBRD)
Iron Curtain
Near Abroad (of Russia)

nomenklatura
North Atlantic Treaty Organization (NATO)
oligarchs
Solidarity

Study Questions

1. In what specific sense was the old USSR economically unsustainable?
2. Compare and contrast trends in GDP per capita in countries that have made a decisive break with communism and countries that have not done so.
3. Why has life expectancy been going down in Russia?
4. Is demographic contraction avoidable in Eastern Europe and the Former Soviet Union?
5. Compare and contrast the role of agriculture in China's economic development since the late 1970s and the place of agriculture in the Russian economy since the fall of communism.
6. Why has the use of fertilizer, machinery, and other agricultural inputs declined in much of Eastern Europe and the Former Soviet Union?
7. Has food security improved or grown worse in the region since communism came to an end?

15

Sub-Saharan Africa

The challenges of development vary considerably from one part of the world to another. Age-old problems of socioeconomic inequality persist in Latin America and the Caribbean, for example. The Middle East and North Africa demonstrate that a generous endowment of natural resources does not really compensate for the absence of institutions needed for broad-based economic progress. Russia and its neighbors are still endeavoring to clear away the wreckage of Marxist central planning.

Notwithstanding challenges yet to be overcome, prospects for each of these regions are far from discouraging. High inflation of the sort experienced during the 1980s and early 1990s no longer threatens in Latin America and the Caribbean. There is hope for representative government and the rule of law in the Middle East and North Africa. Likewise, the worst of the postcommunist transition appears to be over in the Former Soviet Union. At the same time, the outlook for Asia continues to be bright.

Optimism is harder to justify in Sub-Saharan Africa, which has the lowest per capita gross domestic product (GDP) in the world and where impediments to development are legion. Geography is part of the problem. Much of the continent is of ancient geological origin, so most soils are highly weathered and therefore infertile. Other than in settings with torrential rainfall, such as the Congo River Basin, precipitation that is sufficient for nonirrigated agriculture is more the exception than the rule, with droughts a frequent occurrence. In addition, tropical diseases are a chronic threat to people and domesticated livestock. Likewise, crops are under constant attack from insects and other pests.

Even the region's great geographical expanse is not entirely advantageous. It has 45 independent states—about the same as Asia, which is the only continent that is larger and more populous—and many Sub-Saharan nations are small, sparsely populated, or both.[1] Fifteen of these nations are landlocked as well. Just as Bolivia is poorer than the rest of South America and Afghans and the Nepalese have lower average incomes than other Asians, international trade is inhibited and living standards are below regional norms in most African countries with impeded access to overseas markets because they lack seaports of their own.

Another consequence of the continent's ample territorial dimensions is that population densities are low in many places. Where the countryside is inhabited by relatively few people, almost all of whom are destitute, the costs of roads and other infrastructure required for trade and development are well beyond local financing capacities. All too often, then, public works are not built. Even when construction occurs, usually thanks to foreign aid, local authorities often cannot cover operation and maintenance costs, as exemplified by decaying roads, crumbling schools and clinics, and other monuments to previous donor largesse that dot the landscape.

Sub-Saharan geography is daunting enough, yet the challenges of development have been compounded by the misdeeds of rulers. Colonized long after the Western Hemisphere and several decades after Europeans seized much of Asia, Africa was carved up during the second half of the nineteenth century. The imperial era created a lasting linguistic legacy: Anglophone, Francophone, and Lusophone. However, structures required for effective governance were not put in place. Also, colonial boundaries, which later became national frontiers, were drawn with little or no regard for ethnic identifications and cleavages. The result once independence was achieved was country after country characterized by an absence of statecraft in the face of dysfunctional ethnic and regional groupings (Bates, 2000; Herbst, 2000; Salih, 2001).

In many places, this inauspicious start was followed by civil strife. In Burundi and Rwanda, a pair of tiny states with the highest population densities in the entire continent, Hutus and Tutsis have fought often. Likewise, clan warfare has put an end to national government in Somalia. In large countries, population clusters in remote areas that are hostile to ascendant groups in capital cities have rebelled from time to time (Herbst, 2000, pp. 145–152). This was true of Nigeria in the 1960s and of Angola, Ethiopia, and Mozambique in the 1970s and 1980s. War no longer rages

---

[1] In West Africa, only Mali, Niger, and Nigeria have extensive territories. East Africa has just two large countries: Sudan and Ethiopia. The Democratic Republic of the Congo, in the central part of the continent, is sizable, as are Angola, Mozambique, and South Africa, farther south. Of these nine nations, Angola, Mali, Mozambique, Niger, and Sudan have small populations.

between Arabs in northern Sudan and Dinka, Nuer, and other Africans in the southern part of the country, although government-backed gangsters have perpetrated genocide this century in the western state of Darfur. There has been little interruption in the fighting that erupted in the **Democratic Republic of the Congo (DRC)**[2] after the death of its long-time dictator, Mobutu Sese Seko, who had prevailed in the civil war that broke out in the early 1960s when independence was achieved from Belgium. Thus far, millions of Congolese have lost their lives.

During the Cold War, **proxy wars** were fought throughout the continent, with factions allied with the United States and other Western nations pitted against Soviet-supported opponents. These days, the stakes involved often relate to mineral wealth: diamonds, for instance, in Liberia and Sierra Leone, which in the 1990s experienced social breakdown similar to that of Somalia, and gold and diamonds in the eastern DRC. By and large, the outside world has had no direct interest other than the avoidance of state failure, which as the case of Taliban-controlled Afghanistan made clear can have serious global repercussions if ignored.

Aside from its enormous and tragic human toll, warfare has crippled a number of African economies. Even where armed conflict has been avoided, policies that discourage trade and investment frequently have been followed, thereby compounding difficulties related to geography (Collier and Gunning, 1999). Tiny GDPs in much of Sub-Saharan Africa reflect the region's limited participation in global markets. South Africa is the only economy of appreciable size. Its GDP is more than two-thirds greater than that of Nigeria, which is Africa's most populous nation and leading oil producer and has the region's second largest economy. A scattering of other countries export fossil fuels, copper, and other natural resources, with few rewards trickling down to local populations. It is telling that GDP for the entire region at the turn of the twenty-first century was comparable to the annual economic output of Belgium and that median national GDP is approximately $2 billion, which is about the same as the annual output of a city of 60,000 people in Western Europe or the United States (World Bank, 2000, p. 7).

In 1958, when Ghana (formerly the Gold Coast) ceased being a British colony, there were just two self-ruling states between South Africa, at the continent's antipodal tip, and the Arab nations of North Africa. One of the two was Liberia, which was founded in the 1800s by emancipated slaves from the United States. The other was Ethiopia, which had been

---

[2] The DRC was formerly known as Zaire.

governed by the ancestors of Emperor Haile Sellassie for centuries. By the middle 1970s, when the Portuguese (who were the first Europeans with colonies south of the Sahara) completed their withdrawal from Angola, Mozambique, and other possessions, virtually the entire continent was independent, although white minorities ruled in South Africa, Namibia, and Rhodesia (now Zimbabwe). Three, four, or five decades is precious little time to develop the habits of modern statecraft almost from scratch, particularly in light of difficult natural conditions and ethnic fragmentation. So it is hardly remarkable that African governments have been weak, propped up by foreign aid totaling 10 to 20 percent of GDP in many countries and as much as 25 to 30 percent of GDP in a number of others (World Bank, 2009b).

Yet there are glimmers of hope. Nelson Mandela adroitly managed the transition from apartheid to majority rule in South Africa. Uganda has made progress in the fight against the acquired immune deficiency syndrome (AIDS). Long dominated by military strongmen, Nigeria now has a democratically elected government. Positive developments such as these hint at the possibilities for the region as a whole.

15.1 Trends in GDP per Capita

Although its average income is much lower than GDP per capita in Latin America and the Caribbean, Sub-Saharan Africa went through economic travails during the 1980s similar to those experienced on the other side of the Atlantic Ocean (Chapter 12).

Some of the causes of poor economic performance were the same as well. Global demand was weak for commodities exported by the region. In many nations, a combination of inflation and a fixed exchange rate resulted in currency overvaluation, which in turn created import bias (Chapter 6). Price controls were widespread. In addition, key economic sectors were dominated by parastatal monopolies, which along with overstaffed bureaucracies and insolvent government-owned banks marginalized the private sector while simultaneously driving up fiscal deficits (World Bank, 2000, pp. 18–28; Collier and Gunning, 1999).

The burden was especially severe for rural producers, who lacked the political clout of urban elites and workers. State-run marketing boards, which were authorized to purchase all commodity output, paid low prices to farmers, just as had happened during the early years of the Soviet Union (Chapter 7). Another element of the anti-rural bias in public policy was minimal spending on education, public health, and transportation infrastructure in the countryside and on agricultural research and development.

Aside from distorted policies, which discouraged exports and investment (Table 15.1, columns 2 and 4), drought, war, and dictatorship added to the misery of many African nations in the 1980s. Incomes declined in more than half of the nations listed in Table 15.1 and went up at a snail's pace in many of the rest. There were only three small countries where GDP per capita (column 1) grew by more than 2 percent per annum: Botswana (a lightly populated place that is blessed with enormous diamond deposits), Mauritius (an Indian Ocean archipelago), and Swaziland (which is nearly surrounded by South Africa).

As happened in Latin America and the Caribbean, the last decade of the twentieth century brought a measure of relief south of the Sahara. Commodity exports were buoyed by sustained expansion of the global economy. In addition, the disappearance of the USSR in 1991 ended the Cold War, which in turn caused proxy conflicts to wind down and obliged a number of former Soviet allies to abandon Marxist economic practices. Pro-market reform was also pursued elsewhere—often as part of a recommendation from the International Monetary Fund (IMF) that structural adjustment (Chapter 8) be undertaken. State-owned enterprises were privatized and prices were deregulated. Steps toward trade liberalization were taken as well.

In Eritrea (which achieved independence from Ethiopia), Ethiopia, and Mozambique, GDP grew appreciably faster than human numbers during the 1990s because governmental control of the economy diminished and civil warfare abated (Table 15.1, column 5). The greatest improvement in living standards was registered in Uganda, which along with Mozambique was an enthusiastic convert to market-friendly policies. Exports grew faster in the former nation than in any other part of the region (column 6) and, with inflation barely in the double digits (column 7), investment took place (column 8).

However, per capita GDP shrank during the waning years of the twentieth century in nearly half of the nations listed in Table 15.1, with the worst contraction coinciding with armed conflict in Burundi and Rwanda, Liberia and Sierra Leone, as well as the DRC. Few African nations have come close to duplicating the success of Botswana. As already noted, this small country, which is located just north of South Africa, has sizable diamond deposits. In addition, Botswana has just one ethnic group, which means there is no prospect of tribal conflict. Democratic governance has been sustained for decades and, thanks to mining, living standards have improved steadily.

Table 15.1 Economic trends in Sub-Saharan Africa from 1980 to 2007

Country (ranked by column 1)	A. Average annual rate of growth, 1980–90 (%)				Country (ranked by column 5)	B. Average annual rate of growth, 1990–99 (%)				Country (ranked by column 9)	C. Average annual rate of growth, 2000–07 (%)				Remittances ÷ exports (%), 2004–2007	Freedom status 2008*
	GDP per capita (1)	Exports (2)	Inflation (3)	Gross Investment (4)		GDP per capita (5)	Exports (6)	Inflation (7)	Gross Investment (8)		GDP per capita (9)	Exports (10)	Inflation (11)		(12)	(13)
Botswana	8.2	10.6	13.6	—	Uganda	4.2	16.3	13.7	9.9	Angola	8.6	—	155.4	—	NF
Mauritius	5.3	10.4	9.5	10.2	Mozambique	4.1	13.4	36.4	13.1	Sierra Leone	6.7	—	10.0	42.0	PF
Swaziland	3.8	14.4	13.8	—	Mauritius	3.9	—	—	—	Chad	6.0	32.1	6.1	—	NF
Burundi	1.6	3.4	4.4	4.5	Eritrea	2.5	0.5	9.7	—	Mozambique	5.2	16.7	8.3	1.0	PF
Lesotho	1.6	4.1	13.8	6.9	Lesotho	2.2	11.3	9.6	2.3	Sudan	4.5	26.2	9.5	19.0	NF
Chad	1.4	6.5	2.9	19.0	Ethiopia	2.0	9.3	7.4	13.4	Ethiopia	4.3	13.4	6.7	14.3	PF
Burkina Faso	1.1	−0.4	3.3	8.6	Benin	1.9	1.9	9.4	5.3	Botswana	3.5	5.7	8.2	1.4	F
Angola	1.0	3.7	5.9	−6.8	Botswana	1.9	2.5	10.0	−1.3	Tanzania	3.5	9.3	8.7	0.3	PF
Kenya	0.8	4.3	9.1	0.8	Sudan	1.7	—	—	—	Uganda	3.2	8.0	5.2	42.6	PF
Uganda	0.7	1.8	104.0	9.6	Ghana	1.6	10.8	27.2	4.2	Nigeria	3.1	—	18.7	26.9	PF
Cameroon	0.5	5.9	5.6	−2.7	Guinea	1.6	4.7	6.2	2.4	Rwanda	3.1	9.5	6.2	8.5	NF
Rep. of Congo	0.5	5.1	0.5	11.9	Burkina Faso	1.4	0.4	6.2	4.8	Namibia	2.8	5.6	6.3	0.2	F
Mali	0.4	4.8	4.5	5.4	Malawi	1.4	4.9	33.5	−7.5	Mauritius	2.8	3.7	5.4	5.1	F
Senegal	0.3	3.7	6.5	3.9	Mauritania	1.3	1.6	6.1	6.8	Zambia	2.8	24.4	19.8	1.2	PF
Benin	0.1	−2.4	1.3	−6.2	Ivory Coast	1.1	4.7	8.0	17.6	Ghana	2.7	4.3	20.5	2.3	F
Mozambique	0.1	−6.8	38.3	−2.5	Mali	0.8	9.6	8.5	−0.8	South Africa	2.6	4.0	7.1	0.9	F
Zimbabwe	0.1	4.3	11.6	1.3	Namibia	0.8	4.3	9.8	2.5	Lesotho	2.5	12.1	6.4	1.5	F
Ghana	−0.3	2.5	42.1	4.5	Senegal	0.6	2.6	5.2	3.1	Mali	2.2	9.0	3.3	10.2	F
Guinea	−0.5	—	—	—	Swaziland	0.2	—	—	—	Burkina Faso	2.1	7.0	2.2	7.5	PF
Rwanda	−0.5	3.4	4.0	3.7	Tanzania	0.2	9.5	23.2	−1.7	Mauritania	1.6	13.8	8.7	0.1	PF

(continued)

Table 15.1 (Continued)

Country (ranked by column 1)	A. Average annual rate of growth, 1980–90 (%)			Country (ranked by column 5)	B. Average annual rate of growth, 1990–99 (%)				Country (ranked by column 9)	C. Average annual rate of growth, 2000–07 (%)			Remittances ÷ exports (%), 2004–2007	Freedom status 2008*
	GDP per capita (1)	Exports (2)	Inflation (3)	Gross Investment (4)		GDP per capita (5)	Exports (6)	Inflation (7)	Gross Investment (8)		GDP per capita (9)	Exports (10)	Inflation (11)	(12)	(13)
Ethiopia	−0.8	2.4	4.6	3.5	Zimbabwe	0.2	11.0	23.8	−0.7	Kenya	1.6	7.5	6.0	10.2	PF
Mauritania	−0.9	3.6	8.4	−4.1	Gabon	−0.1	—	—	—	Senegal	1.4	1.7	2.2	34.8	PF
Malawi	−1.0	2.5	14.6	−2.8	South Africa	−0.1	5.3	10.2	3.0	DRC	1.3	10.8	160.5	—	NF
South Africa	−1.0	1.9	14.9	−4.8	C. Af. Rep.	−0.3	6.7	4.9	−1.7	Cameroon	1.2	0.6	2.3	3.4	NF
Sudan	−1.2	0.9	31.3	—	Nigeria	−0.4	2.5	34.8	5.8	Rep. of Congo	1.2	—	11.6	0.2	NF
Tanzania	−1.2	−1.5	25.1	—	Kenya	−0.5	0.4	14.8	4.9	Swaziland	1.0	8.1	10.4	0.1	NF
Namibia	−1.4	−0.1	13.9	11.9	Chad	−0.6	5.0	7.6	4.4	Guinea	0.8	2.2	15.3	0.9	NF
Nigeria	−1.4	−0.3	16.7	−8.6	Togo	−0.8	1.5	8.3	11.6	Niger	0.5	−10.4	2.5	8.9	PF
Togo	−1.4	0.1	4.8	2.9	Niger	−0.9	1.7	6.4	5.4	Gabon	0.4	−3.0	7.7	0.0	PF
Sierra Leone	−1.5	2.1	64.0	−6.5	Madagascar	−1.2	3.6	20.6	0.9	Madagascar	0.3	6.5	10.6	0.1	PF
DRC	−1.7	9.6	62.9	—	Cameroon	−1.4	2.7	5.5	0.0	Malawi	0.2	−8.6	25.9	0.1	PF
Madagascar	−1.8	−1.7	17.1	4.9	Zambia	−1.7	1.8	56.9	11.3	Togo	−0.4	4.0	0.7	21.3	PF
C. Af. Rep.	−2.0	−1.2	7.9	1.8	Rep. of Congo	−1.9	4.3	7.1	4.7	Benin	−0.6	1.8	3.0	—	F
Zambia	−2.2	−3.4	42.2	−2.7	Angola	−2.4	8.2	813.8	12.9	Burundi	−0.6	—	9.3	0.0	PF
Gabon	−2.7	3.0	1.9	−4.6	Rwanda	−3.5	−6.0	16.3	2.1	C. Af. Rep.	−1.1	−0.1	2.2	—	PF
Ivory Coast	−2.9	1.9	2.8	−28.8	Burundi	−5.1	2.4	11.7	−12.4	Ivory Coast	−1.2	6.1	2.6	0.0	NF
Niger	−3.2	−2.9	1.9	−5.9	Sierra Leone	−7.2	−12.2	31.2	−10.3	Eritrea	−1.2	−0.7	14.1	3.1	NF
					DRC	−8.3	−5.5	1423.0	−3.5	Zimbabwe	NA	−2.9	188.3	—	NF

Sources: World Bank (2001), pp. 278–279 and 294–295; World Bank (2009b); and Freedom House (2008).

*F = free; PF = partly free; NF = not free.

Interethnic struggles have created economic chaos more recently in the Ivory Coast and Eritrea. Also, Zimbabwe's agony has been prolonged as octogenarian Robert Mugabe and his henchmen have clung to power. But in most of the continent, living standards have improved since 2000. As noted above, annual growth in GDP per capita exceeded 2 percent during the 1980s in just three of the nations listed in Table 15.1. During the 1990s, this threshold was passed in five countries. From 2000 to 2007, though, GDP per capita went up by more than 3 percent per annum in 11 countries (Table 15.1, column 9), including Angola, Sierra Leone, and a number of others that had suffered much before the turn of the twenty-first century.

The nascent turnaround south of the Sahara is partly a long-term consequence of policy reform during the 1990s, which has encouraged exports and slowed inflation (Table 15.1, columns 10 and 11). Also, the forgiveness of debts has improved the financial standing of a number of governments. Foreign investment also has occurred, including by the South Africans and Chinese (Box 15.1).

Box 15.1 South African and Chinese investors range across Africa

As the investment environment has improved south of the Sahara, banks, mobile phone companies, mining firms, and farmers from Africa's leading economy have parlayed their regional knowledge into successful new businesses. South African construction companies have built airports and roads in various countries. Likewise, entrepreneurs from places like Johannesburg and Cape Town are largely responsible for burgeoning agricultural exports to Europe as well as the development of extractive industries.

The Chinese have a major presence as well. Aside from being active in Siberia and Latin America, energy and mining companies from the world's most populous nation have penetrated Sub-Saharan Africa, as exemplified by the fact that Angola is now China's leading source of petroleum. Chinese companies have succeeded partly because they are willing to operate in hostile environments, such as southern Sudan. In addition, they often win oil and mining contracts in exchange for the construction of roads, ports, and other infrastructure (Ghazvinian, 2007, pp. 276–277).

(continued)

> (*continued*)
> There are concerns about Chinese investment in the region, including among Africans themselves. In addition, Chinese firms are reluctant to hire local labor, which recently has prevented them from signing some oil deals (Faucon and Swartz, 2009).

As in other parts of the developing world, remittances (Table 15.1, column 12) have become sizable in many countries relative to the foreign exchange earned from exports. Francophone states in West Africa, such as Mali and Senegal, receive euros from immigrants living in France. Likewise, Nigerians and other English-speakers from the same part of the continent have settled in Great Britain and send some of their earnings home. Similar patterns are observed in East Africa, although most of Sudan's remittances come from the Middle East.

Political and civil liberties (Table 15.1, column 13) are broadly correlated with economic performance. Of the 18 nations with the greatest improvement in GDP per capita since 2000, seven are categorized as free, another seven partly so, and only four not free. In contrast, eight of the 20 countries with the worst economic record afford their citizens few rights, 11 are partly free, and only one is fully free. Another parallel with other parts of the world is that oil development does not seem to mix with political and civil liberties. Angola, Cameroon, Chad, the Congo Republic, and Sudan, which all export petroleum, are categorized as not free. Another two, Gabon and Nigeria, are partly so. It is hoped that the political and civil liberties that Ghana has enjoyed for several years will survive the recent discovery of oil off the country's coast.

15.2 Demographic Trends

Progress in modern times against mortal illness is taken for granted in most of the world, affluent and developing alike. By the same token, there is a widespread trend toward increased life expectancies at birth. Setbacks, such as those that Russia has experienced (Chapter 14), are uncommon in most of the world.

In contrast, reduced life expectances are not at all rare in Sub-Saharan Africa. Communicable diseases, such as malaria, were never brought fully under control in the region, and in fact are now resurging. AIDS also claims many lives, which as discussed below has created a severe demographic crisis in a number of countries.

Human Fertility

South Africa and its Anglophone neighbors suffer more from AIDS than the rest of the region, and probably document infection and mortality rates at least as well. The same part of the continent also has experienced significant reductions in human fertility since 1980, as have Ghana and the Ivory Coast (Table 15.2, columns 2 and 3). In most other countries, households with five or six children are still typical, with Niger (a landlocked nation to the north of Nigeria) having the world's highest total fertility rate (TFR): 7.0 births per woman in 2007.

The reasons for elevated fertility are readily discerned. Aside from the insular nation of Mauritius, Gabon and a few other oil-exporters, and most of Anglophone Southern Africa, income per capita is below $2,000 (Table 15.2, column 1). In addition, urbanization (column 4) is limited in the region, with 70 to 80 percent of the population living in rural areas in a number of countries. Yet another factor related to family size is educational access for women. Where GDP per capita is above $2,000, differences between male and female illiteracy rates (columns 5 and 6) are modest, having all but disappeared in Botswana, Namibia, South Africa, and Swaziland. But elsewhere in the region, illiteracy is widespread, especially for women.

High TFRs sometimes have been reinforced by political interests, religious traditions, and public policy. In countries with ethnic divisions, political leaders worry about slower natural increase within their own constituencies, so encourage procreation. In Francophone nations inhabited by Muslims, Roman Catholics, or both, there were colonial-era edicts against birth control, motivated by a desire to foster the peopling of sparsely inhabited settings. To this day, no French-speaking nation in Africa actively promotes the use of contraceptives—as happens routinely in Anglophone countries, where higher living standards and other conditions correlate with the wider acceptance of birth control (Becker, 2002).

Large families have predominated south of the Sahara even though labor force participation by African women, which is usually a good indicator of female economic empowerment, was uniformly high in 1980 and has changed little since then (Table 15.2, columns 7 and 8). While elevated female participation in the labor force helps to explain why TFRs are low in Eastern Europe and the Former Soviet Union, human fertility in Sub-Saharan Africa is little affected by the fact that 40 percent or more of all

Table 15.2 Human fertility and its determinants in Sub-Saharan Africa from 1980 to 2007

Country (ranked by column 1)	2007 GDP per capita, PPP (constant 2005 international $) (1)	Total fertility rate		Urban % of total population in 2007 (4)	Adult illiteracy in early 2000s (%)		Female % of labor force		Contraceptive prevalence rate in 2000–2007 (% women 15–49) (9)
		1980 (2)	2007 (3)		Males (5)	Females (6)	1980 (7)	2007 (8)	
Gabon	14,323	5.0	3.1	85	10	18	43	43.8	33
Botswana	12,847	6.1	2.9	59	17	17	45	43.7	48
Mauritius	10,668	2.7	1.7	42	10	15	37	36.1	76
South Africa	9,215	4.6	2.7	60	11	13	43	45.2	60
Angola	5,085	7.2	6.5	56	17	46	46	46.6	6
Namibia	4,868	6.5	3.6	36	11	13	49	46.3	55
Swaziland	4,522	6.2	3.6	25	19	22	37	50.1	51
Rep. Congo	3,316	6.1	4.5	61	—	—	42	41.1	21
Cameroon	2,009	6.4	4.3	56	23	40	41	41.3	29
Sudan	1,970	6.4	4.2	43	29	48	27	30.4	8
Nigeria	1,859	6.9	5.3	48	20	36	37	35.9	13
Mauritania	1,820	6.3	4.4	41	37	52	41	42.9	8
Ivory Coast	1,596	7.4	4.5	48	39	61	29	31.0	13
Senegal	1,573	7.0	5.1	42	48	67	40	42.2	12
Kenya	1,456	7.4	5.0	21	22	30	46	46.4	39
Lesotho	1,456	5.5	3.4	25	26	10	46	52.4	37
Chad	1,395	6.7	6.2	26	57	79	44	48.7	3
Zambia	1,283	7.1	5.2	35	19	39	44	43.2	34

(continued)

Table 15.2 (*Continued*)

Country (ranked by column 1)	2007 GDP per capita, PPP (constant 2005 international $) (1)	Total fertility rate		Urban % of total population in 2007 (4)	Adult illiteracy in early 2000s (%)		Female % of labor force		Contraceptive prevalence rate in 2000–2007 (% women 15–49) (9)
		1980 (2)	2007 (3)		Males (5)	Females (6)	1980 (7)	2007 (8)	
Ghana	1,260	6.6	3.8	49	28	42	49	48.9	17
Benin	1,239	7.1	5.4	41	—	—	41	40.5	17
Tanzania	1,141	6.6	5.2	25	21	34	50	49.8	26
Guinea	1,076	7.0	5.4	34	57	82	48	47.1	9
Burkina Faso	1,061	7.7	6.0	19	63	78	47	46.8	17
Mali	1,023	7.6	6.5	32	65	82	46	38.3	8
Uganda	1,000	7.1	6.7	13	18	34	48	47.8	24
Madagascar	881	6.5	4.8	29	23	35	50	48.8	27
Rwanda	818	8.5	5.9	18	29	40	52	52.9	17
Mozambique	758	6.5	5.1	36	43	67	51	56.1	17
Togo	744	7.0	4.8	41	31	62	39	38.3	17
Ethiopia	736	6.8	5.3	17	50	77	45	47.3	15
Malawi	719	7.5	5.6	18	21	35	51	50.1	42
C. Af. Rep.	674	5.8	4.6	38	35	67	47	45.5	19
Sierra Leone	641	6.5	6.5	37	50	73	38	50.3	5
Niger	592	8.1	7.0	16	57	85	42	30.7	11
Eritrea	591	6.5	5.0	20	35	60	41	41.0	8
Burundi	322	6.8	6.8	10	33	48	53	51.4	9
DRC	282	6.7	6.3	33	19	46	42	38.6	21
Zimbabwe	—	7.0	3.7	37	6	12	49	43.3	60

Source: World Bank (2009b).

the region's workers are female. The reason for this is that most of the jobs that women hold are agricultural or in the service sector and, with few exceptions, the work they do is menial and low paid. Accordingly, contraceptive prevalence (column 9) is low, as can be seen in Burkina Faso (in the interior of West Africa), Ethiopia, Mozambique, Tanzania, and many other places where grinding rural poverty and large families are the norm and where children labor alongside their mothers to feed the family.

Finally, human fertility has been affected somewhat by economic crisis and AIDS. People are marrying at a later age, in the hope of finding work first. Likewise, spousal separation, often resulting from employment-related migration by men, is becoming more common. Behavioral changes such as these hinder reproduction. Nevertheless, the trend toward lower human fertility is weaker south of the Sahara than anywhere else in the world.

Natural Increase

With fertility rates greater than or equal to five births per woman in 21 of the countries listed in Table 5.2, including in three of the region's four most populous nations (the DRC, Ethiopia, and Nigeria), the demographic transition in this, the poorest part of the developing world, is not far advanced. Generally more than 30 per thousand, crude birth rates (CBRs) above 40 per thousand are not unusual (Table 15.3, column 3). AIDS and other diseases have prevented death rates from falling below 10 per thousand almost everywhere (column 2). However, births still outnumber deaths, so natural increase is elevated, greater than 20 per thousand in many places.

CBRs in South Africa and other Anglophone nations are lower than rates in the rest of the continent, yet crude death rates (CDRs) are about the same. Southern Africa has the highest rate of human immunodeficiency virus (HIV) infection in the region (Table 15.3, column 5). Indeed, no part of the world faces a worse demographic crisis due to this sexually transmitted disease. Other Sub-Saharan nations also suffer. Kenya certainly has not been spared, even though the incidence of infection is zero according to official figures (Table 15.3, column 6). In West Africa, 3 percent of Nigeria's adult population is infected, as is 4 percent of the Ivory Coast's.

Formerly well above 10 percent of the adult population, the prevalence in Uganda has been brought down thanks to an innovative approach (Box 15.2). But regrettably, few of Uganda's neighbors are following its lead, which means that rising health care expenditures and perhaps even labor scarcity can be expected in the years to come. Other impacts of the

Table 15.3 Population trends in Sub-Saharan Africa

Country (ranked by column 1)	2007 GDP per capita, PPP (constant 2005 international $) (1)	Crude death rate in 2007 (per 1,000 population) (2)	Crude birth rate in 2007 (per 1,000 population) (3)	Average annual population growth in 2006–2007 (%) (4)	Percentage of adults with HIV/AIDS in 2007 (5)
Gabon	14,323	12	26	0.7	6
Botswana	12,847	14	25	0.6	24
Mauritius	10,668	7	14	0.3	2
South Africa	9,215	17	22	0.5	18
Angola	5,085	21	47	1.2	2
Namibia	4,868	12	26	0.8	15
Swaziland	4,522	21	29	0.4	26
Rep. Congo	3,316	11	35	1.0	4
Cameroon	2,009	14	35	1.0	5
Sudan	1,970	10	32	1.1	1
Nigeria	1,859	17	40	1.1	3
Mauritania	1,820	8	32	1.3	1
Ivory Coast	1,596	15	35	0.9	4
Senegal	1,573	9	35	1.4	1
Kenya	1,456	12	39	1.3	0
Lesotho	1,456	19	29	0.3	23
Chad	1,395	15	45	1.4	4
Zambia	1,283	19	39	1.0	15
Ghana	1,260	9	30	1.0	2
Benin	1,239	11	41	1.5	1
Tanzania	1,141	13	39	1.2	6
Guinea	1,076	12	40	1.1	2
Burkina Faso	1,061	14	44	1.4	2
Mali	1,023	15	48	1.5	2
Uganda	1,000	13	47	1.7	5
Madagascar	881	10	36	1.3	0
Rwanda	818	17	44	1.4	3
Mozambique	758	20	39	1.0	12
Togo	744	10	37	1.3	3
Ethiopia	736	13	38	1.2	2
Malawi	719	15	41	1.3	12
C. Af. Rep.	674	18	36	0.9	6
Sierra Leone	641	22	46	0.9	2
Niger	592	14	49	1.7	1
Eritrea	591	9	39	1.6	1
Burundi	322	16	47	2.0	2
DRC	282	18	50	1.4	—
Zimbabwe	—	18	28	0.7	15

Source: World Bank (2009b).

demographic crisis could include diminished productivity and profits, not to mention lower tax revenues (needed to pay for health care, among other things) and GDP decline. As countries like South Africa experience all these effects, democratic institutions are sure to feel the strain. In other parts of the continent, demographic collapse induced by AIDS could eliminate all chances of rising out of poverty and establishing decent democratic governments (Lamptey et al., 2002).

Box 15.2 HIV/AIDS prevention in Uganda

Almost alone in Sub-Saharan Africa, Uganda has significantly lowered the incidence of HIV/AIDS. The country's **ABC Program** deserves much of the credit for this accomplishment.

The ABC Program has three features: (A) abstinence education focused on building up the self-esteem of preadolescent and adolescent girls, many of whom face pressure from older males to engage in sexual intercourse; (B) the encouragement of couples to be faithful; and (C) education about condoms.

The effectiveness of this combined approach has been documented by international agencies (WHO, n.d.). In contrast, countries that have tried to combat the disease mainly or entirely by promoting the use of condoms have little to show for their efforts (Green, 2005).

15.3 Agricultural Development

In terms of economic performance as well as demographic trends, the general picture in Sub-Saharan Africa has been discouraging, notwithstanding a scattering of positive signs here and there. The same pattern holds for agricultural development. Positive changes have occurred in a few places, although overall results have been disappointing. Moreover, agriculture's future outlook, like everything else in the region, is clouded by AIDS and other illnesses.

As indicated at the beginning of the chapter, limited agricultural development south of the Sahara is partly the result of unpromising geography. Since much of the region is of ancient geological origin, fertile soils eroded away long ago, so that subsoil and even basement rock are all that remain. Also, much of what remains near the surface contains aluminum in high enough concentrations that plant growth is stunted. The contrast is dramatic with the foothills of the Himalayas, which are newer geological formations, and downstream river valleys in South Asia (Voortman, Sonneveld, and Keyzer, 2000).

Just as good land is in short supply, precipitation tends to be deficient. Two-thirds of Sub-Saharan Africa is dry, with droughts occurring often in the Sahel (just south of the Sahara Desert) and the eastern and southern parts of the continent. At the same time, pests such as tsetse flies prey on livestock, just as labor productivity is sapped by malaria and other diseases. Moreover, actions by government have diminished the incentives for agricultural production.[3]

Exceptions to the broad Sub-Saharan pattern of agricultural stagnation have arisen because of favorable geography as well as public policies that, at the least, do not discriminate against economic activity in the countryside. All too unusual in the region, this combination of circumstances has been present in South Africa, where farms established by the white minority and employing up-to-date technology (made available by agricultural research and extension) produce sugar, citrus and deciduous fruits, wine, and sunflower oil for international markets. Cut-flower enterprises and tobacco farms enjoyed similar success in Zimbabwe until the late 1990s, when commercial agriculture was largely destroyed by rural invasions instigated by the Mugabe dictatorship. Closer to the equator, cocoa producers in Ghana and the Ivory Coast are internationally competitive. So are farms yielding pineapples and other horticultural products in the same two nations and Kenya. Additional examples of commercial agriculture linked to global markets include coffee, tea, and oil palm estates in a number of countries.

Otherwise, Africa's farmers struggle. Facing difficult environmental conditions as well as neglect or worse by national governments, they generally avoid the use of purchased inputs, and thus are barely able to feed themselves and their families.

Factor Use

Trends in fertilizer applications south of the Sahara contrast sharply with trends in the rest of the developing world (Chapters 5 and 9). Already low 30 years ago, application rates have failed to rise since then in many African nations and have fallen in quite a few (Table 15.4, columns 2 and 3). Henao

---

[3] For example, Delgado (1998) relates postindependence policy initiatives in the region to the changing fashions of foreign-aid providers. Binswanger and Townsend (2000) underscore colonial legacies and postcolonial realities bearing on agricultural development. Sahn, Dorosh, and Younger (1997) highlight ways that structural adjustment programs have had a positive impact on agricultural development, which is contrary to what many believe. Finally, Anderson and Masters (2009) analyze distorted agricultural incentives south of the Sahara during the past half century.

Table 15.4 Agricultural inputs in Sub-Saharan Africa, 1979 to 2005

Country (ranked by GDP per capita in 2007)	Rural population density (rural inhabitants per sq. km of arable land) 2005	Fertilizer use (kg per ha of arable land)		Irrigated land (% of cropland)		Agricultural machinery (tractors per sq. km of arable land)	
		1979–1981	2003–2005	1979–1981	2001–2005	1979–1981	2003–2005
	(1)	(2)	(3)	(4)	(5)	(6)	(7)
Gabon	65	2	4	1	1.4	0.4	0.3
Botswana	208	3	12	0	0.3	0.5	1.6
Mauritius	717	255	239	15	20.1	0.3	0.5
South Africa	129	87	50	8	9.5	1.4	0.4
Angola	224	5	3	2	2.3	0.4	0.3
Namibia	161	—	2	1	1.0	0.4	0.4
Swaziland	482	107	39	23	26.0	1.8	2.2
Rep. Congo	290	1	9	0	0.4	0.1	0.1
Cameroon	136	6	9	0	0.4	0.0	0.0
Sudan	112	5	4	14	10.9	0.1	0.1
Nigeria	238	6	6	1	0.8	0.0	0.1
Mauritania	353	8	6	23	9.8	0.1	0.1
Ivory Coast	282	26	22	1	1.1	0.2	0.3
Senegal	270	10	13	3	4.6	0.0	0.0
Kenya	536	16	34	1	1.7	0.2	0.3
Lesotho	460	15	34	0	0.9	0.5	0.6
Chad	180	1	5	0	0.8	0.0	0.0
Zambia	142	14	7	0	2.8	0.1	0.1
Ghana	281	10	7	1	0.5	0.2	0.1
Benin	185	1	6	1	0.4	0.0	0.0
Tanzania	317	4	5	1	1.8	0.1	0.2
Guinea	503	2	3	8	5.6	0.0	0.5
Burkina Faso	235	3	5	0	0.5	0.0	0.0
Mali	168	6	9	3	5.0	0.0	0.1
Uganda	469	0	1	0	0.1	0.1	0.1
Madagascar	452	3	3	21	30.6	0.1	0.0
Rwanda	635	0	0	0	0.7	0.0	0.0
Mozambique	306	11	6	2	2.7	0.2	0.1
Togo	151	1	6	0	0.3	0.0	0.0
Ethiopia	481	—	13	1	2.6	0.0	0.0
Malawi	421	20	21	1	2.3	0.1	0.1
C. Af. Rep.	134	1	0	—	0.1	0.0	0.0

(*continued*)

Table 15.4 (*Continued*)

Country (ranked by GDP per capita in 2007)	Rural population density (rural inhabitants per sq. km of arable land) 2005 (1)	Fertilizer use (kg. per ha of arable land)		Irrigated land (% of cropland)		Agricultural machinery (tractors per sq. km of arable land)	
		1979–1981 (2)	2003–2005 (3)	1979–1981 (4)	2001–2005 (5)	1979–1981 (6)	2003–2005 (7)
Sierra Leone	588	6	1	4	5.0	0.1	0.0
Niger	77	0	0	0	0.5	0.0	0.0
Eritrea	573	—	4	—	3.6	—	0.1
Burundi	732	1	2	1	1.6	0.0	0.0
DRC	595	1	0	0	0.1	0.0	0.0
Zimbabwe	261	61	37	3	5.2	0.7	0.7

Source: World Bank (2009b).

and Baanante (2006) estimate that fertilizer use is below levels needed to maintain soil fertility on 85 percent of the region's farmland. Where farmers are, in effect, mining nutrients, land degradation is the typical result (Chapter 5).

Declining fertilizer use is often blamed on reduced input subsidies, which in turn are linked to the structural adjustment that insolvent governments have had to undergo. However, this analysis ignores other aspects of structural adjustment, including currency devaluation, which strengthen production incentives and increase the use of purchased inputs in the countryside. That fertilizer application rates are well below 30 kilograms per hectare in all but a handful of Sub-Saharan nations undoubtedly has a lot to do with a general lack of transportation infrastructure, which is a great barrier to specialization and trade of all sorts.

Limited irrigation also has much to do with financing challenges arising in places that are poor, sparsely populated, or both. In addition, many parts of Africa lack groundwater reserves and good reservoir sites. The latter limitation is common where rainfall is sparse and the topography is generally flat, as opposed to a landscape featuring narrow gorges that can be dammed easily as well as rivers passing through those gorges that can fill up newly created reservoirs in short order.

The challenge of agricultural water development in Sub-Saharan Africa is indicated by the fact that just five of the countries listed in Table 15.4

irrigated more than 10 percent of their cropland three decades ago (column 4) and, because of a major increase in rain-fed farmland in the Sahelian nation of Mauritania, only four did so after 2000 (column 5). Two of the four are insular: Madagascar (a large island off Africa's southeast coast, where rice is grown in paddies) and Mauritius. Swaziland benefits from excellent conditions for the irrigated production of sugar and other crops. Irrigated farming of fertile land alongside the Nile River, which Egyptians have been doing for millennia (Chapter 13), also takes place upstream in Sudan, with cotton and groundnuts being major crops. Outside of these four nations and a couple of others, at least nineteen out of every twenty cultivated hectares are rain-fed.

As with fertilizer use, Sub-Saharan irrigation is paltry relative to what one finds in Asia. However, one must be cautious about this interregional comparison, avoiding in particular the idea that Green Revolution technology from places like India can be transferred without modification to Africa. Aside from accounting for differences in rural population densities, which are much higher in agricultural settings in Asia, intensification has to fit with local environmental realities. For example, new crop varieties suited to African soils and weather conditions must be developed, along with the right guidelines for fertilization and other kinds of land treatment (Voortman, Sonneveld, and Keyzer, 2000). To carry out such tasks, agricultural research and extension networks will need considerable strengthening. In addition, biotechnology will have to be harnessed. South Africa is doing this. However, few other countries are following suit, in part because of pressure applied by opponents of genetically modified organisms (GMOs) who are from outside the region (Box 15.3).

Box 15.3 Zambia's rejection of donated GMOs

In 2002, 3 million Zambians were starving and the United States offered 35,000 tons of food free of charge. Yet authorities in Lusaka, the national capital, rejected the donation since it would have included genetically modified products.

A variety of reasons were provided for this decision, including the possibility that human health might be affected or that genetically altered varieties might not be consumed and instead start to grow locally. In addition, European activists opposed to agricultural biotechnology let Zambian leaders know that acceptance of the food aid could jeopardize farm exports from the country in the future (Lewin, 2007).

> The same pressure has been applied elsewhere in Africa. Aside from keeping donated food away from hungry people, activists have prevented Sub-Saharan farmers from benefiting from biotechnological advances (Aerni, 2008; Paarlberg, 2008, pp. 134–138).

Finally, just as agricultural intensification has proven to be a great challenge, few Sub-Saharan nations have experienced significant mechanization. In Botswana, where GDP per capita is high by regional standards and the ratio of rural population to arable land is modest (Table 15.4, column 1), tractor use has risen since 1980 (columns 6 and 7). Also, production of sugar and other commercial crops is somewhat mechanized in Swaziland. The ratio of tractors to arable land has declined in South Africa, mainly because the country has reduced agricultural subsidies (captured entirely by white producers) in order to qualify for membership in the World Trade Organization (WTO). Elsewhere, mechanization is negligible. Even animal traction, which has the side benefit of organic fertilization, is limited, both because many farmers are too poor to afford draft livestock and the forage grasses they consume and because of tsetse flies and other pests.

Intensification, Extensification, and Trends in Output

With commercial fertilizer used sparingly and irrigation unheard of in many places, Sub-Saharan food supplies vary primarily because of changes in land use. For each of the crops listed in Table 15.5, annual growth in production from the 1970s through the 1990s (column 4) has resulted more from increases in planted area (column 2) than from improved yields (column 3). Annual yield increases have equaled or exceeded 1.0 percent for just two commodity groups: roots and tubers and maize. In contrast, yields of millet and pulses (important sources of nutrition for the rural poor) have changed little over the years, while fruit production per hectare has not varied at all.

Agricultural output per hectare has grown slowly. Indeed, average cereal yields in 15 of the 38 countries listed in Table 15.6 (columns 1 and 2) were either lower in 2005 to 2007 than they had been 25 years earlier or had not changed appreciably. Per hectare output grew by 50 percent or more in 13 nations. Whereas grain yields rose above 1.5 tons per hectare decades ago in many other parts of the world, this threshold has been surpassed in just eight of the countries listed in the table.

Of the 38 Sub-Saharan nations listed in Table 15.6, agricultural land use held steady or declined between 1980 and 2005 in just three: Botswana,

Table 15.5 Trends in crop area, yield, and output in Sub-Saharan Africa, 1970 to 2000

	2000 Production (million tons) (1)	Average annual change, 1970 to 2000 (%)		
		Area (2)	Yield (3)	Production (4)
A. Crops				
Roots & Tubers	154	1.7	1.0	2.8
Fruits	47	1.6	0.0	1.6
Maize	38	1.5	1.2	2.7
Vegetables	22	1.9	0.8	2.6
Sorghum	18	1.2	0.5	1.6
Millet	14	1.4	0.4	1.8
Rice	11	2.4	0.6	2.9
Pulses	7	1.6	0.2	1.9
Oil Crops	6	0.9	0.7	1.6
B. Animal Products				
Total Milk	19	—	—	1.8
Total Meat	8	—	—	2.0
Total Eggs	1	—	—	3.7
Cattle Hides	0.5	—	—	1.7

Source: Dixon et al. (2001), pp. 44–45.

Mauritius, Swaziland (columns 3 and 4). During the same quarter century, increases of 50 percent or more occurred in 10 countries, including Burkina Faso, Mozambique, and Sudan. Within this group, yields grew slowly between 1980 and 2005, were still low in 2005, or both.

The rapid expansion of farmed area combined with modest intensification has had a positive impact on food production. However, the gains reported in columns 5 and 6 of Table 15.6 compare poorly with contemporaneous increases in food production in Asia (Chapter 11). Furthermore, human numbers were growing considerably faster in Africa, so that per capita production actually declined in various places. There were several exceptions to this trend in West Africa, such as Benin and Ghana. Elsewhere, sluggish output growth has resulted from the full range of impediments to agricultural development reviewed in this chapter, from inhospitable environments to policy-induced market distortions to civil conflict.

The same impediments to development have held down the productivity of rural labor in most of the continent. Reflecting per capita GDPs that

Table 15.6 Trends in cereal yields, farmed area, food production, value added per agricultural worker, and agricultural protection

Country (ranked by GDP per capita 2007)	Cereal yield (kg per ha)		Arable and permanent cropland (1,000 hectares)		Food production index (1999–2001 = 100)		Value added per agricultural worker (2000 $)		Nominal rate of assistance*	
	1979–1981 (1)	2005–2007 (2)	1980 (3)	2005 (4)	1979–1981 (5)	2003–2005 (6)	1979–1981 (7)	2003–2005 (8)	1980–1984 (9)	2000–2004 (10)
Gabon	1,718	1,648	290	325	79	101.7	1,814	1,592	—	—
Botswana	203	508	402	377	87	106.1	630	390	—	—
Mauritius	2,536	7,667	100	100	90	106.5	3,087	5,011	—	—
South Africa	2,105	3,081	12,440	14,753	93	107.7	2,899	2,495	22.9	-0.1
Angola	526	525	2,900	3,300	90	136.3	—	174	—	—
Namibia	377	424	655	815	107	130.8	919	103	—	—
Swaziland	1,345	1,196	183	178	80	104.6	1,671	1,330	—	—
Rep. Congo	838	788	488	495	82	108.8	385	—	—	—
Cameroon	849	1,355	5,910	5,960	80	108.3	834	648	-11.2	-0.1
Sudan	645	681	12,360	19,434	105	112.0	—	667	-29.3	-11.9
Nigeria	1,265	1,460	27,850	32,000	57	105.2	414	—	9.4	-5.4
Mauritania	384	762	210	500	87	107.1	299	356	—	—
Ivory Coast	865	1,785	1,955	3,500	71	101.7	1,074	795	-32.2	-24.5
Senegal	690	967	2,341	2,550	74	90.1	336	215	-20.5	-7.5
Kenya	1,364	1,723	3,800	5,264	68	106.4	262	332	-18.6	9.3
Lesotho	977	546	292	330	89	102.8	723	423	—	—
Chad	587	907	3,137	4,200	80	110.7	155	215	—	—
Zambia	1,676	1,697	5,094	5,260	73	105.3	196	204	-2.7	-28.5
Ghana	807	1,365	1,900	4,185	69	118.3	670	320	-21.2	-1.4
Benin	698	1,173	1,500	2,750	63	119.7	311	519	—	—

(continued)

Table 15.6 (Continued)

Country (ranked by GDP per capita 2007)	Cereal yield (kg per ha)		Arable and permanent cropland (1,000 hectares)		Food production index (1999–2001 = 100)		Value added per agricultural worker (2000 $)		Nominal rate of assistance*	
	1979–1981 (1)	2005–2007 (2)	1980 (3)	2005 (4)	1979–1981 (5)	2003–2005 (6)	1979–1981 (7)	2003–2005 (8)	1980–1984 (9)	2000–2004 (10)
Tanzania	1,063	1,162	8,000	9,200	77	106.9	—	295	−56.3	−12.4
Guinea	958	1,434	702	1,200	96	113.8	—	190	—	—
Burkina Faso	575	1,131	2,745	4,840	63	114.7	134	173	—	—
Mali	804	1,110	2,010	4,800	77	113.6	241	241	—	—
Uganda	1,555	1,527	4,080	5,400	70	107.8	—	175	−6.2	0.4
Madagascar	1,664	2,493	2,540	2,950	84	104.6	197	174	−38.8	1.0
Rwanda	1,134	1,117	760	1,200	85	116.4	371	182	—	—
Mozambique	603	949	2,870	4,400	101	106.5	—	148	−25.2	12.4
Togo	729	1,151	1,950	2,490	77	108.2	345	347	—	—
Ethiopia	—	1,557	13,000	13,115	—	110.9	—	158	−17.5	−11.2
Malawi	1,161	1,416	1,518	2,600	93	95.7	109	116	—	—
C. Af. Rep.	529	1,100	1,870	1,930	80	108.8	377	381	—	—
Sierra Leone	1,249	1,023	450	600	85	114.5	367	—	—	—
Niger	440	437	10,212	14,482	98	105.2	222	157	—	—
Eritrea	—	436	—	637	—	90.9	—	71	—	—
Burundi	1,081	1,316	930	971	80	105.3	177	70	—	—
DRC	807	772	6,620	6,700	72	97.1	241	149	—	—
Zimbabwe	1,359	674	2,505	3,220	83	84.7	307	222	−24.0	−38.7

Sources: World Bank (2009b); FAO (1982), pp. 45–56; FAO (2002a), pp. 3–13; Anderson and Masters (2009), pp. 20–26.
*NRA = relative effect (%) of government policies on gross returns to crop and livestock production.

are high by regional norms, agricultural value added (Table 15.6, columns 7 and 8) is above $1,000 per worker in Gabon, Mauritius, Namibia, South Africa, and Swaziland. Elsewhere, value added amounts to no more than a few hundred dollars for everyone who farms.

Finally, explicit and implicit taxation of agriculture was at least as severe in Sub-Saharan Africa as in any other part of the world during the early 1980s (Table 15.6, column 9). But reforms during the ensuing two decades changed the picture considerably. Immediately after the turn of the century, there was no penalization to speak of in seven of the 15 countries listed in the table that Anderson and Masters (2009) have analyzed. This penalization remained severe only in the Ivory Coast, Zambia, and Zimbabwe (column 10).

15.4 Consumption Trends and Food Security

Going along with declining per capita food production in many Sub-Saharan nations has been a deterioration in human nutrition. Applying Rask's (1991) methodology for the analysis of dietary change (Chapter 10), one finds that per capita consumption of plant products, which is not highly sensitive to changes in income, varied just a little in most of the region during the last two decades of the twentieth century (Table 15.7, column 1). Thanks to good weather and policy reform, substantial increases occurred in a number of West African nations. In Burkina Faso and Chad, average intake of milk, eggs, and meat (column 2) grew along with per capita consumption of plant products. But in Benin, Ghana, and Mali, agricultural development and limited poverty alleviation allowed people to improve their diets not by eating more animal products, but instead by consuming plant products in greater quantities. In Mauritania and Nigeria, living standards did not improve and combined food intake per capita (column 3) went down, with plant products being substituted for animal products.

Income elasticities of demand are higher for eggs, meat, and milk than for other foods. Also, most Africans are poor, so per capita consumption of animal products went up significantly in just three of the countries listed in Table 15.7: Botswana, Chad, and Mauritius. In many more Sub-Saharan nations, however, this consumption went down, often without any compensating increase in the consumption of plant products. This was the trend during the 1980s and 1990s in Gabon, South Africa, and a few other places where the average person still consumes more than 500 to 600 cereal-equivalent kilograms per annum. But there are other places where per capita consumption was at or below this threshold around 1980 and yet diets subsequently deteriorated. Zimbabwe is one such place. So is Uganda.

Table 15.7 Food consumption trends in Sub-Saharan Africa, 1979 through 1999

Country (ranked by income per capita 2000)	Years	Plant product consumption per capita (in cereal equivalent tons) (1)	Animal product consumption per capita (in cereal equivalent tons) (2)	Total food consumption per capita (in cereal equivalent tons) (3)	Total food self-sufficiency* (4)
Mauritius	1979–1981	0.28	0.44	0.72	1.22
	1989–1991	0.29	0.57	0.85	1.01
	1997–1999	0.30	0.60	0.90	0.88
South Africa	1979–1981	0.28	0.78	1.05	1.13
	1989–1991	0.29	0.73	1.02	1.00
	1997–1999	0.29	0.60	0.88	0.94
Botswana	1979–1981	0.21	0.55	0.75	1.75
	1989–1991	0.22	0.77	0.99	1.16
	1997–1999	0.22	0.67	0.89	1.02
Namibia	1979–1981	0.22	0.96	1.18	1.31
	1989–1991	0.22	0.63	0.85	1.41
	1997–1999	0.22	0.42	0.64	1.35
Gabon	1979–1981	0.24	0.92	1.16	0.55
	1989–1991	0.25	0.84	1.09	0.51
	1997–1999	0.25	0.72	0.97	0.58
Swaziland	1979–1981	0.25	0.83	1.07	1.31
	1989–1991	0.27	0.66	0.92	1.02
	1997–1999	0.26	0.68	0.93	0.95
Lesotho	1979–1981	0.24	0.37	0.61	0.78
	1989–1991	0.25	0.35	0.59	0.71
	1997–1999	0.25	0.28	0.53	0.66
Zimbabwe	1979–1981	0.24	0.33	0.56	1.21
	1989–1991	0.22	0.29	0.51	1.10
	1997–1999	0.23	0.23	0.45	1.04
Guinea	1979–1981	0.25	0.13	0.37	0.87
	1989–1991	0.22	0.11	0.34	0.76
	1997–1999	0.25	0.14	0.39	0.80
Ghana	1979–1981	0.19	0.22	0.41	0.72
	1989–1991	0.22	0.24	0.46	0.71
	1997–1999	0.28	0.19	0.48	0.76
Mauritania	1979–1981	0.18	0.89	1.07	0.82
	1989–1991	0.24	0.82	1.06	0.80
	1997–1999	0.26	0.64	0.91	0.74
Cameroon	1979–1981	0.26	0.29	0.54	0.82
	1989–1991	0.24	0.33	0.56	0.78
	1997–1999	0.25	0.30	0.55	0.80
Sudan	1979–1981	0.21	0.62	0.83	0.99
	1989–1991	0.21	0.51	0.72	0.91
	1997–1999	0.22	0.62	0.85	0.87
Ivory Coast	1979–1981	0.31	0.34	0.65	0.77
	1989–1991	0.27	0.25	0.52	0.79
	1997–1999	0.29	0.17	0.46	0.83

(continued)

Table 15.7 (*Continued*)

Country (ranked by income per capita 2000)	Years	Plant product consumption per capita (in cereal equivalent tons) (1)	Animal product consumption per capita (in cereal equivalent tons) (2)	Total food consumption per capita (in cereal equivalent tons) (3)	Total food self-sufficiency[*] (4)
Senegal	1979–1981	0.24	0.35	0.60	0.67
	1989–1991	0.24	0.40	0.64	0.67
	1997–1999	0.24	0.42	0.66	0.69
Togo	1979–1981	0.26	0.15	0.40	0.85
	1989–1991	0.26	0.20	0.45	0.82
	1997–1999	0.28	0.19	0.46	0.79
Uganda	1979–1981	0.23	0.30	0.53	0.98
	1989–1991	0.25	0.25	0.50	0.99
	1997–1999	0.24	0.22	0.45	0.97
Angola	1979–1981	0.22	0.37	0.59	0.71
	1989–1991	0.18	0.35	0.53	0.72
	1997–1999	0.20	0.28	0.48	0.78
Central African Republic	1979–1981	0.25	0.31	0.57	0.98
	1989–1991	0.20	0.44	0.64	0.94
	1997–1999	0.21	0.46	0.66	0.96
Kenya	1979–1981	0.23	0.42	0.65	0.94
	1989–1991	0.19	0.45	0.64	0.97
	1997–1999	0.20	0.38	0.58	0.90
Benin	1979–1981	0.23	0.22	0.45	0.97
	1989–1991	0.26	0.21	0.47	0.93
	1997–1999	0.28	0.21	0.49	0.95
Burkina Faso	1979–1981	0.18	0.18	0.36	0.92
	1989–1991	0.23	0.23	0.46	0.96
	1997–1999	0.25	0.24	0.49	0.95
Eritrea	1997–1999	0.19	0.22	0.41	0.80
Rwanda	1979–1981	0.26	0.11	0.37	0.73
	1989–1991	0.23	0.10	0.34	0.67
	1997–1999	0.23	0.11	0.34	0.62
Chad	1979–1981	0.17	0.34	0.51	0.96
	1989–1991	0.18	0.40	0.58	0.96
	1997–1999	0.23	0.38	0.61	0.98
Madagascar	1979–1981	0.25	0.48	0.72	0.97
	1989–1991	0.22	0.42	0.64	0.98
	1997–1999	0.21	0.37	0.58	0.97
Mozambique	1979–1981	0.21	0.12	0.33	0.91
	1989–1991	0.20	0.11	0.31	0.84
	1997–1999	0.22	0.09	0.31	0.89
Nigeria	1979–1981	0.22	0.27	0.49	0.69
	1989–1991	0.27	0.19	0.46	0.66
	1997–1999	0.32	0.18	0.50	0.64
Mali	1979–1981	0.18	0.45	0.62	0.97
	1989–1991	0.24	0.43	0.68	0.97
	1997–1999	0.23	0.44	0.67	0.97

(*continued*)

Table 15.7 (*Continued*)

Country (ranked by income per capita 2000)	Years	Plant product consumption per capita (in cereal equivalent tons) (1)	Animal product consumption per capita (in cereal equivalent tons) (2)	Total food consumption per capita (in cereal equivalent tons) (3)	Total food self-sufficiency* (4)
Zambia	1979–1981	0.25	0.26	0.50	0.88
	1989–1991	0.22	0.22	0.44	0.99
	1997–1999	0.21	0.18	0.40	0.84
Niger	1979–1981	0.23	0.41	0.63	0.73
	1989–1991	0.23	0.27	0.49	0.60
	1997–1999	0.22	0.26	0.48	0.63
Ethiopia	1979–1981	0.20	0.30	0.50	1.00
	1989–1991	0.18	0.25	0.43	0.95
	1997–1999	0.20	0.23	0.42	0.98
Malawi	1979–1981	0.25	0.12	0.37	1.01
	1989–1991	0.22	0.10	0.32	0.93
	1997–1999	0.24	0.09	0.33	0.93
Burundi	1979–1981	0.23	0.12	0.35	0.91
	1989–1991	0.21	0.11	0.31	0.92
	1997–1999	0.18	0.08	0.27	0.91
Republic of the Congo	1979–1981	0.24	0.30	0.54	0.77
	1989–1991	0.23	0.34	0.57	0.71
	1997–1999	0.23	0.26	0.50	0.67
Tanzania	1979–1981	0.25	0.27	0.51	0.99
	1989–1991	0.23	0.29	0.52	1.00
	1997–1999	0.21	0.25	0.46	0.95
Sierra Leone	1979–1981	0.23	0.18	0.41	0.85
	1989–1991	0.22	0.12	0.34	0.85
	1997–1999	0.23	0.13	0.37	0.76
Democratic Republic of the Congo	1979–1981	0.23	0.12	0.36	0.90
	1989–1991	0.24	0.11	0.36	0.90
	1997–1999	0.19	0.09	0.28	0.90

Notes: Total food self sufficiency equals total domestic food production divided by total domestic food consumption.

Food consumption is particularly deficient in the poorest nations in Central, Eastern, and Southern Africa. Armed struggle has taken a severe toll on human diets in Ethiopia and Eritrea, in Lusophone Angola and Mozambique, and the DRC, as well as Sierra Leone in West Africa. Acute resource scarcity and violent ethnic conflict have combined in Burundi and Rwanda to create an appreciable decline not in per capita consumption of meat, eggs, and milk, which has always been negligible, but rather in the average intake of plant products. Warfare has been avoided in Tanzania as well as Malawi and Zambia, which are the two poorest

Anglophone nations in Southern Africa. The citizens of these lands have little food because of occasional droughts and also because the challenges of development have not been addressed satisfactorily.

Many of the poorest countries in the region (and the entire world) actually have relatively high indices of food self-sufficiency (Table 15.7, column 4). However, this is no mark of success. Instead, national food production is close to domestic consumption because the money needed to purchase imports is lacking—because not enough hard currency is being earned from exports, to be specific. By the same token, a falling ratio of domestic production over national consumption does not necessarily indicate failure. For example, South Africa had a high ratio 25 years ago, in part because of an apartheid-era emphasis on self-sufficiency. Insofar as the decline since then has resulted from increased specialization and trade, with the country importing more food as it exports more nonagricultural goods in which it holds a comparative advantage, this change is welcome. As in the Middle East and North Africa, indices of food self-sufficiency well under 1.00 in Angola, Gabon, and Nigeria reflect Dutch Disease, with petroleum exports driving up domestic currency values. Likewise, exchange rates can be distorted to the detriment of farmers and others whose output is tradable if a country is a major exporter of diamonds and other mining products.

Overall food consumption is low, stagnating, or both in so many countries south of the Sahara, so standard indicators of food security are alarming. Outside the region, infant mortality rates above 100 per thousand live births have become rare. But 11 of the 38 nations listed in Table 15.8 are in this category (column 2), mainly because so many infants are poorly fed. Childhood malnutrition, as measured by the incidence of children who are abnormally underweight and short (columns 3 and 4), is common in the region, although percentages in South Asia are at least as high (Chapter 11).

No part of the world has a higher percentage of the population living on less than $1.25/day (Table 15.8, column 5), although estimates of this indicator are lacking for many places. Analysis of economic inequality is similarly spotty. There are some countries, such as Burundi, where GDP per capita and Gini coefficients (column 6) are both low. However, there are many more where poverty as bad as anything encountered in South Asia coincides with income inequality of Latin American proportions. Angola is one such place. Zimbabwe is another. Gini coefficients for each of these nations and several others are about equal to that of South Africa, where average earnings for the white minority continue to be an order of magnitude above mean incomes for the rest of the population.

Table 15.8 Food insecurity indicators in Sub-Saharan Africa

Country (ranked by column 1)	Total food consumption per capita 1997–99* (in cereal equivalent tons) (1)	Infant mortality rate in 2007 (per 1,000 live births) (2)	Malnutrition prevalence		% of population living on $1.25/day or less in late 1990s or early 2000s (5)	Gini index late 1990s/early 2000s (6)
			Percent of abnormally short 5-year-olds in late 1990s or early 2000s (3)	Percent of abnormally underweight 5-year-olds in late 1990s or early 2000s (4)		
Mauritius	0.97	12.60	—	—	—	—
South Africa	0.93	46.00	—	—	8.2	58
Botswana	0.91	33.00	29.10	10.70	—	—
Namibia	0.90	47.00	29.50	18.00	—	—
Gabon	0.89	60.00	26.30	8.80	0.9	41
Swaziland	0.88	66.00	36.60	9.10	29.4	51
Lesotho	0.85	68.00	45.20	16.60	20.8	52
Zimbabwe	0.67	59.00	35.80	14.00	34.9	50
Ghana	0.66	73.00	35.60	18.80	10.5	43
Guinea	0.66	93.00	39.30	22.00	32.2	43
Mauritania	0.64	75.00	39.50	30.40	5.7	39
Cameroon	0.61	86.80	35.40	15.10	10.2	45
Sudan	0.58	69.00	47.60	38.40	—	—
Ivory Coast	0.58	89.00	40.00	17.00	6.8	48
Senegal	0.55	59.00	20.10	14.00	10.8	39
Togo	0.53	65.00	29.80	23.20	11.4	34
Angola	0.50	116.00	50.80	27.50	29.9	59
Uganda	0.50	82.00	44.80	19.00	19.1	43
Kenya	0.49	80.00	35.80	16.00	6.1	48
C. Af. Rep.	0.49	113.00	44.60	21.80	28.3	44
Benin	0.48	78.00	39.10	21.50	15.7	39

(continued)

Table 15.8 (Continued)

Country (ranked by column 1)	Total food consumption per capita 1997–99* (in cereal equivalent tons) (1)	Infant mortality rate in 2007 (per 1,000 live births) (2)	Malnutrition prevalence		% of population living on $1.25/day or less in late 1990s or early 2000s (5)	Gini index late 1990s/ early 2000s (6)
			Percent of abnormally short 5-year-olds in late 1990s or early 2000s (3)	Percent of abnormally underweight 5-year-olds in late 1990s or early 2000s (4)		
Burkina Faso	0.48	104.00	43.10	35.20	20.3	40
Eritrea	0.48	46.00	43.70	34.00	—	—
Madagascar	0.46	70.00	52.80	36.80	26.5	47
Rwanda	0.46	109.00	51.70	18.00	38.2	47
Chad	0.46	124.25	44.80	33.90	25.6	40
Nigeria	0.45	97.00	43.00	27.20	29.6	43
Mozambique	0.45	115.00	47.00	21.20	35.4	47
Mali	0.42	117.00	38.00	28.00	18.8	39
Zambia	0.41	103.00	52.00	23.30	32.8	51
Niger	0.40	83.00	54.80	39.90	28.1	44
Ethiopia	0.39	75.00	50.70	34.60	9.6	30
Malawi	0.37	71.00	52.00	18.40	32.3	39
Burundi	0.34	108.00	63.10	38.90	36.4	33
Rep. Congo	0.33	79.45	31.20	11.80	22.8	47
Tanzania	0.31	73.00	44.40	16.70	46.8	35
Sierra Leone	0.28	155.00	47.00	28.00	20.3	43
DRC	0.27	108.00	44.40	33.60	25.3	44

Source: World Bank (2009b).
*From Table 15.7, column (3).

15.5 Summary

Africa was the cradle of humankind. No matter where we or our immediate ancestors were born or moved from, each and every one of us has a family tree with roots south of the Sahara. Yet African environments, with some exceptions, are not especially hospitable to people. As a rule, soils are infertile and rainfall sparse. Where precipitation is heavy, disease-bearing organisms proliferate. Past the middle of the twentieth century, human settlement reflected these conditions and population densities tended to be low.

Problematical geography was bound to impede economic development south of the Sahara—a region that was no poorer 50 years ago than some other parts of the developing world, including Asia. However, the prolonged slide in per capita GDP rankings that has happened in Africa since then has resulted as much from the mistakes of government as from the limitations of nature. Poorly prepared for independence, some nations embraced Marxist totalitarianism, with disastrous results. Elsewhere, warfare broke out, often with different sides receiving external assistance during the Cold War and occasionally with genocidal results. Even where autocracy, central planning, and armed struggle were avoided, governments interfered regularly with market forces. Far more often than not, this interference was detrimental to agriculture, which has been the dominant economic sector in numerous countries.

Africa's poor economic record is reflected in a demographic transition that is not far advanced. Other than in Anglophone nations at the southern tip of the continent, human fertility remains well above the 2.1 births per woman required for one generation to be as large as the next. As a result, natural increase is elevated, significantly exceeding levels in other parts of the world. During the next few years, about the only countries where birth rates could align with death rates may be those suffering a calamitous upswing in mortality because of AIDS.

Signs of agricultural underdevelopment are unmistakable in the countryside. Even where governments have refrained from pricing and other policies that harm rural producers, there has been little investment either in infrastructure for irrigation and transportation or in the science and technology base for farming. Most rural households engage in subsistence production. Using few if any purchased inputs, they depend entirely on what they are able to wrestle from the unforgiving ground with their own sweat. They are, in effect, mining nature, at great cost to themselves and producing barely enough to nourish their families. Per capita food consumption is miserably low in most of the continent, as are standard indicators of food security.

Except in Sudan, the DRC, and Somalia, armed conflict is at an ebb in Africa and the continent's worst experiments with autocracy and market suppression are fading into the past. Since the middle 1990s, democracy has spread and GDP has grown faster than human numbers in a number of countries. There is a chance that the standard model (Chapter 8) can be applied, as will be needed if poverty, hunger, and disease are to be alleviated.

Key Words and Terms

ABC Program
Democratic Republic of the
 Congo (DRC)

proxy war

Study Questions

1. In what ways do environmental realities add to the challenge of economic development in Sub-Saharan Africa?
2. In what ways has the legacy of colonialism held back the region's development?
3. How have public policies and government action (or inaction) affected Sub-Saharan agriculture and its contributions to overall development?
4. Where are rates of HIV infection particularly high and what are the likely demographic impacts of AIDS in these places?
5. Explain why human fertility is higher in Sub-Saharan Africa than anywhere else in the world.
6. Why has Africa not experienced a Green Revolution similar to Asia's?
7. Why is there little mechanization of agriculture south of the Sahara?
8. Why are ratios of food self-sufficiency not that low in a number of African nations with acute food insecurity?

16

The Global Food Economy in the Twenty-First Century

Population growth and food availability, both age-old concerns, posed unusual challenges during the second half of the twentieth century. In absolute as well as relative terms, demographic expansion was unprecedented, with human numbers more than doubling to surpass 6.0 billion in 2000. Adding to the growth in food demand were pronounced improvements in living standards, especially in China and other Asian nations where large segments of the population could not afford an adequate diet little more than a generation ago. Sizable though the increases in food demand were, however, there was more growth in agricultural output. As a result, average food consumption rose and the undernourished portion of the human race shrank.

As Tweeten (1998) observes, humankind has dodged a Malthusian bullet. Moreover, this achievement has come about almost entirely because of productivity gains, as reflected in higher yields. Per hectare output of cereals, which make up at least 60 percent of what people eat if feed grains consumed by livestock are factored in, has gone from a little under 1.5 metric tons worldwide in the early 1960s to 3.5 tons at present. Food consequently has become less scarce. We know this because real food prices are much lower today than they used to be.

The Green Revolution underlying much of the increase in worldwide yields and output had its greatest impacts where the threat of hunger was worst in the 1960s and 1970s. During those years, alarm over Asia's prospects reached a crescendo, with many convinced that mass starvation was unavoidable. But at the same time, the region's farmers were beginning to sow new varieties of rice and wheat, developed by scientists to yield more grain in response to irrigation and fertilization (Chapter 3).

Since then, no other part of the world has made greater strides toward food security (Chapter 8).

Hundreds of millions of people remain poor and hungry in the world's most populous continent, especially in and around India. However, undernourishment is now worse south of the Sahara—especially in rural areas, where people feed themselves as best they can by relying on their own muscle power and whatever fertility the environment provides. Few national governments possess the means and inclination to address pressing needs. Foreigners send emergency rations when mundane deprivation is punctuated by warfare or a natural disaster. Otherwise, efforts to raise living standards and reduce hunger meet with little success.

No one should be indifferent to Africa's plight. Diseases not contained in one setting inevitably spread because international travel, either for business or pleasure, has become routine. Likewise, people who are desperate because of war or famine do not stay put, but instead move to places where they can work or at least avoid starvation. Still relatively modest, migration across the Mediterranean to Europe could swell if Africa's decline is not reversed. Furthermore, Afghanistan's experience in the late twentieth century demonstrates that the hoodlums who seize control as nations collapse do not just bully the impoverished locals. Rather, they can strike anywhere, so human agony south of the Sahara could easily lead to more international terrorism.

Having evaluated poverty, hunger, and other problems in Africa, Asia, and other parts of the world, an expert panel calling itself the **Copenhagen Consensus** (2004) has recommended a limited number of high-priority initiatives. One of these is freer international trade. Since many of the panel's members are economists, including four Nobel laureates, the call for unencumbered commercial exchange, which has been repeated often since the days of Adam Smith and David Ricardo, is to be expected. But more to the point, poverty cannot be alleviated if producers in places like the African countryside are denied full access to markets.

Aside from championing free commerce, which allows maximum living standards to be derived from available resources, the same group of experts advocates the enhancement of productive capacity in impoverished settings. Much of this enhancement has to do with human health. For example, anemia currently prevents hundreds of millions of people, mainly women and children, from leading active and productive lives. A good way to deal with this problem would be to spend $12 billion or so on the distribution of iron tablets. More money is proposed for the containment of the human immunodeficiency virus (HIV) and alleviation of the symptoms of acquired immune deficiency syndrome (AIDS). Controlling malaria, which kills more than 1 million people every year, is also urgent (Copenhagen Consensus, 2004).

Additionally, the Copenhagen Consensus (2004) highlights the need for more agricultural research and development in the tropics and subtropics, where practically all the world's food-insecure live. Among those who would agree is Gordon Conway, who led the Rockefeller Foundation from 1998 through 2004. However, reenacting the Green Revolution, which the same foundation supported from its inception, is not viable. This is mainly because water is acutely scarce in many of the world's hungriest places, where agriculture already uses 75 percent or more of the available hydrologic resources (Chapter 3). In addition, better agricultural technology is needed not only in places well suited to crop production, which have been served in the past, but also in the less-hospitable and neglected areas inhabited by the world's poorest farmers (Conway, 1997, pp. 134–135).

Conway (1997) calls for a **Doubly Green Revolution**, one that is scientifically grounded in two ways (p. 72). First, sound ecological principles guide the management of land, water, and biological resources. Second, biotechnology is harnessed. With respect to the latter, no one denies the importance of monitoring and containing potential risks, such as unchecked proliferation of wild relatives of genetically modified organisms (GMOs). But at the same time, opportunities to raise agricultural yields must be exploited fully as long as human numbers and food demand continue to grow. Weighing the risks and gains, the former president of the Rockefeller Foundation sees no alternative to bettering crop and livestock production through biotechnology.

Myriad benefits flow from sustainable agricultural development, of the sort envisioned by the Copenhagen Consensus and Gordon Conway. One is to arrest the conversion of forests and other habitats into cropland and pasture, which is hard to avoid if production per hectare remains unchanged in the face of increasing food demand. In addition, food security depends on advances in agriculture. This is because the greatest reductions in hunger occur if broad-based economic growth, which raises earnings, is coupled with lasting reductions in food prices, which are achieved with technologically driven improvements in food availability.

16.1 Victims of Our Own Success?

No one objects to combating anemia, malaria, and HIV/AIDS. The sums required are tiny relative to global economic output, and the benefits of disease control—consisting of increases in gross domestic product (GDP), not to mention diminished human suffering—far outweigh the costs. Likewise, the sustainable improvement of agriculture is widely supported. But as our ability to produce food improves, new issues emerge, including some that are entirely novel in the broad sweep of human experience. The increased incidence of obesity is one of these issues.

When addressing the problem of obesity, one must keep in mind that the desire to consume a lot of calories is etched in our genes. Our distant ancestors faced food shortages regularly, which meant that an ability to store calories as fat was advantageous. Having this ability rarely led to weight gain because hard physical labor kept everyone trim. Also, human beings who stored fat readily were in a better position to reproduce, thereby passing along their genetic traits. However, a consequence of this selection from long ago is that many of us today have an urge to overeat "somewhere between the propensity to breathe and the propensity to have sex" according to Stephen Bloom, the director of metabolic medicine at the University of London's Imperial College. Breaking the habit of overeating, he adds, is much harder than giving up smoking (Associated Press, 2004).

Aside from being a consequence of the urge to overeat, the incidence of obesity has been driven up by changing technology and economic realities. Few jobs these days require much physical exertion and many can be accomplished while sitting comfortably at a desk. Also, many people now shop online, rather than walking or even driving to a store, and send e-mails or use cellular telephones, instead of dropping by their friends or associates for a visit. Barring declines in caloric intake, developments such as these cause people to gain weight.

Besides making our lives more sedentary, technological change has given us more food that is affordable, fast, and (quite often) fattening. In a statistical analysis, Lakdawalla and Philipson (2002) find that cheaper food, which is mainly a consequence of improved agricultural productivity, has been responsible for approximately two-fifths of the increases in body mass indexes (BMIs)[1] that have occurred in the United States since 1980. Also having an impact are innovations such as vacuum packing,

---

[1] A person's BMI equals his or her weight in kilograms divided by the square of his or her height in meters, or 704.5 multiplied by weight in pounds divided by the square of height in inches (Chapter 10).

microwave ovens, and improved food preservation, which "have enabled food manufacturers to cook food centrally and ship it to consumers for rapid consumption" (Cutler, Glaeser, and Shapiro, 2003, p. 94). Thanks to these innovations, we now eat more frequently during the day, snacking on potato chips, candy, and so forth. Research reveals a linkage between international differences in the incidence of obesity and the availability of these mass-produced items (Cutler, Glaeser, and Shapiro, 2003).

Obesity and its causes are not confined to the United States and other prosperous settings. To the contrary, new technology for food preparation has had similar impacts in emerging economies. Supermarkets and other mass outlets have proliferated recently in Latin America (Chapter 12), thereby putting convenience foods in the hands (and mouths) of many. Changes of the same sort are also occurring everywhere else in the developing world with a growing middle class. As a result, U.S.-style eating habits are spreading, as are waistlines.

Africans, Asians, and Latin Americans who suffered chronic hunger as infants or even while in the womb may be particularly inclined to overeat. Take China for example. Millions starved to death there in the late 1950s and early 1960s, all of them the direct or indirect victims of the agricultural collectivization and forced industrialization ordered by Mao Zedong (Chapter 7). Now approaching or in their fifties, Chinese born during this period appear to be unusually prone to gain weight. Moreover, people who went hungry during Mao's **Great Leap Forward** tend to be anxious about the nutritional status of their offspring and grandchildren, and so feed them too much if they have the means to do so. An indicator of this is that one-third of Beijing's residents, whose incomes exceed earnings in the rest of the country, are overweight (Duncan, 2003).

That excess weight is, indeed, a global phenomenon is underscored by data compiled by the **International Obesity Task Force (IOTF)**, which was created by the **World Health Organization (WHO)**. To be sure, the incidence of obesity is highest in wealthy nations. In the United States, for example, the portion of adult males with a BMI at or above 30 (i.e., the obesity cutoff) rose from 10.4 percent in the early 1960s to 19.9 percent in the early 1990s, as the rate for adult females went from 15.1 percent to 24.9 percent. However, the obese share of the population also has risen outside the Organization for Economic Cooperation and Development (OECD). For example, the portion of Brazilian men with a BMI of 30 or more increased from 3.1 percent in 1975 to 5.9 percent in 1989. During the same 14 years, the incidence for women rose from 8.2 percent to 13.3

percent. As of 1990, 8 percent of black men and 44 percent of black women in South Africa's Cape Peninsula were above the same BMI threshold (WHO, 2000, p. 21).

Obesity is unhealthy in various ways. People who are extremely overweight are more susceptible to type 2 diabetes. They also run greater risks of cardiovascular disease and kidney failure. Drawing on recent research, the IOTF provides estimates of the expense of diagnosing and treating illnesses related to obesity, reporting for example that this expense in the United States was nearly $46 billion per annum (6.8 percent of total health care costs) in the early 1990s (Wolf and Colditz, 1994, cited in WHO, 2000, p. 83). In the developing world, more than 115 million people suffer from obesity-related ailments. For example, the economic cost of chronic diseases related to diet currently exceeds 2 percent of GDP in China, which is higher than the economic cost of undernourishment in that country (Burslem, 2004).

For the sake of limiting morbidity, mortality, and health care costs, governments could do various things to reduce obesity. Education about the causes of weight gain gives people ideas about how to avoid the problem, and disseminating more information about health consequences should help motivate people to eat right and exercise. Fiscal instruments are also an option. Health insurance premiums could be raised for heavy people, for example. Alternatively, foods rich in fat, oil, and sugar could be taxed, just as is done with tobacco products. Fiscal approaches such as these might be justified if people who get sick because they are overweight do not pay all their treatment costs.

However, applying fiscal measures to correct for the non-internalization of medical costs seems impolitic. Much of the inconvenience and cost of obesity is shouldered by overweight people themselves, rather than society as a whole. Furthermore, taxation of fattening products may be inequitable if these items are favored by poorer segments of the population. Conceding the obstacles to combating obesity with taxes, Rashad and Grossman (2004) recommend subsidies for exercise as an alternative. However, this approach also seems unpromising, in light of the trouble and expense of monitoring people who have agreed to exercise in return for a financial reward. Another option would be to pay people who lose weight and do not gain it back, although here again monitoring costs would be problematic.

So no policy measure seems adequate. More might well be gained from advances in food science that yield healthy and tasty alternatives to the fattening products that many of us crave.

16.2 The New Food Economy

Problems such as fattening diets are sometimes blamed on large agribusinesses eager to process and sell huge volumes of edible output (Pollan, 2006, pp. 52–53). However, this view is simplistic and misleading, not least because it neglects the importance of a basic fact of modern economic life. We refer here to the rising value of people's time, including the implicit expense of cooking meals from scratch rather than simply warming up, thawing, or hydrating processed and prepared foods.

In his presidential address to the **American Agricultural Economics Association (AAEA)** in 1999, John Antle focused on this determinant of the **new food economy**, as he calls it. Time's value goes up, he observed, as living standards improve. Also having an impact is female economic empowerment, which raises the opportunity cost of however many hours women spend in the kitchen. Combined with technological advances that lower the cost of transmitting information, increases in the value of time leave no part of the food economy unaffected. One impact is to increase the importance, even dominance, of services, as reflected in the growing number and variety of edible goods (Box 16.1).

Box 16.1 The new food economy

"The increasing opportunity cost of time goes hand in hand with an increasing specialization of labor, and together these factors drive the transition from the industrialized to the service- and information-based economy. The opportunity cost of time and the specialization of labor explain the growth in demand for an array of increasingly specialized goods and services, including the diversity of manufactured foods and food services that substitute for time spent preparing foods and raw ingredients in the home. The increasing opportunity cost of time also explains the explosive growth in the utilization of information technology that allows all kinds of repetitive tasks to be carried out by machines rather than people, further reinforcing the continued specialization of labor that leads to yet higher incomes. Just as significantly, the transition to a service- and information-based economy is associated with a rapidly declining cost of information that has a profound impact on the structure and efficiency of markets" (Antle, 1999, p. 994).

Not so long ago, it was expensive for suppliers to discern consumers' preferences about nutritional content, convenience attributes such as ease of preparation, and so forth. Likewise, suppliers found it difficult to respond to heterogeneous preferences, by producing and delivering a diverse set of quality-differentiated goods. Accordingly, a limited number of uniform commodities were available in the food economy, with prices being the main focus of competition in the marketplace. No more! Today's information technology allows for easy tracking of preferences as well as the fine-tuning of production processes. The result is a seemingly endless variety of products, each with its own price and characteristics.

Combined with lower commodity prices, the rising importance of services explains why the value of unprocessed farm products is declining over time relative to what people spend on food. In the United States, for example, this share has fallen from 41 percent in 1950 to 15 percent today (ERS-USDA, 2009a). Furthermore, demand for services of all sorts shows every sign of continuing to increase. This certainly is true, for example, of the services that people pay for so they can spend less time preparing their own meals.

In places like the United States, fast food appears to be subject to Engel's Law (Chapter 2). That is, income elasticities of demand for cheeseburgers, fried chicken, and other low-priced fare are a fraction of 1.0 and fall as GDP per capita goes up, especially if that fare is fattening or otherwise nutritionally undesirable. However, Engel's Law appears less applicable to full-service restaurants. In the United States, real sales revenues for these establishments, which offer entertainment of a sort along with meals cooked by trained chefs, are expected to be 18 percent higher in 2020 than in 2000. This growth is comparable to increases in GDP per capita in the same period. It is also well above the 6 percent increase in inflation-adjusted revenues projected for fast-food eateries during the same two decades (Stewart et al., 2004).

Demographic change has something to do with the rising demand for food consumed outside the home. For example, the number of domiciles with either a single person or multiple adults but no children is increasing faster than the number of households with children. The former group eats out more than the latter group. So do the elderly, who comprise a growing segment of the population. Nevertheless, the primary reason why people are using their own kitchens less is income growth, which drives up the opportunity cost of time. In the United States, for example, five-sixths of the aforementioned rise of 18 percent in the sales revenues of full-service restaurants between 2000 and 2020 is expected to occur because of higher earnings (Stewart et al., 2004).

Throughout the OECD as well as in other parts of the world with an urban middle class, farmers have ceased to dominate the food economy in all respects other than their utilization of land and water resources. Specialization and trade are far advanced, with the value of inputs and post-harvest services provided by agribusinesses greatly exceeding the value that farmers contribute to food output. Already sizable, the value of preparation services provided by restaurants and other enterprises is growing. Looking to the future, expansion of services such as these is certain to be a major source of growth in the food economy.

16.3 The Changing Role of Government

Aside from becoming an ever more important part of the food economy, post-harvest services are characterized by substantial economies of size. A typical consequence is that processing, preparation, and related lines of commerce are concentrated, with a few firms accounting for a sizable portion of total service output. The supermarket business, for instance, is currently dominated by a limited number of large chains and is likely to become even more concentrated in the years to come, as some businesses merge and others cease operating (Duncan, 2003).

While retail prices are pulled down as economies of size are captured, monopolistic behavior, which raises prices, is a possibility in any market served by a handful of firms. Antitrust and other policies can be applied to counteract monopolization. However, the economic case for this kind of corrective intervention by government perhaps is not as compelling as one might suppose. This is the general conclusion Antle (1999) reaches after surveying a number of empirical studies. By and large, these studies indicate that long-term trends in prices are explained more by developments such as increased product differentiation than monopolistic behavior. Apparently, competition among firms that together account for a large share of sales in selected markets is vigorous enough to keep prices close to efficient levels—the prices households pay for retail products as well as the prices growers receive for output.

If the economic case for aggressive governmental action to combat monopolization in the food economy is not really overpowering, the economic rationale is extraordinarily weak for price supports and deficiency payments aimed at propping up agricultural incomes. The average annual return on assets used in the commercial production of crops and livestock has held fairly steady at around 10 percent for many years, which is about what one expects of a competitive industry in long-run equilibrium

(Chapter 4). Furthermore, the beneficiaries of arrangements such as the U.S. Farm Bill and Europe's Common Agricultural Policy (CAP) tend to be more affluent than other growers, let alone most of their fellow citizens.

Arrangements such as price supports and deficiency payments are likely to diminish over time. True, farmers sometimes find it easier to organize politically for "entitlements" from the nonagricultural majority of taxpayers as their numbers shrink. However, agriculture is now a prime focus of multilateral trade negotiations, with protectionism and subsidies in the OECD coming in for pointed criticism (Chapter 6). At the same time, expansion of the European Union (EU), which has made farmers east of the old Iron Curtain eligible for CAP largesse, as well as soaring U.S. budget deficits have bolstered the case for an agricultural sector that depends less on taxpayers and trade barriers (Chapter 10).

Any support that remains is likely to come with strings attached. Some of these strings are environmental. In every part of the world, not least in wealthy nations, farms are a major source of water pollution (Chapter 5). Recognizing this, the nonagricultural majority and its political representatives are bound to insist on sound management of natural resources in return for any income assistance provided to growers.

Support for agriculture also hinges on food safety, the demand for which is being driven up for various reasons. Improvements in our ability to detect harmful components have had an effect. Also, consumption of fresh fish and meat, which are subject to spoilage, has increased and the marketing channels connecting ranches and ports to consumers' tables have grown more complex, which can introduce or spread hazards. Even the rising importance of food preparation raises questions about safety, since the responsibility for adequate cooking has shifted from households to businesses (Unnevehr, 2004).

Consumers cannot be faulted for paying attention to food safety—as they have done in recent years after outbreaks of **Bovine Spongiform Encephalopathy (BSE)**, or "mad cow" disease, in Great Britain and North America, for example. However, their worries can be exploited for protectionist ends. This exploitation is supposed to be limited by the **Codex Alimentarius**, which the Food and Agriculture Organization (FAO) and WHO established in 1963 to develop food standards that protect consumers' health while simultaneously ensuring "fair trade practices" (Codex Alimentarius Commission, 2004). Accepting the Codex approach, the World Trade Organization (WTO) allows trade restrictions based on sound science.

The conundrum is that countries do not always agree about what sound science is, not to mention the right safety standards for edible products. For example, the United States has complained about the EU's restrictions on GMOs, contending that these restrictions have little objective basis and are instead political and protectionist (Miller and Conko, 2004, pp. 195–196). But even in the United States, worries about food safety, though not backed up by scientific evidence, sometimes impede technological improvement. For example, McDonalds, fearing consumer resistance, announced in 2000 that it would not use any genetically modified products. This has reduced plantings of potatoes containing the *Bacillus thuringiensis* (Bt) bacterium, which resist pests and have other beneficial properties (Unnevehr, 2004).

Harmonization of food-safety rules, which is the purpose of the Codex Alimentarius, should help to preempt protectionism that masquerades as something else. Provided, of course, that the rules have a demonstrable scientific basis, harmonization also overcomes the barriers to technological progress that arise if standards in some places permit new products (e.g., those resulting from the use of biotechnology) while other jurisdictions' standards are too restrictive. At the same time, uniform and reasonable rules governing the global food economy should facilitate the specialization, trade, and productivity growth needed for a growing population to be fed.

16.4 Back to the Future Food Economy?

Our thinking about food is bound to evolve as people eat better and fewer of us go hungry. A minor worry as long as sustenance was hard to come by, obesity is now the subject of widespread debate and concern. Likewise, affluent people and countries tend to be exacting about what they consume.

Their preoccupations do not relate only to microbes and chemicals. Local produce is also preferred over food trucked in from hundreds of kilometers away or flown in from other continents. Demanded as well are products containing none of the inputs of a modern food economy: commercial fertilizer, pesticides, and genetically modified seeds, in particular. More than a few consumers are troubled by their being fed thanks to the toil and sweat of impoverished farmworkers. Also provoking concern are measures such as the raising of hogs, chickens, and other livestock in confined conditions, which keeps meat, eggs, and other animal products cheap.

As any economist is bound to point out, these preferences cannot be accommodated without affecting the affordability of food. Due to the discouragement of agricultural biotechnology, crop yields in Europe have been reduced by approximately 15 percent (Collier, 2008), thereby driving up prices. Agricultural self-sufficiency, motivated by a desire to reduce **food miles**, has the same effect. For example, energy use and prices of horticultural products would increase in Great Britain if imports from Kenya were curtailed (Box 16.2). There would be similar consequences if Chicagoans had to rely exclusively on local greenhouses for items such as fresh tomatoes and lettuce. In fact, the economic advantages of farmers in warmer and more distant settings frequently outweigh the expense of fuel and other inputs required to transport produce to customers in colder places.[2]

Box 16.2 Are intercontinental food miles a problem? The case of Kenya and Great Britain

During the early years of the twenty-first century, Kenya registered impressive growth in a range of horticultural exports. By 2004, its foreign sales of fresh vegetables and other produce, mainly to the United Kingdom, totaled $161 million. The East African nation also exported $470 million in cut flowers, bulbs, and so forth, which made it the leading supplier of these products to the EU (Desrochers and Shimizu, 2008).

This success drew the attention of anti-food-mile campaigners in Great Britain, even though the carbon released into the atmosphere as horticultural goods are flown in from Kenya is less than 0.1 percent of total carbon emissions for the importing nation. One factor containing these emissions is that 60 to 80 percent of the roses, green beans, and other lightweight items grown in Kenya arrive in the holds of passenger aircraft, occupying what otherwise would be empty space. In addition, the energy needed to produce, say, cut flowers in Kenya is little more than one-sixth of the energy requirements of the same crop in European greenhouses (Williams, 2007, cited by Desrochers and Shimizu, 2008).

In light of this evidence, British activists who have continued to oppose Kenyan imports have had to shift to other arguments.

---

[2] The advantages of farmers in Florida and the southwestern part of the United States explain why the production of horticultural goods in midwestern and northeastern greenhouses all but disappeared after the completion of the interstate highway system, which President Dwight Eisenhower pushed for in the 1950s in order to lower shipping costs and therefore promote cross-country commerce.

The point here is not to condemn food that is local, organic, free-range, and so on. Vegetables from a nearby farm where the operator applies manure and weeds by hand rather than using chemical inputs often is tastier than the fare provided by the conventional food economy. The same can be said of chicken, for example, not raised in a concentrated animal feeding operation (CAFO). These alternative products are costlier as well, yet they are purchased by consumers with enough earnings to indulge their discerning palates.

However, there is no such thing as an inexpensive lunch outside the conventional food economy. Someone who recognizes this is Michael Pollan, a professor of journalism at the University of California. Although he is a sharp critic of agribusinesses and the mass production of food, Pollan shares some edifying lessons in the final chapter of a recent book, in which he recounts trying to harvest a meal of local game, seafood, and greens on his own. Pollan is not above using implements that a genuine hermit would never possess: a rifle to bring down a wild boar, an automobile to reach hunting and fishing grounds, and so forth. But even with these implements, the experience leaves the professor exhausted, as he admits (Pollan, 2006, pp. 391–411).

At least a few activists are heedless of what can be learned from Pollan's brief experiment with food self-sufficiency. More than that, they are not content with free commerce, which in the final analysis is driven by the independent choices made by individual consumers. Instead, the activists prefer collective action, provided of course that the outcome is acceptable to them. Specific proposals are lacking, although creating the food economy they seem to have in mind would involve harnessing governmental power to reverse much of the specialization and trade and many of the technological strides that have been achieved in recent decades.

The activists have had their greatest successes in their campaigns against GMOs. The reasons for this, it must be said, are easy to understand. Many people avoid ingesting things that owe their existence to biotechnology, not nature, even if the chances of harmful consequences are remote. This reluctance is reinforced by a general aversion to risk, which most of the population shares, and by the perception that the harm resulting from the consumption of GMOs is novel or unfathomable.[3]

---

[3] Activists cultivate this impression by insisting that foods with genetically modified ingredients be so labeled. The alternative is to label foods that are free of these ingredients, so consumers desiring these items can find them easily. This alternative also has the virtue of not stoking unwarranted fears about GMOs.

In addition, unwillingness to try genetically modified products is pronounced in affluent settings where food purchases comprise a small portion of household earnings and budgets, such as Europe and Japan. In these settings, the costs of not using GMOs, in terms of higher food prices, are barely noticeable.

Much more is at stake in the developing world, where hundreds of millions of poor households spend much of their meager earnings on food. Regulatory repression of agricultural biotechnology in wealthy nations (Miller and Conko, 2004, pp. 102–108 and 185–187) diminishes the number of advances that could raise yields in Africa, Asia, and Latin America. So does the adoption by poor countries of regulatory regimes that reflect extreme caution about farmers' use of GMOs (Paarlberg, 2008, pp. 121–122). In addition, agricultural biotechnology has been discouraged south of the Sahara because African nations are unwilling to jeopardize farm exports to Europe, either now or in the future (Aerni, 2008; Paarlberg, 2008, pp. 134–138).

Four, five, or six decades from now, it may be possible to indulge a desire to go back to the future in the world food economy, forgoing the efficiencies of specialization and trade across great distances and not improving technology for crop and livestock production. By that time, the human population will be stabilizing and the demand for food will be growing at a leisurely pace, driven by modest increases in consumption per capita.

But in the meantime, human numbers and food demand will increase. Just because a Malthusian bullet was dodged between 1950 and 2000 does not mean that we will never have to do so in the future. If Malthusian bullets start flying again, the consequences of going back prematurely to the sort of food economy envisioned by those who protest against food miles and GMOs would be devastating. Norman Borlaug, who won the Nobel Peace Prize in 1970 for his contributions to the Green Revolution (Chapter 3), sounded a warning toward the end of his illustrious life and career: "If our (Green Revolution) varieties had been subjected to the kinds of regulatory strictures and requirements that are being inflicted upon the new biotechnology, they would never have become available" (Miller and Conko, 2004, p. x).

In truth, resistance to the scientific advances underpinning agricultural development appears to be waning—dramatically so in much of the world. For the time being, GMOs remain proscribed in Europe and a handful of other rich places. Long accepted in North America, genetically modified crops are now being sown and reaped in Argentina, Brazil,

China, India, and many other tropical and subtropical lands. Moreover, African nations are turning to China and other places for advances in biotechnology that Europe has chosen to forgo (*The Economist*, 2010b).

Thanks to developments such as these, there is no reason for supply not to keep up with demand in the future. As it has done in the past, the world food economy is entirely capable of feeding the entire human race while simultaneously contributing to overall economic progress. Each and every one of us has a vital stake in this outcome.

Key Words and Terms

American Agricultural Economics Association (AAEA)
Bovine Spongiform Encephalopathy (BSE)
Codex Alimentarius
Copenhagen Consensus
Doubly Green Revolution
food miles
Great Leap Forward
International Obesity Task Force (IOTF)
new food economy
World Health Organization (WHO)

Study Questions

1. Summarize and explain the main recommendations of the Copenhagen Consensus.
2. Describe the Doubly Green Revolution, as envisioned by Gordon Conway.
3. Analyze the causes of obesity and evaluate the various proposals that have been made to deal with the problem.
4. Relate the growing emphasis on processing and preparation services in the food economy to time values and improvements in information technology.
5. Does a strong rationale exist for applying antitrust laws in response to the growing concentration of grocery stores and other agribusinesses?
6. Explain the function of the Codex Alimentarius.
7. Is it always economically rational to reduce food miles?
8. Will biotechnology contribute more or less to food supplies in the future? Should it?

Abbreviations and Acronyms

AAA	Agricultural Adjustment Act of the United States
AAEA	American Agricultural Economics Association
ABC	Uganda's program for the prevention of HIV-AIDS
AHDR	Arab Human Development Report
AIDS	acquired immune deficiency syndrome
BMI	body mass index
BRIC	Brazil, Russia, India, and China
BSE	Bovine Spongiform Encephalopathy ("Mad Cow")
Bt	*Bacillus thuringiensis*
CAFO	concentrated animal feeding operation
CAP	Common Agricultural Policy of the European Union
CBR	crude birth rate
CDC	U.S. Centers for Disease Control and Prevention
CDR	crude death rate
CGIAR	Consultative Group on International Agricultural Research
CIMMYT	International Maize and Wheat Improvement Center
CRP	Conservation Reserve Program of the United States
CS	consumers' surplus
DRC	Democratic Republic of Congo, formerly Zaire
EBRD	European Bank for Reconstruction and Development
EKC	Environmental Kuznets Curve
EMBRAPA	Empresa Brasileira de Pesquisa Agrícola
EU	European Union
FAO	Food and Agriculture Organization of the United Nations
FDI	foreign direct investment
G10	Group of Ten
G20	Group of Twenty
G33	Group of Thirty-Three
GATT	General Agreement on Tariffs and Trade
GDP	gross domestic product
GHG	greenhouse gas

GLASOD	Global Land Assessment of Degradation
GMO	genetically modified organism
HDI	human development index
HIV	human immunodeficiency virus
HRS	Household Responsibility System (of post-Maoist China)
IFPRI	International Food Policy Research Institute
IITA	International Institute of Tropical Agriculture
IMF	International Monetary Fund
IOTF	International Obesity Task Force of the World Health Organization
IPCC	Intergovernmental Panel on Climate Change
IPR	intellectual property right
IRRI	International Rice Research Institute
ISI	Import-Substituting Industrialization
MC	marginal cost
MDG	Millennium Development Goal
MITI	Japan's Ministry of International Trade and Investment
MP	marginal product
MU	marginal utility
MV	marginal value
NAFTA	North American Free Trade Agreement
NATO	North Atlantic Treaty Organization
NEP	New Economic Policy (of the early Soviet Union)
NEV	net economic value
NPC	nominal protection coefficient
NRA	nominal rate of assistance
OECD	Organization for Economic Cooperation and Development
OPEC	Organization of Petroleum Exporting Countries
PPF	production possibilities frontier
PPP	purchasing power parity
PS	producers' surplus
PSE	producer subsidy equivalent
SOE	state-owned enterprise
TFP	total factor productivity
TFR	total fertility rate
UNDP	United Nations Development Program
UNPD	United Nations Population Division
USLE	Universal Soil Loss Equation
WHO	World Health Organization
WTO	World Trade Organization
WTP	willingness-to-pay

Map Annex

Map Annex

Africa and Southwest Asia: Political

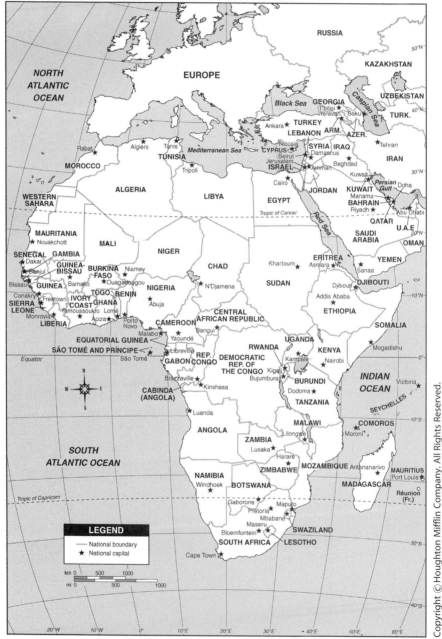

Asia and the South Pacific: Political

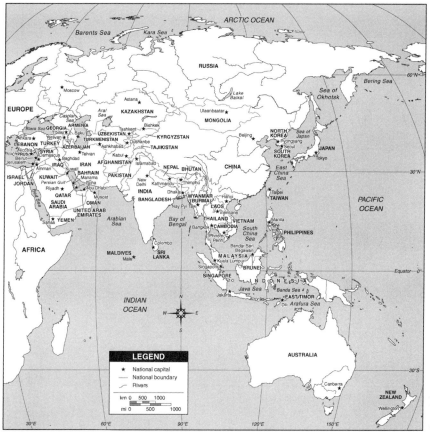

Education Place: http://www.eduplace.com

European Countries

Education Place: http://www.eduplace.com

North America: Countries

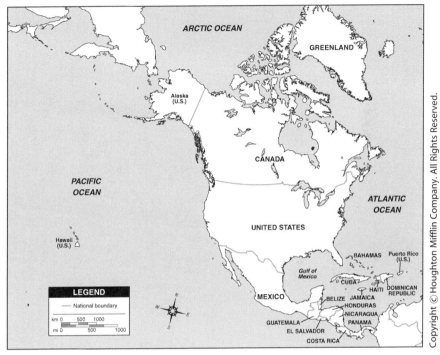

South America: Political

References

Aerni, Philipp. (2008). "A New Approach to Deal with the Global Food Crisis." *African Technology Development Forum Journal*, 5: 16–31.
Ahmad, Sameena. (2004, March 20). "Behind the Mask: A Survey of Business in China." *The Economist*, 370: 1–20.
Ahuluwalia, Montek S. (2002). "Economic Reforms in India Since 1991: Has Gradualism Worked?" *Journal of Economic Perspectives*, 16: 67–88.
Alexandratos, Nikos, ed. (1995). *World Agriculture towards 2010.* Rome: Food and Agriculture Organization of the United Nations.
Allina-Pisano, Jessica. (2008). "Seeds of Discontent: Russia's Food Woes." *Current History*, 107: 330–335.
Alston, Julian M., and Philip G. Pardey. (1996). *Making Science Pay: The Economics of Agricultural R&D Policy.* Washington: AEI Press.
Ames, Bruce, and Lois Gold. (1989). "Pesticides, Risk, and Apple Sauce." *Science*, 244: 755–757.
Anand, Greta. (2010, February 23). "Green Revolution in India Wilts as Subsidies Backfire." *Wall Street Journal*, 245: A1 and A16.
Anderson, Jock R., Robert W. Herdt, Grant M. Scobie, Carl E. Pray, and Hans E. Jahnke. (1985). *International Agricultural Research Centers: A Study of Achievements and Potential.* Bulletin Number 32, Department of Agricultural Economics and Business Management, University of New England, Armidale.
Anderson, Kym. (2009). *Distortions of Agricultural Incentives: A Global Perspective 1955–2007.* Washington: World Bank.
———, and Will Martin. (2006). "Agriculture, Trade Reform, and the Doha Agenda," in Kym Anderson and Will Martin (eds.), *Agricultural Trade Reform and the Doha Development Agenda.* Washington: World Bank.
———, and William A. Masters. (2009). *Distortions to Agricultural Incentives in Africa.* Washington: World Bank.
———, and Johan Swinnen. (2008). *Distortions to Agricultural Incentives in Europe's Transition Economies.* Washington: World Bank.
———, and Alberto Valdés. (2008). *Distortions to Agricultural Incentives in Latin America.* Washington: World Bank.
Anderson, Terry, and P. J. Hill. (1975). "The Evolution of Property Rights." *Journal of Law and Economics*, 18: 163–179.
Antle, John M. (1999). "The New Economics of Agriculture." *American Journal of Agricultural Economics*, 81: 993–1010.
Askari, Hossein, and John Cummings. (1976). *Agricultural Supply Response: A Survey of the Empirical Evidence.* New York: Praeger.

Aslund, Anders. (2002). *Building Capitalism: The Transformation of the Former Soviet Union*. Cambridge: Cambridge University Press.

Associated Press. (2004, May 8). "Obesity Becoming Major Global Problem." *USA Today* (http://www.usatoday.com/news/health/2004-05-08-global-obesity_x.htm?POE=click-refer). Accessed on April 13, 2006.

Babinard, Julie, and Per Pinstrup-Andersen. (2001). *Shaping Globalization for Poverty Alleviation and Food Security: Nutrition*. Washington: International Food Policy Research Institute.

Ballard, Charles L., John B. Shoven, and John Whalley. (1985). "General Equilibrium and Computations of the Marginal Welfare Costs of Taxes in the United States." *American Economic Review*, 75: 128–138.

Ban, Ki-Moon. (2008, June 3). "Address by the Secretary-General to the High-Level Conference on World Food Security." U.N. Food and Agriculture Organization, Rome.

Barnett, Harold J., and Chandler Morse. (1963). *Scarcity and Growth: The Economics of Natural Resource Availability*. Baltimore: Johns Hopkins University Press.

Barraclough, Solon. (1991). *An End to Hunger?* London: Zed Books.

Bates, Robert H. (2000). "Ethnicity and Development in Africa: A Reappraisal." *American Economic Review*, 90: 399–404.

Beaumont, Peter. (1998). "Restructuring of Water Usage in the Tigris-Euphrates Basin: The Impact of Modern Water Management Policies," in Jeff Albert, Magnus Bernhardsson, and Roger Kenna (eds.), *Transformations of Middle Eastern Natural Environments: Legacies and Lessons* (Bulletin 103). New Haven: Yale School of Forestry and Environmental Studies.

Beck, Barbara. (2009, July 27). "A Slow-Burning Fuse: A Special Report on Ageing Populations." *The Economist*, 392: 1–16.

Becker, Charles M. (2002). "Fertility Decline in Sub-Saharan Africa: Introduction." *Journal of African Policy Studies*, 7(2/3): 1–16.

Becker, Gary, and H. Gregg Lewis. (1976). "On the Interaction between the Quantity and Quality of Children." *Journal of Political Economy*, 81: S279–S288.

Bennett, Adam. (2003). "Failed Legacies." *Finance and Development*, 40: 22–25.

Binswanger, Hans, and Robert Townsend. (2000). "The Growth Performance of Agriculture in Sub-Saharan Africa." *American Journal of Agricultural Economics*, 82: 1075–1086.

Bishop, Joshua, and Jennifer Allen. (1989). "The On-Site Costs of Soil Erosion in Mali" (Environment Working Paper Number 21), World Bank, Washington.

Bourguignon, Françoise, and Christian Morrisson. (2002). "Inequality among World Citizens, 1880–1992." *American Economic Review*, 92: 727–744.

Brainerd, Elizabeth, and David M. Cutler. (2005). "Autopsy on an Empire: Understanding Mortality in Russia and the Former Soviet Union." *Journal of Economic Perspectives*, 19: 107–130.

British Broadcasting Corporation (BBC). (2009, April 14). "Germany Bans Monsanto's GM Maize." *BBC News* (http://news.bbc.co.uk/2/hi/europe/7998181.stm). Accessed on April 15, 2009.

Brown, Lester. (1994, August 28). "How Could China Starve the World? Its Boom Is Consuming Global Food Supplies." *Washington Post*, 708: C01.

Burslem, Chris. (2004). "The Changing Face of Malnutrition." *IFPRI Forum*, October: 1 and 9–12.

Carr, Geoffrey. (2003, March 29). "Climbing the Helical Staircase: A Survey of Biotechnology." *The Economist*, 366: 1–24.

Chandran, Rina. (2010). "Debate over GM Eggplant Consumes India." *Reuters* (http://www.reuters.com/article/idUSSGE6070CP20100216). Accessed on February 18, 2010.
Chapman, Peter. (2007). *Bananas: How the United Fruit Company Shaped the World*. Edinburgh: Cannongate Books Ltd.
Chenery, Hollis B., and Moises Syrquin. (1975). *Patterns of Development, 1950–1970*. London: Oxford University Press.
Cipolla, Carlo M. (1965). *The Economic History of World Population*. Baltimore: Penguin Books.
Clark, Colin. (1967). *Population Growth and Land Use*. London: Macmillan.
Coase, Ronald. (1960). "The Problem of Social Cost." *Journal of Law and Economics*, 3: 1–44.
Cochrane, Willard. (1965). *The City Man's Guide to the Farm Problem*. Minneapolis: University of Minnesota Press.
———. (1979). *The Development of U.S. Agriculture: A Historical Analysis*. Minneapolis: University of Minnesota Press.
Codex Alimentarius Commission. (2004). *FAO/WHO Food Standards: Codex Alimentarius*. Rome: U.N. Food and Agriculture Organization.
Coelli, Tim J., and D. S. Prasada Rao. (2003). "Total Factor Productivity Growth in Agriculture: A Malmquist Index Analysis of 93 Countries, 1980–2000" (Working Paper No. 02/2003). Centre for Efficiency and Productivity Analysis, University of Queensland, St. Lucia.
Collier, Paul. (2008). "The Politics of Hunger: How Illusion and Greed Fan the Food Crisis." *Foreign Affairs*, 87: 67–79.
———, and Jan Willem Gunning. (1999). "Why Has Africa Grown Slowly?" *Journal of Economic Perspectives*, 13: 23–40.
Commission of the European Communities. (2006). "Communication from the Commission: An EU Strategy for Biofuels" (SEC-2006-142). Brussels.
Conquest, Robert. (1986). *The Harvest of Sorrow: Soviet Collectivization and the Terror-Famine*. Oxford: Oxford University Press.
Conway, Gordon. (1997). *The Doubly Green Revolution: Food for All in the 21st Century*. London: Penguin Books.
Copenhagen Consensus. (2004). *Copenhagen Consensus 2004: Today's Challenge—Tomorrow's Opportunity* (www.copenhagenconsensus.com). Accessed on March 14, 2006.
Council for Economic Planning and Development, Republic of China. (1999). *Taiwan Statistical Data Book*. Taipei.
Craig, Barbara J., Klaus W. Deininger, and Philip G. Pardey. (1994). "An Embodiment Approach to Measuring Productivity Growth in U.S. Agriculture." Department of Agricultural and Applied Economics, University of Minnesota, St. Paul.
Crook, Clive. (2001, September 29). "A Survey of Globalization." *The Economist*, 360: 1–32.
Crosson, Pierre, and Jock R. Anderson. (1992). "Resources and Global Food Prospects: Supply and Demand for Cereals to 2030" (Technical Paper Number 184), World Bank, Washington.
Cutler, David M., Edward L. Glaeser, and Jesse M. Shapiro. (2003). "Why Have Americans Become More Obese?" *Journal of Economic Perspectives*, 17: 93–118.
Dalrymple, Dana G. (1985). "The Development and Adoption of High-Yielding Varieties of Wheat and Rice in Developing Countries." *American Journal of Agricultural Economics*, 67: 1067–1073.

Dasgupta, Susmita, Benoit Laplante, Hua Wang, and David Wheeler. (2002). "Confronting the Environmental Kuznets Curve." *Journal of Economic Perspectives*, 16: 147–168.

Datt, Gaurav, and Martin Ravallion. (2002). "Is India's Economic Growth Leaving the Poor Behind?" *Journal of Economic Perspectives*, 16: 89–108.

Davis, Otto A., and Andrew B. Whinston. (1965). "Piecemeal Policy in the Theory of Second Best." *Review of Economic Studies*, 34: 323–331.

Deevey, Edward S. (1960). "The Human Population." *Scientific American*, 203: 194–204.

Deininger, Klaus, and Lyn Squire. (1997). "Income Growth and Income Inequality." *Finance and Development*, 34(1): 38–41.

Delgado, Christopher L. (1998). "Africa's Changing Agricultural Development Strategies: Past and Present Paradigms as a Guide to the Future." *Brown Journal of World Affairs*, 5(1): 175–214.

de Soto, Hernando. (2000). *The Mystery of Capital: Why Capitalism Triumphs in the West and Fails Everywhere Else*. New York: Basic Books.

Desrochers, Pierre, and Hiroko Shimizu. (2008). "Yes, We Have No Bananas: A Critique of the 'Food Miles' Perspective" (Policy Primer Number 8). Mercatus Center, George Mason University, Arlington.

de Vivo, G. (1987). "Ricardo, David," in John Eatwell, Murray Milgate, and Peter Newman (eds.), *The New Palgrave: A Dictionary of Economics*. London: Macmillan Press.

Diamond, Jared. (1997). *Guns, Germs, and Steel*. New York: W.W. Norton.

Diao, Xinshen, Agapi Somwaru, and Terry Roe. (2001). "A Global Analysis of Agricultural Trade Reform in WTO Member Countries," in *Background for Agricultural Trade Reform in the WTO: The Road Ahead*. Washington: Economics Research Service, U.S. Department of Agriculture.

Dixon, John, Aiden Gulliver with David Gibbon. (2001). *Farming Systems and Poverty: Improving Farmers' Livelihoods in a Changing World*. Rome: U.N. Food and Agriculture Organization.

Dixon, John and Kirk Hamilton. (1996). "Expanding the Measure of Wealth." *Finance and Development*, 33(4): 15–17.

Doornbosch, R., and R. Steeblick. (2007). *Biofuels: Is the Cure Worse than the Disease?* Paris: Organization for Economic Cooperation and Development.

Duncan, Emma. (2003, December 13). "Spoilt for Choice: A Survey of Food." *The Economist*, 369: 1–16.

Easterly, William. (2001). *The Elusive Quest for Growth*. Cambridge: Massachusetts Institute of Technology Press.

Eberstadt, Nicholas. (1995). "Population, Food, and Income: Global Trends in the Twentieth Century," in Ronald Bailey (ed.), *The True State of the Planet*. New York: The Free Press.

———. (2009a). "Drunken Nation: Russia's Depopulation Bomb." *World Affairs*, 171: 51–62.

———. (2009b). "Will China Continue to Rise?" in Gary J. Schmitt (ed.), *The Rise of China: Essays on the Future Competition*. New York: Encounter Books.

Economic Research Service of the U.S. Department of Agriculture (ERS-USDA). (2001). *Agricultural Outlook: June/July 2001*. Washington.

———. (2002). *Agricultural Productivity in the United States* (http://www.ers.usda.gov/data/agproductivity). Accessed on August 12, 2009.

———. (2003). *International Food Consumption Patterns* (http://www.ers.usda.gov/data/InternationalFoodDemand/). Accessed on July 28, 2009.

———. (2009a). *Agricultural Outlook: Statistical Indicators* (http://www.ers.usda.gov/Publications/AgOutlook/AOTables/). Accessed on August 14, 2009.

———. (2009b). *Food Security in the United States* (http://www.ers.usda.gov/Briefing/FoodSecurity/). Accessed on November 27, 2009.

The Economist. (2009a, October 31). "Go Forth and Multiply a Lot Less." 393: 29–31.

———. (2009b, May 23). "Outsourcing's Third Wave." 391: 61–63.

———. (2009c, September 12). "When the Rains Fail." 392: 27–29.

———. (2010a, August 28). "Brazilian Agriculture: The Miracle of the Cerrado." 396: 58–60.

———. (2010b, February 27). "Taking Root: The Spread of GM Crops." 394: 70–72.

Eggen, Dan. (2008, May 2). "Bush Seeks $770 Million More in World Food Aid." *Washington Post* (http://www.washingtonpost.com/wp-dyn/content/article/2008/05/01/AR2008050101677.html?nav=emailpage). Accessed on November 30, 2009.

Ehrlich, Paul R. (1968). *The Population Bomb*. San Francisco: Sierra Club.

Ellman, Michael, and Vladimir Kontorovich. (1998). *The Destruction of the Soviet Economic System: An Insiders' History*. Armonk: M.E. Sharpe.

Engerman, Stanley L., and Kenneth L. Sokoloff. (1997). "Factor Endowments, Institutions, and Differential Paths of Growth among New World Economies: A View from Economic Historians of the United States," in Stephen Haber (ed.), *How Latin America Fell Behind: Essays on the Economic Histories of Brazil and Mexico*. Stanford: Stanford University Press.

Estudillo, Jonna P., and Keijiro Otsuka. (2010). "Rural Poverty and Income Dynamics in Southeast Asia." *Handbook of Agricultural Economics*, 4: 3435–3468.

European Bank for Reconstruction and Development (EBRD). (2002). *Transition Report: Agriculture and Rural Transition*. London.

Faucon, Benoit, and Spencer Swartz. (2009, September 30). "Africa Pressures China's Oil Deals." *Wall Street Journal*, 254: A14.

Faulconbridge, Guy. (2010). "Russia Says Population Up for First Year Since 1995." *Reuters* (http://in.reuters.com/articlePrint?articleId=INTRE6012KM20100119). Accessed on January 20, 2010.

Feldstein, Martin. (2009, June 22). "The Exploding Carbon Tax." *The Weekly Standard*, 14: 22–24.

Ferguson, Niall. (2008). *The Ascent of Money: A Financial History of the World*. New York: Penguin Press.

Fischer, Günther, and Gerhard K. Heilig. (1997). "Population Momentum and the Demand on Land and Water Resources." *Philosophical Transactions of the Royal Society, Series B*, 352: 869–889.

Fischer, Stanley. (2001). "Exchange Rate Regimes: Is the Polar View Correct?" *Journal of Economic Perspectives* 15: 3–24.

Fogel, Robert. (2004a). *The Escape from Hunger and Premature Death, 1700–2100: Europe, America, and the Third World*. Cambridge: Cambridge University Press.

———. (2004b). "Health, Nutrition, and Economic Growth." *Economic Development and Cultural Change*, 52: 643–658.

———, and Dora Costa. (1997). "A Theory of Technophysio Evolution, with Some Implications for Forecasting Population, Health Care Costs, and Pension Costs." *Demography*, 34: 49–66.

Food and Agriculture Organization (FAO) of the United Nations. (1982). *Food Production Yearbook*. Rome.

———. (1992). *Water for Sustainable Food Production and Rural Development—The UNSED Agenda: Targets and Cost Estimates*. Rome.

———. (1997). *Food Security: Some Macroeconomic Dimensions*. Rome.

———. (2002a). *Food Production Yearbook*. Rome.

———. (2002b). *State of Food Security in the World 2001*. Rome.

———. (2009a). *FAOSTAT: Online Database* (http://faostat.fao.org/). Rome. Accessed on August 13, 2009.

———. (2009b). *Food Security Statistics* (http://fao.org/economic/ess/food-security-statistics/en/). Rome: FAO. Accessed on November 5, 2009.

———. (2009c). *State of the World's Forests 2009*. Rome.

Foster, Phillips, and Howard D. Leathers. (1999). *The World Food Problem*. 2nd ed. Boulder, CO: Lynne Rienner Publishers.

Fraga, Arminio. (2004). "Latin America Since the 1990s: Rising from the Sickbed?" *Journal of Economic Perspectives*, 18: 89–106.

Freedom House. (2008). *Combined Average Ratings: Independent Countries, 2008* (http://www.freedomhouse.org/template.cfm?page=410&year=2008). Accessed on August 3–10, 2009.

Gardner, Bruce L. (2002). *American Agriculture in the Twentieth Century: How It Flourished and What It Cost*. Cambridge: Harvard University Press.

———. (2005). "Golitsino Papers: Summary of Findings and Implications." *Comparative Economic Studies*, 47: 224–228.

Ghazvinian, John. (2007). *Untapped: The Scramble for Africa's Oil*. Orlando: Harcourt.

Gibson, Paul, John Wainio, Daniel Whitley, and Mary Bohman. (2001). "Profiles of Tariffs in Foreign Agricultural Markets" (Agricultural Economic Report Number 796), Economics Research Service, U.S. Department of Agriculture, Economic Research Service, Washington.

Gillam, Carey. (2009, August 18). "Monsanto, Dupont Square off in Crop Seed Turf War," *Reuters* (http://www.reuters.com/article/rbssIndustryMaterialsUtilitiesNews/idUSN1843575620090818). Accessed on August 19, 2009.

Good, Keith. (2007, August 21). *Farmpolicy* (http://www.farmpolicy.com). Accessed on October 22, 2009.

Gordon, H. Scott. (1954). "Economic Theory of a Common Property Resource: The Fishery." *Journal of Political Economy*, 62: 124–143.

Green, Edward C. (2005). "AIDS in Africa—A Betrayal." *The Weekly Standard*, 10 (19): 27–29.

Gregory, Paul, and Kate Zhou. (2010). "How China Won and Russia Lost." *Policy Review*, 158: 35–50.

Griliches, Zvi. (1958). "Research Costs and Social Returns: Hybrid Corn and Related Innovations." *Journal of Political Economy*, 66: 419–431.

Guriev, Sergei, and Andrei Rachinsky. (2005). "The Role of Oligarchs in Russian Capitalism." *Journal of Economic Perspectives*, 19: 131–150.

Hardin, Garrett. (1968). "The Tragedy of the Commons." *Science*, 162: 1243–1248.

Hayami, Yujiro, and Vernon W. Ruttan. (1970). "Factor Prices and Technical Change in Agricultural Development: The United States and Japan, 1880–1960." *Journal of Political Economy*, 78: 1115–1141.

———. (1985). *Agricultural Development*, rev. ed. Baltimore: Johns Hopkins University Press.

Heath, John, and Hans Binswanger. (1996). "Natural Resource Degradation Effects of Poverty and Population Growth Are Largely Policy Induced: The Case of Colombia." *Environment and Development Economics*, 1: 65–84.

Henao, Julio, and Carlos Baanante. (2006). "Agricultural Production and Soil Nutrient Mining in Africa" (Summary). International Fertilizer Development Center, Muscle Shoals.

Hendrickson, Mary, and William Heffernan. (2007). "Concentration of Agricultural Markets." Department of Rural Sociology, University of Missouri.

Henneberry, David, Shida Rastegari, and Luther Tweeten. (1991). "A Review of International Agricultural Supply Response." *Journal of International Food and Agribusiness Marketing*, 2: 49–68.

Herbst, Jeffrey. (2000). *States and Power in Africa: Comparative Lessons in Authority and Control*. Princeton: Princeton University Press.

Hertel, Thomas, Bernard Hoekman, and Will Martin. (2002). "Developing Countries and a New Round of WTO Negotiations." *World Bank Research Observer*, 17: 113–140.

Hopkins, Jeffrey, and Mitchell Morehart. (2002). "An Empirical Analysis of the Farm Problem: Comparability of Rates of Return," in Luther Tweeten and Stanley R. Thompson (eds.), *Agricultural Policy for the 21st Century*. Ames: Iowa State University Press.

Huang, Jikun, Scout Rozelle, and Mark W. Rosegrant. (2000). "China's Food Economy to the Twenty-First Century: Supply, Demand, and Trade." *Economic Development and Cultural Change*, 47: 737–766.

Huff, H. Bruce, Ekaterina Krivonos, and Dominique van der Mensbrugghe. (2007). "Review and Synthesis of Empirical Results of Studies of World Trade Organization Agricultural Trade Reform," in Alex McCalla and John Nash (eds.), *Reforming Agricultural Trade for Developing Countries*. Washington: World Bank.

Huffman, Wallace E., and Robert E. Evenson. (1989). "Supply and Demand Functions for Multiproduct U.S. Cash Grain Farms: Biases Caused by Research and Other Policies." *American Journal of Agricultural Economics*, 71: 761–773.

Intergovernmental Panel on Climate Change (IPCC). (2008). *Climate Change 2007: Synthesis Report* (http://www.ipcc.ch/pdf/assessment-report/ar4/syr/ar4_syr.pdf.). Accessed on July 2, 2009.

Johnson, D. Gale. (2000). "Population, Food, and Knowledge." *American Economic Review*, 90: 1–14.

Johnson, Glenn, and C. Leroy Quance. (1972). *The Overproduction Trap in U.S. Agriculture*. Baltimore: Johns Hopkins University Press.

Johnston, Bruce F. (1966). "Agriculture and Economic Development." *Food Research Institute Studies*, 6: 251–312.

———, and John W. Mellor. (1961). "The Role of Agriculture in Economic Development." *American Economic Review*, 51: 566–593.

Kerr, Richard A. (1998). "Acid Rain Control: Success on the Cheap." *Science*, 282: 1024–1027.

Knutson, R., C. R. Taylor, J. Penson, and E. Smith. (1989). "*Economic Impacts of Reduced Chemical Use*," Knutson and Associates, College Station.

Kotkin, Stephen. (2001). *Armageddon Averted: The Soviet Collapse, 1970–2000*. Oxford: Oxford University Press.

Kremer, Michael. (2002). "Pharmaceuticals and the Developing World." *Journal of Economics Perspectives*, 16: 67–90.

Kuran, Timur. (2004). "Why the Middle East Is Economically Underdeveloped: Historical Mechanisms of Institutional Stagnation." *Journal of Economic Perspectives*, 18: 71–90.

Kuznets, Simon. (1955). "Economic Growth and Income Inequality." *American Economic Review*, 45: 1–28.

———. (1965). *Economic Growth and Structure*. New York: W.W. Norton.

Kynge, James. (2002, December 12). "World Report—China." *The Financial Times*, 598: I–IV.

Lakdawalla, Darius, and Tomas Philipson. (2002). "The Growth of Obesity and Technological Change: A Theoretical and Empirical Examination" (working paper W8946), National Bureau of Economic Research, Chicago.

Lamptey, Peter, Merywen Wigley, Dara Carr, and Yvette Collymore. (2002). "Facing the HIV/AIDS Pandemic." *Population Bulletin*, 57: 1–40.

Landes, David S. (1998). *The Wealth and Poverty of Nations*. New York: W.W. Norton.

Lee, Ronald. (2003). "The Demographic Transition: Three Centuries of Fundamental Change." *Journal of Economic Perspectives*, 17: 167–190.

Leon, David A., Lyudmila Saburova, Susannah Tomkins, Evgueny Andreev, Nikolay Kiryanov, Martin McKee, and Vladimir M. Shkolnikov. (2007). "Hazardous Alcohol Drinking and Premature Mortality in Russia: A Population Based Case-Control Study." *The Lancet*, 369: 2001–2009.

Lewin, Alexandra C. (2007). "Zambia and Genetically Modified Food Aid." Cornell University (http://cip.cornell.edu/DPubs/Repository/1.0/Disseminate?view=body& id=pdf_1&handle=dns.gfs/1200428165). Accessed on January 14, 2010.

Lewis, Bernard. (2002). *What Went Wrong? Western Impact and Middle Eastern Response*. New York: Oxford University Press.

Lomborg, Bjørn. (2001). *The Skeptical Environmentalist: Measuring the True State of the World*. Cambridge: Cambridge University Press.

Long, Simon. (2004, February 21). "India's Shining Hopes: A Survey of India." *The Economist*, 370: 1–20.

Lora, Eduardo. (2001). "Structural Reform in Latin America: What Has Been Reformed and How to Measure It" (working paper number 466), Research Department, Inter-American Development Bank, Washington.

Maddison, Angus. (1995). *Monitoring the World Economy, 1820–1992*. Paris: Organization for Economic Cooperation and Development.

Mahar, Dennis. (1989). *Government Policies and Deforestation in Brazil's Amazon Region*. Washington: World Bank.

Malthus, Thomas. (1963). *An Essay on the Principle of Population*. Homewood: R.D. Irwin.

Mankiw, N. Gregory. (2001). *Principles of Economics*. 2nd ed. Fort Worth: Harcourt College Publishers.

Matras, Judah. (1977). *Introduction to Population: A Sociological Approach*. Englewood Cliffs: Prentice-Hall, Incorporated.

McInerney, John. (1976). "The Simple Analytics of Natural Resource Economics." *Journal of Agricultural Economics*, 27: 31–52.

McKinnon, Ronald I. (1991). *The Order of Market Liberalization: Financial Control in the Transition to a Market Economy*. Baltimore: Johns Hopkins University Press.

Meadows, Donella H., Dennis L. Meadows, Jørgen Randers, and William W. Behrens III. (1972). *The Limits to Growth*. New York: Universe Books.

Mellor, John W. (ed.). (1995). *Agriculture on the Road to Industrialization*. Baltimore: Johns Hopkins University Press.

Mendelsohn, Robert. (2006). "The Role of Markets and Governments in Helping Society Adapt to a Changing Climate." *Climatic Change*, 78: 203–215.
Merkel-Hess, Kate. (2009). "The Revolution Will Be Digitized." *Current History*, 108: 291–292.
Miller, Henry I., and Gregory Conko. (2004). *The Frankenfood Myth: How Protest and Politics Threaten the Biotech Revolution*. Westport: Praeger.
Mitchell, Donald O., Merlinda D. Ingco, and Ronald C. Duncan. (1997). *The World Food Outlook*. Cambridge: Cambridge University Press.
Morgan, Dan. (1979). *Merchants of Grain*. New York: Viking Press.
Moyo, Dambisa. (2009) *Dead Aid*. New York: Farrar, Straus, and Giroux.
Myrskylä, Mikko, Hans-Peter Kohler, and Francesco C. Billari. (2009). "Advances in Development Reverse Fertility Decline." *Nature*, 460: 741–743.
Naughton, Barry. (1995). *Growing out of the Plan: Chinese Economic Reform, 1978–1993*. Cambridge: Cambridge University Press.
———. (2009). "In China's Economy, the State's Hand Grows Heavier." *Current History*, 108: 277–283.
Neely, Christopher. (1999). "How Big is Japan's Debt." *International Economic Trends* (http://research.stlouisfed.org/publications/iet/19990201/cover.pdf). Accessed on November 15, 2005.
Nelson, Gerald C., Mark W. Rosegrant, Jawoo Koo, Richard Robertson, Timothy Sulser, Tingju Zhu, Claudia Ringler, Siwa Msangi, Amanda Palazzo, Miroslav Batka, Marilia Magalhaes, Rowena Valmonte-Santos, Mandy Ewing, and David Lee. (2009). *Climate Change: Impact on Agriculture and Costs of Adaptation*. Washington: International Food Policy Research Institute.
Nordhaus, William. (2008). *A Question of Balance: Weighing the Options on Global Warming Policies*. New Haven: Yale University Press.
Normile, Mary Anne, Anne B.W. Effland, and C. Edwin Young. (2004). "U.S and E.U. Farm Policy—How Similar?" in *U.S.-E.U. Food and Agriculture Comparisons*. Washington: Economic Research Service, U.S. Department of Agriculture.
North, Douglass. (1973). *Institutions, Institutional Change, and Economic Performance*. Cambridge: Cambridge University Press.
Ocampo, José Antonio. (2004). "Latin America's Growth and Equity Frustrations during Structural Reforms." *Journal of Economic Perspectives*, 18: 67–88.
Oldeman, L. R., R. T. A. Hakkeling, and W. G. Sombroek. (1991). *World Map of the Status of Human-Induced Soil Degradation: An Explanatory Note*. 2nd rev. ed. Wageningen: International Soil Reference and Information Centre.
Organization for Economic Cooperation and Development (OECD). 2008. *Agricultural Policies of the OEDC Countries at a Glance* (http://browse.oecdbookshop.org/oecd/pdfs/browseit/5108041E.PDF). Accessed on December 16, 2009.
O'Rourke, P. J. (1994). *All the Trouble in the World*. New York: Atlantic Monthly Press.
Paarlberg, Robert. (2008). *Starved for Science: How Biotechnology Is Being Kept out of Africa*. Cambridge: Harvard University Press.
Pardey, Philip G., Julian M. Alston, and Jennifer S. James. (2008). "Agricultural R&D Policy: A Tragedy of the International Commons" (Staff Paper P08-08), Department of Applied Economics, University of Minnesota, St. Paul.
———, and Nienke M. Beintema. (2001). *Slow Magic: Agricultural R&D a Century after Mendel*. Washington: International Food Policy Research Institute.
Perkins, Dwight H., Steven Radelet, and Steven Lindauer. (2006). *Economic Development*, 6th ed. New York: W.W. Norton and Company.

Peterson, E. Wesley F. (2009). *A Billion Dollars a Day: The Economics and Politics of Agricultural Subsidies*. Malden: Blackwell Publishing.

Pingali, Prabhu L., Cynthia B. Marquez, Agnes C. Rola, and Florencia G. Palis. (1995). "The Impact of Long-Term Pesticide Exposure on Farmer Health: A Medical and Economic Analysis in the Philippines," in Prabhu L Pingali and Pierre A. Roger (eds.), *Impact of Pesticides on Farmer Health and the Rice Environment*. Boston: Kluwer Academic Publishers.

Pinstrup-Andersen, Per. (2002). *More Research and Better Policies Are Essential for Achieving the World Food Summit Goal*. Washington: International Food Policy Research Institute.

Pollan, Michael. (2002, November 10). "The Unnatural Idea of Animal Rights." *New York Times Magazine*, 92: 58–64, 100–102.

———. (2006). *Omnivore's Dilemma: A Natural History of Four Meals*. New York: Penguin Press.

Porzecanski, Arturo. (2009). "Latin America: The Missing Financial Crisis," U.N. Economic Commission for Latin America and the Caribbean, Washington.

Pullen, J. M. (1987). "Malthus, Thomas Robert," in John Eatwell, Murray Milgate, and Peter Newman (eds.), *The New Palgrave: A Dictionary of Economics*. London: Macmillan Press.

Rashad, Inas, and Michael Grossman. (2004). "The Economics of Obesity." *The Public Interest*, 156: 104–112.

Rask, Kolleen, and Norman Rask. (2004). "Reaching Turning Points in Economic Transition: Adjustments to Distortions in Resource-Based Consumption of Food." *Comparative Economic Studies*, 46: 542–569.

Rask, Norman. (1991). "Dynamics of Self-Sufficiency and Income Growth," in Fred J. Ruppel and Earl D. Kellogg (eds.), *National and Regional Self-Sufficiency Goals: Implications for International Agriculture*. Boulder: Lynne Rienner Publishers.

Rauch, Jonathan. (2003). "Will Frankenfood Save the Planet?" *Atlantic Monthly*, 292 (October): 103–108.

Reardon, Thomas, and Julio A. Berdegué. (2002). "The Rapid Rise of Supermarkets in Latin America: Challenges and Opportunities for Development." *Development Policy Review*, 20: 371–388.

Reed, Thomas C. (2004). *At the Abyss: An Insider's History of the Cold War*. New York: Ballantine Books.

Reid, Michael. (2008, November 11). "The Party's Over: A Special Report on Spain." *The Economist*, 389: 1–16.

Reuters. (2008, July 21). "Excess Finance Plague of 21st Century: Tremonti" (press release).

Ricardo, David. (1965). *The Principles of Political Economy and Taxation*. London: Dent and Sons, Limited.

Richardson, Stanley Dennis. (1990). *Forests and Forestry in China: Changing Patterns of Resource Development*. Washington: Island Press.

Righter, Rosemary. (2002, February 12). "The Rot behind China's Bank Scandals." *The Times of London*, 67372: 31.

Rosegrant, Mark, Ximing Cai, and Sarah Cline. (2002). "Global Water Outlook to 2025," International Food Policy Research Institute, Washington.

Roy, Amit H. (2003). "Fertilizer Needs to Enhance Production—Challenges Facing India," in Rattan Lal, David Hansen, Norman Uphoff, and Steven Slack (eds.), *Food Security and Environmental Quality in the Developing World*. Boca Raton: Lewis Publishers.

Runge, C. Ford, Benjamin Senauer, Philip G. Pardey, and Mark W. Rosegrant. (2003). *Ending Hunger in our Lifetime: Food Security and Globalization.* Baltimore: Johns Hopkins University Press.

Ruttan, Vernon W. (2002). "Productivity Growth in World Agriculture: Sources and Constraints." *Journal of Economic Perspectives,* 16: 161–184.

Sachs, Jeffrey. (1997, July 14). "The Limits of Convergence." *The Economist,* 344: 19–22.

Sahn, David, Paul Dorosh, and Stephen D. Younger. (1997). *Structural Adjustment Reconsidered: Economic Policy and Poverty in Africa.* Cambridge: Cambridge University Press.

Salih, Mohamed Abdul Rahim M. (2001). *African Democracies and African Politics.* London: Pluto Press.

Samuelson, Paul A. (1969). "The Way of an Economist," in Paul A. Samuelson (ed.), *International Economic Relations: Proceedings of the Third Congress of the International Economic Association.* London: Macmillan.

Sánchez, Pedro A., and Terry J. Logan. (1992). "Myths and Science about the Chemistry and Fertility of Soils in the Tropics," in Rattan Lal and Pedro A. Sánchez (eds.), *Myths and Science of Soils of the Tropics.* Madison: Soil Science Society of America.

Schepf, Randall, Erik Dohlman, and Christine Bolling. (2001). "Agriculture in Brazil and Argentina: Developments and Prospects for Major Field Crops" (Agriculture and Trade Report WRS-01-3), Economic Research Service, U.S. Department of Agriculture, Washington.

Scherr, Sara J., and Satya Yadav. (1997). "Land Degradation in the Developing World: Issues and Policy Options for 2020" (2020 Brief Number 44), International Food Policy Research Institute, Washington.

Schneider, Robert. (1992). "Brazil: An Analysis of Environmental Problems in the Amazon" (Report 9104-Br), World Bank, Washington.

———. (1995) "Government and the Economy on the Amazon Frontier" (Environment Paper 11), World Bank, Washington.

Schultz, Theodore W. (1960). "Value of U.S. Farm Surpluses to Underdeveloped Countries." *Journal of Farm Economics,* 42: 1019–1030.

———. (1961). "Investment in Human Capital." *American Economic Review,* 51: 1–17.

———. (1964). *Transforming Traditional Agriculture.* New Haven: Yale University Press.

Schweizer, Peter. (1994). *Victory.* New York: Atlantic Monthly Press.

Sen, Amartya. (1981). *Poverty and Famines: An Essay on Entitlement and Deprivation.* Oxford: Clarendon Press.

———. (2000). "Culture and Development," (mimeo), World Bank Meeting, Tokyo.

Senauer, Benjamin, and Mona Sur. (2001). "Ending Global Hunger in the 21st Century: Projections of the Number of Food Insecure People." *Review of Agricultural Economics,* 25: 68–81.

Short, Philip. (1999). *Mao: A Life.* New York: Henry Holt.

Simpson, David, Roger Sedjo, and John Reid. (1996). "Valuing Biodiversity: An Application to Genetic Prospecting." *Journal of Political Economy,* 104: 163–185.

Skidelsky, Robert. (1996). *The Road from Serfdom: The Economic and Political Consequences of the End of Communism.* New York: Penguin Books.

Smith, Adam. (1964). *The Wealth of Nations.* London: J.M. Dent and Sons.

Smith, Lisa, and Lawrence Haddad. (2000). "Overcoming Child Nutrition in Developing Countries" (Discussion Paper 30), International Food Policy Research Institute. Washington.

Sohngen, Brent, and Robert Mendelsohn. (2003). "An Optimal Control Model of Forest Carbon Sequestration." *American Journal of Agricultural Economics*, 85: 448–457.

Solow, Robert. (1957). "Technical Change and the Aggregate Production Function." *Review of Economics and Statistics*, 39: 312–320.

Song, Sora. (2004, March 22). "We're Fat. Now What?" *Time*, 163: 73.

Southgate, Douglas. (1994). "The Causes of Tropical Deforestation and Agricultural Development in Latin America," in Katrina Brown and David W. Pearce (eds.), *The Causes of Tropical Deforestation: The Economic and Statistical Analysis of Factors Giving Rise to the Loss of Tropical Forests*. London: University College London Press.

———, and Morris Whitaker. (1994). *Economic Progress and the Environment: One Developing Country's Policy Crisis*. New York: Oxford University Press.

Sowell, Thomas. (1998). *Conquests and Cultures*. New York: Basic Books.

Stelzer, Irwin M. (2005). "The Axis of Oil." *Weekly Standard*, 10(20): 25–28.

Stewart, Bobby, Rattan Lal, and Samir El-Swaify. (1991). "Sustaining the Resource Base of an Expanding World Agriculture," in Rattan Lal and Frans Pierce (eds.), *Soil Management for Sustainability*. Ankeny: Soil and Water Conservation Society.

Stewart, Hayden, Noel Blisard, Sanjib Bhuyan, and Rodolfo M. Nayga. (2004). "The Demand for Food away from Home: Full Service or Fast Food?" (Agricultural Economic Report Number 829), Economic Research Service, U.S. Department of Agriculture, Washington.

Tawney, R. H. (1966). "Religion and the Rise of Capitalism," in Stanley Cohen and Forrest Hill (eds.), *American Economic History*. New York: Lippincott.

Thurow, Lester. (1992). *Head to Head: The Coming Economic Battle among Japan, Europe, and America*. New York: William Morrow.

Tietenberg, Thomas. (2003). *Environmental and Natural Resource Economics*. 6th ed. Chicago: Addison Wesley Publishing.

Timmer, C. Peter, Walter P. Falcon, and Scott R. Pearson. (1983). *Food Policy Analysis*. Baltimore: Johns Hopkins University Press.

Tolstoy, Leo. (1992). *Anna Karenina*. New York: Barnes and Noble.

Truth about Trade and Technology. (2004). "The Truth about Organic Food" (http://www.truthabouttrade.org/print.asp?id=2384). Des Moines. Accessed on February 23, 2006.

Tweeten, Luther. (1998). "Dodging a Malthusian Bullet in the 21st Century." *Agribusiness*, 14: 15–32.

———. (1999). "The Economics of Global Food Security." *Review of Agricultural Economics*, 21: 473–488.

———. (2007). *Prescription for a Successful Economy: The Standard Economic Model*. Lincoln: iUniverse.

———, and George Brinkman. (1976). *Micropolitan Development*. Ames: Iowa State University Press.

———, and Jeffrey Hopkins. (2003). "Economies of Size, Government Payments, and Land Costs," in Charles Moss and Andrew Schmitz (eds.), *Government Policy and Farmland Markets*. Ames: Iowa State Press.

———, and Donald McClelland (eds.). (1997). *Promoting Third-World Agricultural Development and Food Security*. Westport: Praeger Publishers.

———, and Stanley R. Thompson (eds.). (2002). *Agricultural Policy for the 21st Century*. Ames: Iowa State University Press.

United Nations Conference on Trade and Development (UNCTAD). (2008). *Handbook of Statistics* (http://www.unctad.org/Templates/webflyer.asp?docid=10193&intItemID=1397&lang=1). Accessed on November 6, 2009.

United Nations Development Programme (UNDP). (2002a). *Arab Human Development Report 2002: Creating Opportunities for Future Generations*. New York: United Nations Development Programme, Regional Bureau for Arab States.
———. (2002b). *Human Development Report 2002*. New York: Oxford University Press.
———. (2009). *Millennium Development Goals* (http://www.undp.org/mdg/basics.shtml). Accessed on November 17, 2009.
United Nations, Population Division (UNPD). (1993). *World Population Prospects: The 1992 Revision*. New York: UNPD.
———. (2003). *World Population Prospects: The 2002 Revision* (www.un.org/esa/population/unpop.htm). New York. Accessed on October 17, 2005.
———. (2009). *World Population Prospects: The 2008 Revision, Highlights* (Working Paper No. ESA/WP.210). New York.
U.S. Agency for International Development (USAID). (2006). "Protecting and Promoting Property Rights" (http://www.usaid.gov/stories/ukraine/ss_ukraine_land.html). Accessed on November 12, 2008.
U.S. Environmental Protection Agency (USEPA). (2000). *National Water Quality Inventory, 2000 Report*. Washington.
———. (2005). *Greenhouse Gas Mitigation Potential in U.S. Forestry and Agriculture* (EPA 430-R-05-006). Washington.
Unnevehr, Laurian J. (2004). "Mad Cows and Bt Potatoes: Global Public Goods in the Food System." *American Journal of Agricultural Economics*, 86: 1159–1166.
Van Lynden, G. W. J., and L. R. Oldeman. (1997). *The Assessment of the Status of Human-Induced Soil Degradation in South and Southeast Asia*. Wageningen: ISRIC.
Vanzetti, David, and Ramesh Sharma. (2007). "Projecting the Effects of Agricultural Trade Liberalization on Developing Countries," in Alex McCalla and John Nash (eds.), *Reforming Agricultural Trade for Developing Countries*. Washington: World Bank.
von Braun, Joachim. (1989). "Commentary: Commercialization of Smallholder Agriculture—Policy Requirements for Capturing Gains for the Malnourished Poor." *International Food Policy Research Institute Report*, 11: 4.
———. (2008, May 16). "Global Response Needed to Rising Food Prices" (press release), International Food Policy Research Institute.
Von Cramon-Taubadel, Stephan, Oleg Nivyevskiy, Emanuel Elsner von der Malsburg, and Veronika Movchan. (2008). "Ukraine," in Kym Anderson and Johan Swinnen (eds.), *Distortions to Agricultural Incentives in Europe's Transition Economies*. Washington: World Bank.
Voortman, R. L., B. G. J. S. Sonneveld, and M. A. Keyzer. (2000). "African Land Ecology: Opportunities and Constraints for Agricultural Development" (Working Paper Number 37), Center for International Development, Harvard University, Cambridge.
Walder, Andrew G. (2009). "Unruly Stability: Why China's Regime Has Staying Power." *Current History*, 108: 257–263.
Weber, Max. (1930). *The Protestant Ethic and the Spirit of Capitalism*. London: Allen and Unwin.
Westcott, Paul C., and C. Edwin Young. (2000). "U. S. Farm Program Benefits: Links to Planting Decisions and Agricultural Markets." *Agricultural Outlook*, October, AGO-275: 10–13.
Williams, Adrian. (2007). "Comparative Study of Cut Roses for the British Market Produced in Kenya and the Netherlands" (http://www.fairflowers.de/filead

min/flp.de/Redaktion/Dokumente/Studien/Comparative_Study_of_Cut_Roses_Feb_2007.pdf). Accessed on May 7, 2010.

Williams, Brett. (2009). "The WTO Doha Round Draft Text on Agricultural Liberalization." *Farm Policy Journal*, 6: 1–21.

Williams, Jeffrey C., and Brian D. Wright. (1991). *Storage and Commodity Markets*. New York: Cambridge University Press.

Williamson, John. (1990). "What Washington Means by Policy Reform," in John Williamson (ed.), *Latin American Adjustment: How Much Has Happened?* Washington: Institute for International Economics.

———. (2000). "What Should the World Bank Think about the Washington Consensus?" *World Bank Research Observer*, 15: 251–264.

Wischmeier, Walter. (1976). "Use and Misuse of the Universal Soil Loss Equation." *Journal of Soil and Water Conservation*, 31: 5–9.

Wolf, A. M., and G. A. Colditz. (1994). "The Costs of Obesity: The U.S. Perspective." *PharmacoEconomics*, 5: 34–37.

Wood, Stanley, Kate Sebastian, and Sara J. Scherr. (2001). *Pilot Analysis of Global Ecosystems: Agroecosystems*. Washington: International Food Policy Research Institute.

World Bank. (1986). *World Development Report 1986*. New York: Oxford University Press.

———. (2000). *Can Africa Claim the 21st Century?* Washington.

———. (2001). *World Development Report 2000/2001: Attacking Poverty*. Washington.

———. (2003). *Global Economic Prospects: Realizing the Development Promise of the Doha Agenda*. Washington.

———. (2007). *World Development Report 2008: Agriculture for Development*. Washington.

———. (2008). "Double Jeopardy: Responding to High Food and Fuel Prices," *G8 Hokkaido-Toyako Summit* (http://siteresources.worldbank.org/NEWS/Resources/risingfoodprices_chart_apr08.pdf). Accessed on September 9, 2009.

———. (2009a). *Global Economic Prospects: Commodities at the Crossroads*. Washington.

———. (2009b). *World Development Indicators Online* (http://ddp-ext.worldbank.org/ext/DDPQQ). Accessed via Ohio State University Library System on July 27 through August 10, 2009.

——— and U.N. Development Program. (1990). "A Proposal for an Internationally Supported Programme to Enhance Research in Irrigation and Drainage Technology in Developing Countries," Washington.

World Food Program. (2009). *INTERFAIS*, International Food Aid Information System (http://www.wfp.org/fais/). Accessed on July 6, 2009.

World Health Organization (WHO). (2000). *Obesity: Preventing and Managing the Global Epidemic* (Technical Report Series 894). Geneva: World Health Organization.

———. (n.d.). *Uganda Reverses the Tide of HIV/AIDS* (http://www.who.int/inf-new/aids2.htm). Accessed on March 13, 2006.

Yergin, Daniel. (1991). *The Prize: The Epic Quest for Oil, Money, and Power*. New York: Simon and Schuster.

———, and Joseph Stanislaw. (1998). *The Commanding Heights: The Battle between Government and the Marketplace that Is Remaking the Modern World*. New York: Simon and Schuster.

Yousef, Tarik M. (2004). "Development, Growth, and Policy Reform in the Middle East and North Africa since 1950." *Journal of Economic Perspectives*, 18: 91–116.

Canada, 32, 52, 62, 128, 168n, 247, 249, 253, 257, 304
comparative advantage, in agriculture, 253
cap and trade, 127
capital formation, 113, 179, 182, 219, 223, 231, 239, 242, 244
Caribbean, 9, 32, 77, 129, 162, 207, 225, 255, 293, 310, 315, 339, 372, 376
 agricultural development, 302–304
  factor use, 304–307
  intensification and extensification, 307–311
 consumption trends, 311–315
 dietary change, 311–315
 food security, 311–315
 GDP per capita trends, 393–297
 population dynamics, 298–302
  human fertility reduction, 298–300
  natural increase, 301–302
Casey, William, 322
Central African Republic, 397
central planning, 95, 96, 119, 270, 341, 342, 343
Chad, 395
Chávez, Hugo, 297
chemical fertilizers, 51
Chenery, Hollis, 184
Chile, 26, 27, 30, 33, 77, 140, 181, 195, 293, 295, 296, 298, 304, 306, 314, 315
China, 20, 72, 140, 191, 224, 271, 273, 279, 287, 404, 409, 418
 GDP percapita trends, 268–273
 Household Responsibility System (HRS), 191, 271
Clean Air Act, 150
climate change, 75, 123, 127–131
Club of Rome, 15–17, 24, 109
Coase, Ronald, 147
Codex Alimentarius, 413, 414
collectivization, 190, 191, 358, 370, 408
Colombia, 137, 170, 293, 296, 306, 308, 310
commodity programs, 302, 306, 308
 market impacts, 137
Common Agricultural Policy (CAP), 255, 413
communism, 133, 183, 196, 281, 346, 355, 367, 371
 and agriculture, 183, 184, 191
communitarian values in standard model, 217–220
 economic equity, 59, 219
comparative advantage
 illustration, 172–175
 net benefits of trade, distribution of, 154
 theory, 152–153

competitive equilibrium, 97–100, 118, 119, 127
concentrated animal feeding operation (CAFO), 59, 416
concentration ratio, 83
Congo, Republic of, 373n, 374, 380
Consultative Group on International Agricultural Research (CGIAR), 71, 135
consumers' surplus (CS), 97, 98, 154, 176
consumption and production, aligning, 95–113
 agricultural product scarcity
  historical trends, 24, 106
 commodity programs, market impact, 102
 competitive equilibrium, 97–99
 decentralized decision-making, 114–116
 demand and supply, shifts, 116–119
 NEV and maximization, 98, 100, 119–121
 outlook, for twenty-first century, 111–113
contestability of markets, 84
Conway, Gordon, 406
Copenhagen Consensus, 405, 406
Costa Rica, 306, 314, 315
Croatia, 346, 355
crude birth rate (CBR), 22, 233, 247, 349
crude death rate (CDR), 22, 247, 301, 349
currency over-valuation, 157, 158, 192, 194, 224, 296, 375
Czech Republic, 343, 362

decentralized decision-making, 114–116
 demand and supply, shifts, 116–119
 NEV and maximization, 98, 100, 119–121
 production and consumption alignment, 115–116
decoupling, of farm payments, 164
deficiency payment, 104, 105, 162
Deflation, 213
deforestation, 127, 136–140
 Environmental Kuznets Curve, 142–143
 in Brazilian Amazon
  intervention failure, 137
  global warming, 138
 in Latin America, 306n
 in Sub-Saharan Africa, 135
demand for food, 4, 8, 36, 103, 141, 234, 258, 411
 demographic transition, 20–23
  human fertility revolution, 27–31
  human population trends, 31–34
 in developing world, 23–27
 elasticity, 37–40, 44–46, 48, 88, 90, 91, 117, 142 179, 258
 food consumption and income, 34–38
 fundamental economics, 43–44

demand for food (*Continued*)
  fundamental economics, 43–44
    changes, in demand, 47–48
    own-price elasticity, 44–46
  Malthusianism, 12–20
    classical view, 12–14
    criticism, 12–14
    neo-Malthusianism, 15–17
  shifts, 89, 92, 93, 116, 117, 146
  trends and projections, 38–41
Democratic Republic of the Congo (DRC), 373n, 374, 376
demographic transition, 20–23
  crude birth rate (CBR), 22, 233, 247, 302
  crude death rate (CDR), 22–24, 26, 32, 233, 234, 302
  human fertility revolution, 27–31
  human population trends, 10, 15, 33, 64
  in developing world, 23–27
    modern transition, 23–27
  in Eastern Europe and Former Soviet Union, 72, 77, 349
    human fertility, 27, 246, 249, 276
    natural increase, 276–278
  in Sub-Saharan Africa
    human fertility, 381–384
    natural increase, 384–386
  total fertility rate (TFR), 22, 229, 246, 274, 299, 325, 349 381
  see also population dynamics
Denmark, 256
developed world, *see* affluent nations
developing world
  agricultural extensification, 66–68, 74, 75, 85, 140, 235, 279, 281, 284, 307, 308, 331
  agricultural intensification, 70, 81, 95, 133, 239, 257, 279, 306, 332, 391
  agricultural protectionism, 156
  deforestation, 127, 128, 136–140, 143, 235, 237, 240, 306n
  demographic transition, 20–23
  fertilizer application rates, 75, 134, 237, 332, 389
  food insecurity, 9, 76, 78, 199, 201, 203, 204, 209, 218, 221, 239, 245, 287, 340, 370
  foreign exchange earnings, 188
  foreign exchange rates, 214
  governmental intervention, 8, 58, 141, 148, 150, 191–193
  modern transition, 23–27
  population growth, 31–34, 233
  rural population, 68, 74, 79, 235, 271
  taxation, 148, 189, 212, 219
dietary change, 259–262
  and consumption trends, in affluent nations, 261, 262
  obesity problem, 262–264
  Rask's methodology, 259, 260, 285, 335, 362
  in Asia, 285
  in Eastern Europe and Former Soviet Union, 362
  in Latin America and Caribbean, 302–304
  in Middle East and North Africa, 335
dietary energy requirements, 201
diminishing marginal utility, 44, 90, 98
Doha Round of trade negotiations, 165–168
Dominican Republic, 302, 308, 314
doubling time, 11, 223n
Doubly Green Revolution, 406
Dutch Disease, 224, 225, 320, 399
  in Argentina, 310
  in Middle East and North Africa, 328, 329, 330, 339
dynamic benefits of liberalized trade, 163–165

East Asia, 68, 194, 281, 329
Eastern Europe, 9, 77, 185, 221, 234, 347, 355
  agricultural sector, 353–362
    factor use, 358–359
    yields and land use, 359–362
  consumption trends, 362–369
  demographic trends, 349, 380
    human fertility, 325, 349, 381
    natural increase, 233, 276, 301, 353
  dietary change, 259, 285, 335, 362
  economic growth patterns, 342, 343
  food security, 311–315
Eberstadt, Nicholas, 23, 278, 350
economic convergence, 226, 298
  hypothesis, 185, 226, 227, 235
economic growth, 1, 15, 60, 142, 143, 206, 209, 218, 219, 223, 295, 296, 318
  and food prices, 101–103
  income distribution, 184–186, 221–223
    differences and economic convergence, 226
    GDP per capita, regional trends, 223–226
  patterns
    in Eastern Europe and Former Soviet Union, 341–342
  and structure, 188
    standard of living and income distribution, 184–186, 343
  structural transformation, diversity of, 183–84
Ecuador, 11, 27, 39, 140, 297, 304

education, *see* human capital
efficiency, 3, 55, 120, 182, 260, 278, 311, 410
   impeded by monopoly, 100, 105, 120
   impeded by market failure, 124–127
Egypt, 320, 328, 332
Ehrlich, Paul, 1, 17n, 86, 267
El Salvador, 214, 299, 302, 307, 308, 314
elasticity
   demand, 46
      income, 37–40, 43, 48, 142
      own-price, 37, 44–46, 91, 117
   supply, 90–92
energy
   biofuels, 50, 79, 101, 102, 112
   inputs to agriculture, 78–80, 235, 256, 280, 330, 360, 388
   prices of, 79, 80, 92, 101, 297, 347, 349
Engel's Law, 37, 39, 111, 142n, 179, 181, 411
England, 12, 24
environmental deterioration, 132, 133
   in absence of agricultural intensification, 133, 239
      agricultural extensification, 74, 75, 140
      land degradation, 66, 131–136
Environmental Kuznets Curve (EKC), 142–143, 184
epidemiological troubles, in Sub-Saharan Africa, 380
Eritrea, 376, 398
Estonia, 346, 355, 366
Ethiopia, 74, 140, 373, 374, 398
European Bank for Reconstruction and Development (EBRD), 355
European Union (EU), 79, 156, 168n, 244, 322, 324, 343, 413
export cropping, 160
export restrictions, 102, 103, 106, 311, 355
extensification, 49, 62, 64n, 66–70, 74, 75, 85, 95, 112, 139, 140, 235, 279, 281, 284, 307, 331, 333
   agricultural supply, 59, 93–94
   in Asia, 279
   environmental deterioration, 132, 133
   in Latin America and Caribbean, 302–304
   in Middle East and North Africa, 328
   in Sub-Saharan Africa, 395
extension, agricultural, 57, 60, 206, 387

factory farming, 59
fair trade, 160, 161, 413
Farm Bill, 253, 254, 413
fertilization, 51, 55, 64, 72, 74, 87, 88, 93, 132, 133, 134, 135, 145, 237, 238, 279, 329, 331, 359, 390, 391, 404

fertilizer, 49, 51, 55, 56, 61–63, 71, 74, 75, 79, 80, 85, 93, 133, 135, 141, 142, 234, 235, 237, 257, 303, 304, 306–308, 329, 331–334, 359, 371
   application rates, 134, 237, 332, 389
   chemical fertilizers, 51
   commercial fertilizer, 135, 141, 303, 391, 414
   subsidies, 133, 141
Fogel, Robert, 17
food aid, 157, 203–206, 215, 239, 254
Food and Agriculture Organization (FAO) of the United Nations, 1, 67, 199, 209, 259, 413
   declaration on food security, 204
food consumption, 6, 9, 11, 34, 38, 39, 41, 43, 47, 112, 227, 262, 287, 288, 314, 315, 335, 337, 361, 362, 366, 398, 399, 403, 404
   and income, 34–38
   Rask's methodology, 259, 260, 285, 335, 362
   trends
      in affluent nations, 250, 261
      in Asia, 285
      in Eastern Europe and Former Soviet Union, 362
      in Latin America and Caribbean, 302–304
      in Middle East and North Africa, 335
      in Sub-Saharan Africa, 396
food crisis, of 1970, 108
   lessons, 108
food miles, *see* local food
food safety, 36, 250, 290, 413–414
food scarcity, 4, 5, 7, 107, 109, 112, 195
   historical trends, 24, 106
   impacts, 5
food security, 69, 197, 199–202, 204, 205, 207–210, 215, 217–221, 288, 311, 314, 362, 395, 399, 405, 406
   achievement, 23, 71, 115, 204, 249
      economic growth and prices, 206
      food aid, 157, 203–206, 215, 239, 254
   FAO's declaration, 199
   in Asia, 287
   in Eastern Europe and Former Soviet Union, 362
   in Latin America and Caribbean, 311–315
   in Middle East and North Africa, 335
   in Sub-Saharan Africa, 395–401
   in the United States, 201
   insecurity, 201, 203, 239, 335
   synthesis and economic development
      seven-step logical framework, 208
      standard model, 217–220

food self-sufficiency, 160, 166, 340
  and export cropping, 160
  in Middle East and North Africa, 335, 337, 340
  in OECD, 162, 163, 245, 255, 259
  in Sub-Saharan Africa, 399
Ford Foundation, 71, 81
foreign aid, 179, 216, 217, 342, 375
Former Soviet Union, 9, 65, 72, 77, 139, 207, 221, 225–227, 229–231, 234, 235, 237, 239, 341–344, 346, 347, 349, 351, 353, 355, 359, 361, 362, 367, 370, 371, 372, 381
  agricultural sector, 353–358
    factor use, 358–359
    yields and land use, 359–362
  consumption trends, 362–369
  demographic trends, 349–353
    human fertility, 349
    natural increase, 353
  dietary change, 362–369
  economic growth patterns, 344–348
    Russian case study, 348–349
  food security, 362
France, 168n, 244, 246, 254, 258, 260, 380
free trade, 131, 152, 160, 164, 253, 296
  gains, 168
  versus fair trade, 160–161
  see also liberalization
fundamental concepts demand, 7
  demand elasticity, 46
  own-price elasticity, 37, 44–46, 91, 117
  market equilibrium and efficiency, 97, 108, 115, 116, 118, 119
  supply, 90, 92
    agricultural supply, 59–62
    changes, 117
    elasticity, 90–92

Gabon, 380, 381
Gaidar, Yegor, 300
Gambia, 160
García, Alan, 296
GDP per capita trends in affluent nations, economic progress, foreign assistance in, 216–217
  farmers, 85
  in Asia, 268–270
  in China, 271–273
  in India, 273–274
  in Latin America and Caribbean, 293
  in Middle East and North Africa, 274–278
  in Sub-Saharan Africa, 399
  regional trends, 229
    Dutch disease, 225, 310, 320, 329, 399
  see also standards of living

General Agreement on Tariffs and Trade (GATT), 152, 165–168
genetically modified organisms (GMOs)
  environmental benefits of, 6, 390, 406
  public acceptance of, 6
  see also biotechnology, agricultural
Georgia, 347, 349, 353, 359, 366
Germany, 63, 190, 254, 341
Ghana, 374, 380
Gini coefficient, 184, 186, 226, 244, 367
  in affluent countries, 244
  in Asia, 288
  in Eastern Europe and Former Soviet Union, 367
  in Latin America and the Caribbean, 315
  in the Middle East and North Africa, 338
  in Sub-Saharan Africa, 399
  see also income distribution and inequality
global food economy, in twenty-first century, 404–418
  freer international trade, 405
  government, changing role of, 412–414
  new food economy, 410–412
Global Land Assessment of Degradation (GLASOD) methodology, 131, 132
global warming, see climate change
globalization and agriculture, 151–171
  agricultural trade, 156, 160
    freer trade, gains, 163n
    international trade negotiations, 105
  comparative advantage, 55, 74, 152, 153, 156, 167
  net benefits of trade, distribution of, 154
  export cropping, 160
  food self-sufficiency, 160, 166, 287, 314, 340
  free trade versus fair trade, 160
  multinational firms, 158, 159
  trade distortions, net costs of, 154–158
Gorbachev, Mikhail, 348
government's role for, 214, 348
  commodity programs, 348, 349
  in public goods, 195–196
Great Depression, 165, 213n, 254, 297
Green Revolution, 52, 70–72, 74, 75, 81, 82, 86–88, 133, 238, 329
  Doubly Green Revolution, 406
greenhouse gas (GHG), 127, 128, 130, 138
Guatemala, 160, 299, 301–303, 308, 314, 315
Guinea, 385, 388
Gulf of Mexico, 77, 320

Haiti, 295, 299, 301, 302, 308, 314, 315, 318
Hatch Act of 1887, 60
health-care cost, 263, 409
  government's role, 214
  obesity, 409

health cost, 263, 409
Hitler, Adolf, 32
Honduras, 22, 299, 302, 314
Hong Kong, 188, 194, 211, 241, 268, 274, 280, 285, 287, 288
Household Responsibility System (HRS), 191, 271
human capital
  formation, 219, 229
  regional indicators, 230
human fertility
  demographic transition, 27
  female empowerment, 30, 325
  affluent nations, 246, 247
  in Asia, 274–276
  in Caribbean, 298–300
  in Eastern Europe, 349–353
  in Former Soviet Union, 349–353
  in Latin America, 298–300
  in Middle East, 325
  in North Africa, 325
  in Sub-Saharan Africa, 381
  regional trends, 229
  standards of living, 34, 359
human immunodeficiency virus (HIV), 26, 227, 234, 384, 386, 406
human longevity increase and population dynamics, 227
Hungary, 343, 349, 355, 361, 362, 367
hunger, *see* food security
Hussein, Saddam, 226, 321, 322, 324, 333
hybrid crops and seeds, 61n, 62, 82, 103, 107, 109

Import-Substituting Industrialization (ISI), 192, 224, 293
income distribution and inequality, 184, 226
  in affluent countries, 244
  in Asia, 288
  in Eastern Europe and Former Soviet Union, 347
  in Latin America and the Caribbean, 315
  in the Middle East and North Africa, 339
  in Sub-Saharan Africa, 399
  see also Gini coefficient
income elasticity, 37–40, 48, 142, 179, 258, 265, 285, 395, 411
  of food demand, 37, 39, 48, 112, 179
India, 20, 34, 68, 72, 81, 103, 130, 131, 133–135, 140, 159, 168, 186, 202, 223–225, 227, 233, 267, 270, 273, 274, 276, 279, 284, 288, 302, 390, 405
  Bhopal disaster, 159
  fertilizer subsidies, 133

GDP per capita trends, 223
Green Revolution, 267
human longevity, 20, 227
income inequality, 72, 186
intervention failure, 131, 133, 134
liberalization, 168, 273, 274
Indonesia, 140, 180, 181, 183, 185, 267, 277, 279, 284, 285
induced innovation, 63–64, 235, 359
infant mortality, 20, 29, 30, 233, 247, 288, 315, 337, 367, 399
  in Asia, 288
  in the Middle East and North Africa, 337
  in Eastern Europe and Former Soviet Union, 367
  in Latin America and the Caribbean, 315
  in Sub-Saharan Africa, 399
inferior and normal goods, 35–36
inflation, 129n, 157, 180, 212, 213, 224, 225, 242, 244, 245, 268, 293, 295, 296, 346, 347, 372, 375, 379, 411
intellectual property right (IPR), 58, 81
intensification, 68–75, 279, 281, 307, 331, 391
  environmental impacts of, 68, 279
  impacts on agricultural supply, 68–75
  in affluent nations, 257
  in Asia, 279
  in Latin America and Caribbean, 307–311
  in Middle East and North Africa, 331–335
  in Sub-Saharan Africa, 391–395
Intergovernmental Panel on Climate Change (IPCC), 128
internal rate of return, 62, 110
International Food Policy Research Institute (IFPRI), 102, 160, 199
International Maize and Wheat Improvement Center (CIMMYT), 71
International Monetary Fund (IMF), 165n, 211, 295, 376
International Obesity Task Force (IOTF), 408, 409
International Rice Research Institute (IRRI), 71
intervention failure, 123, 128, 131, 133, 134, 137, 141
Iran, 74, 225, 319–321, 323–322, 324, 325, 327, 330–332, 335
Iraq, 226, 321, 322, 324, 328, 330, 332n, 339
irrigation, 51, 52, 54, 64, 71, 72, 74–78, 80, 112, 122, 123, 129, 131–133, 141, 213, 234, 235, 238, 279, 328–334, 359, 389, 390, 391, 404
Israel, 320–322, 330, 331, 334, 335
Italy, 101, 244, 249, 254, 257, 260, 349
Ivory Coast, 379, 381, 384, 387, 395

Jamaica, 295, 302, 307, 308
Japan, 9, 34, 56, 60, 62, 63, 70, 76, 156, 162, 165, 168, 210, 211, 241, 244, 247, 249, 255–258, 260, 262, 265, 267, 272, 278, 281, 284, 341, 417
  financial crisis, 244
  social mobility, 245
John Paul II, 343
Johnson, D. Gale, 17
Jordan, 322, 324, 327, 328, 330, 334, 335

Kazakhstan, 347, 361, 366
Kenya, 22, 160, 384, 387, 415
Kuwait, 241, 321, 324, 325, 328, 334, 335, 339
Kuznets Curve, 142, 143, 184–186
Kyrgyz Republic, 347, 353, 366

land degradation, 8, 66, 82, 132, 133, 135, 136, 389
land grant colleges, 60
Laos, 276, 284
Latin America, 7, 9, 24, 26, 31, 32, 41, 56, 67, 68, 72, 74, 77, 128, 129, 136, 140, 169, 199, 201–203, 207, 217, 221, 223, 225–227, 229, 231, 233–235, 239, 292–318, 325, 331, 366
  agricultural development, 302–311
    factor use, 304–307
    intensification and extensification, 307–311
  GDP per capita trends, 293–298
  population dynamics, 298–302
    human fertility reduction, 298–300
    natural increase, 301–302
Latvia, 346
Lebanon, 321, 322, 325, 327, 330, 334, 335
Lenin, Vladimir, 189
Lesotho, 377, 382, 385, 388, 393, 400
Liberalization
  in China 163, 271
  in Eastern Europe, 355
  in Former Soviet Union, 355
  in India, 273
  in OECD countries, benefits, 163
  trade policy
    standard model, 213, 216
    see also free trade
Liberia, 374, 376
Libya, 324, 329, 330, 332, 333, 335, 339
Lithuania, 346, 367
livestock, 34–36, 36n, 49–52, 56–58, 60, 80, 84, 89, 137, 141, 181, 182, 190, 285, 287
  water pollution, 52
local food, 203, 209, 414–416
Lomé Convention, 169
Lomborg, Bjorn, 76

long-run equilibrium, 110, 412

Macedonia, 346, 355
Macondo, 157, 158
macroeconomic policies, 212, 213
  in affluent countries, 242, 244
  in Asia, 270
  in Latin America and the Caribbean, 295, 296, 298
  in the Middle East and North Africa, 322
  in Eastern Europe and Former Soviet Union, 342, 343
  in Sub-Saharan Africa, 372, 375
Madagascar, 390
Malawi, 398
Malaysia, 195, 274, 277, 281, 284, 287
Mali, 135, 373n, 380, 395
malnutrition, see food security
Malthus, Thomas, 11–12
Malthusianism, 12–20
  erroneous predictions of, 15
  modern variants of, 12–20
  principle of population, 12
  see also Club of Rome, Paul Ehrlich
Mandela, Nelson, 375
Mankiw, Gregory, 153
Mao, Zedong, 191, 271, 272, 408
market economy
  institutional framework, 211
  tendency toward competitive equilibrium, 97–100
market failure, 145–147
  correction, 147–150
  non-internalization and inefficiency, 145–147
Marxism, 74, 96, 215, 372, 376
Mauritania, 320n, 390, 395
Mauritius, 376, 381, 390, 392, 395
mechanization, 17, 63, 64, 75, 78–80, 87, 88, 107, 235, 237, 239, 251, 310, 331, 334
  in Asia, 279, 281
  in Sub-Saharan Africa, 391
Medvedev, Dmitri, 349
Mendel's genetics, 63
Mexico, 51, 71, 74, 77, 193, 292, 293, 296, 297, 299, 302, 303, 306, 308, 310, 314, 320
micronutrient deficiency, 200
Middle East
  agricultural development, 328–335
    factor use, 329–331
    intensification and extensification, 331–335
  consumption trends, 335–339
  dietary change, 335–339
  food security, 335–339
  GDP per capita trends, 320, 322, 323

population dynamics, 324–328
  human fertility, 325
  natural increase, 325–328
minifundios, 303, 304, 306
Mobutu Sese Seko, 374
Moldova, 347, 353, 367
Mongolia, 270, 280, 281
monopoly, 100, 105, 120, 121, 127, 212
Morocco, 320, 324, 325, 328, 330, 333, 335
Morrill Act, 60
mortality, 15, 18, 20–24, 27, 29, 32, 34, 42, 53, 227, 229, 233, 247, 263, 315, 325, 339, 340, 367, 409
  infant, 20, 29, 30, 233, 247, 288, 315, 337, 339, 340, 367, 399
Mozambique, 172–175, 203, 373, 375–377, 392, 398
Mugabe, Robert, 379
multifunctionality (in agriculture), 250
multinational firms, 158–159
Myanmar, 270, 276, 285

Namibia, 375, 381, 395
Nepal, 72, 140, 270, 276, 277, 284, 285
net economic value (NEV), 97, 119–121, 146, 155, 176
  maximization, 119–121
  and trade, 176
The Netherlands, 246, 254, 256–258, 260
New Economic Policy, of the early Soviet Union, 190
new food economy, 410–412
New Zealand, 32, 62, 167, 242, 247, 249, 253, 256, 258, 260, 265, 304
Nicaragua, 297, 314
Niger, 74, 140, 225, 321, 373–375, 380, 381, 384, 395, 399
Nigeria, 74, 135, 140, 225, 321, 373, 374, 375, 380, 381, 384, 395, 399
nomenklatura, 342
nominal protection coefficient (NPC), 252
nominal rate of assistance (NRA), 284, 310, 362
North Africa, 9, 74, 78, 129, 134, 140, 319–340
  agricultural development, 328–335
    extensification, 331–335
    factor use, 329–331
    intensification, 331–335
  consumption trends, 335–339
  dietary change, 335–339
  food security, 335–339
  GDP per capita trends, 320, 322, 323, 324
  population dynamics, 324–328
    human fertility, 325
    natural increase, 324–328

North America, 9, 24, 32, 57, 63, 75, 86, 169, 184, 237, 241, 249, 255, 265, 267, 272, 296, 306, 308, 362, 413, 417
North American Free Trade Agreement (NAFTA), 296
North, Douglass, 126
Norway, 20, 162, 168, 255

obesity, 262, 263, 265, 407–409, 414
  in rich countries, 262
  International Obesity Task Force (IOTF), 408
  overeating, 263, 407
  reduction measures, 409
  repercussions, 425
Oman, 320, 322, 334
opportunity cost, 30, 36, 83, 124, 152, 153, 173–175, 239, 247, 250, 257, 258, 276, 281, 284, 304, 359, 410
organic food, 250–251, 416
Organization for Economic cooperation and Development (OECD), 162, 226, 242, 281, 302, 331, 343, 408
  food self-sufficiency, 260
  liberalization, 255
Organization of Petroleum Exporting Countries (OPEC), 320, 324
Outward-Looking Strategy, 193–195
own-price elasticity, 37, 44–46, 48, 90, 91, 117
O'Rourke, P.J., 29

Pakistan, 72, 74, 140, 267, 277, 281, 284, 287, 288
Panama, 170, 214, 299, 307
Paraguay, 295, 299
per-capita production trends, in agriculture, 237–238
Peru, 295, 296, 298, 299, 303, 304
Philippines, 74, 80, 160, 270, 274, 276, 277, 280, 287, 288
Poland, 27, 181–183, 186, 343, 346, 347, 349, 361, 362, 366
Pollan, Michael, 416
polluter pays principle, 148
pollution, 15, 52, 58, 106, 127, 130, 141–143, 145, 147–150, 159, 161, 263, 413
  agriculture, 127, 130, 141
  impacts, 141, 142
  industrial and municipal, 143
  tax, 127, 148
  water, 52, 127, 141, 142, 145, 413
population dynamics
  affluent nations
    human fertility, 229–233
  in Asia, 267–289
    human fertility reduction, 274–276
    natural increase, 276–278

population dynamics (*Continued*)
  in Latin America and Caribbean, 292–318
    human fertility reduction, 298–300
    natural increase, 301–302
  in Middle East and North Africa
    human fertility, 325
    natural increase of, 325–328
  regional trends in
    human fertility reduction, 229–233
    human longevity increase, 227–229
    natural increase, 233–234
  see also demographic transition
positive-sum game, 153, 174
poverty, 5, 9, 37, 56, 57, 78, 85, 136, 144, 203, 204, 206, 209, 216, 217, 219, 220, 229, 233, 240, 288, 306, 315, 318, 325, 335, 337, 340, 367
  in Asia, 267, 268, 273, 274, 288
  in Eastern Europe and Former Soviet Union, 367
  in Latin America and the Caribbean, 306, 315
  in the Middle East and North Africa, 325, 335, 337
  in Sub-Saharan Africa, 384, 386, 399
poverty trap, 216, 217
price control, 375
price stability, 213
price support, 104, 105, 131, 157, 161, 254, 265, 412, 413
principle of population, 12, 13, 14, 17, 19, 20, 31
producer subsidy equivalent (PSE), 252, 355
producers' surplus (PS), 97, 98, 154, 176
production possibilities frontier (PPF), 172
productivity growth, in agriculture, 8, 9, 110, 196, 258
property rights, 3, 16, 57, 125–127, 147, 148, 171, 178, 212, 217, 295, 342, 346, 355
protectionism, 152, 155, 156, 165–168, 239, 251, 252, 255, 265, 272, 274, 284, 292, 296, 297, 413
public goods, 57, 58, 85, 135, 136, 188, 195, 196, 211, 212, 214, 215, 219
  TFP growth, in U.S. agriculture, 60
public policy, 100–106, 125, 142, 211, 216, 375, 381
purchasing power parity (PPP), 29, 180
Putin, Vladimir, 348, 349

Qatar, 167, 241, 334

Rask's methodology for food consumption analysis, 335

Reagan, Ronald, 242, 322
regional trends, synopsis
  agriculture's response, to demand growth
    input mix, changes, 235–237
    production trends, 237–238
  economic growth and income distribution, 221–226
    GDP per capita, 223–226
    income distribution differences, 221–226
  population dynamics
    human fertility reduction, 229–233
    human longevity increase, 227–229
    natural increase, 233–234
  research and development, agricultural, 214
    economic returns of, 231
    funding of, 57
    in Japan, 60
    in United States, 81
    role of government, 81, 182
    role of private sector, 81, 182
    see also biotechnology, Green Revolution
Ricardo, David, 12, 152, 156, 405
Rockefeller Foundation, 71, 406
Romania, 346, 362
Rosen, Joseph D., 250
rule of 72, *see* doubling time
rule of law, 3, 106, 159, 197, 210, 211, 339, 342
Russia, 31, 32, 40, 52, 65, 66, 139, 168, 223, 224, 226, 227, 235, 271, 296, 341, 342, 347–350, 353, 358, 359, 361, 362, 366, 367, 370, 371
  see also Former Soviet Union Rwanda

Samuelson, Paul, 152
Saudi Arabia, 124, 237, 322, 324, 325, 327, 329, 334, 335, 339
Scandinavia, 24, 25, 26, 53, 219
Schultz, Theodore W., 60
Sen, Amartya, 200
Senegal, 26, 27, 29, 30, 39, 53, 180–183, 185, 380
shifting cultivation, 55, 141
Sierra Leone, 115, 203, 374, 376, 379, 398
Singapore, 188, 194, 211, 241, 268, 272, 274, 280
slash-and-burn farming, *see* shifting cultivation
Slovenia, 346, 359, 362, 366, 367
Smith, Adam, 405
Smith-Lever Act of 1914, 60
social mobility, 245

soils,
  erosion of, 66, 82, 130n, 132, 134, 135, 254
  in temperate latitudes, 52–53
  in tropics and subtropics, 53, 132
Solow, Robert, 60
Somalia, 373, 374, 403
South Africa, 81, 82, 168n, 374–376, 379, 381, 384, 386, 387, 390, 391, 395, 399, 409
South America, 66, 86, 175, 183, 284, 302, 304, 373
South Asia, 5, 53, 68, 74, 78, 113, 133, 140, 162, 202, 203, 207n, 208, 221, 223, 226, 227, 229, 231, 234, 235, 239, 240, 276, 281, 293, 299, 399
South Korea, 9, 36, 161, 168, 170, 194, 211, 268, 274, 281, 284, 285, 287
Southern Mexico, 51, 303
Soviet Union, *see* Former Soviet Union
Spain, 139, 244, 246, 302, 320, 349
specialization and diversity, in agriculture, 54–59
  economic forces and public policies, 54
  environmental variations, 55
speculation, 100, 101
Sri Lanka, 140, 276, 277, 281, 285, 287, 288, 315
state-owned enterprises (SOEs), 210, 271, 295, 348, 376
Stalin, J., 190
standard model
  communitarian values and economic equity, 217–220
  liberal trade policy, 213
  macroeconomic policies, 212
  market economy, institutional framework, 211, 217
  public goods, government's role, 58, 214
standards of living, 8, 34, 65, 179, 219, 242–245, 358
  and economic structure, 180
  and environmental quality, 151
  and income distribution, 184–186
  in Japan
    financial crisis, 244
    social mobility, 245
  in Spain, 256, 258
  in United States
    economic inequality, 293, 315
    GDP per capita growth, 299, 375
    social mobility, 245
  see also GDP per capita trends
static resource allocation gains of liberalized trade, 162, 163
structural adjustment, 216, 295, 298, 376, 387n, 389

structural transformation, 179–186
  see also Import-Substituting Industrialization, Outward-Looking Strategy
Sub-Saharan Africa, 372–403
  agricultural development, 386–395
    factor use, 387–391
    intensification and extensification, 391–395
  consumption trends and food security, 395–401
  demographic trends, 380–386
    human fertility, 381–384
    natural increase, 384–386
  GDP per capita trends, 372–380
  mechanization, 391
subsidies, 57, 64, 78, 79, 86, 102, 105, 123, 127, 130, 131, 133, 135, 137, 141, 142, 144, 158, 162, 164, 217, 235, 237, 247, 251, 252, 255, 257, 258, 284, 290, 328, 329, 334, 339, 355, 358, 359, 361, 362, 370, 391, 409
  agricultural subsidies, 130, 164, 251–255
  and protectionism, 251–255
  communist-era subsidies, 235, 359
  exports, 157
  farm, 142
  fertilizers, 133
  irrigation, 123, 329
    environmental impacts, adverse, 221, 334
  production, 100, 127
  PSE, 252, 253, 255
  removal of, 361
  underwriting, 217
Sudan, 320n, 373n, 374, 379, 380, 390, 392, 403
supply of food, 63, 93, 109, 188, 221
  agriculture
    increase, 59–75
    nature, 50–59
    per-capita production trends, 85–87
  fundamental economics, 89–90
    changes, 92–93
    elasticity, 90–92
    real and imagined characteristics, 93–94
  shifts, 116–119
supply management, 103
support price, 103, 104, 106, 206
Swaziland, 376, 381, 390–392, 395
Sweden, 27, 37, 39, 182–186, 219, 246, 260
Switzerland, 255
Syria, 26, 27, 33, 180, 181, 183, 185, 320–323, 327, 330–334

Taiwan, 194, 211, 268, 272, 281, 290
Tajikistan, 347, 359, 366, 367

Tanzania, 26, 27, 37, 38, 54, 180, 181, 183, 185, 186, 384, 398
tariff, 102, 152, 155, 156n, 162, 164, 166, 167, 168, 254, 273
   GATT, 152
technology development, *see* research and development, agricultural
technology transfer, *see* extension, agricultural
technophysio evolution, 18–19
temperate zones, 52, 241n
Thailand, 27, 29, 30, 33, 36, 74, 156, 169, 180, 181, 183, 185, 195, 268, 276, 278, 280, 281, 287, 288
Thatcher, Margaret, 242
Togo, 378, 385, 394
total factor productivity (TFP), 60, 109, 195, 310
   inflation-adjusted value of U.S. agricultural output, 62
total fertility rate (TFR), 22, 27, 29, 30, 229, 233, 246, 276, 325, 350
totalitarianism, 402
trade barriers, 79, 86, 102, 131, 162, 163, 192, 258, 413
trade distortions, net costs of
   currency over-valuation, 157
   protectionism, 165
tragedy of the commons, 14
Tunisia, 322, 324, 325, 327, 330, 333
Turkey, 74, 168n, 255, 319, 322, 325, 327, 330, 331, 332n, 337, 339, 340
Turkmenistan, 347, 353, 361, 366
Tyers, Rodney, 164

Uganda, 375, 376, 384, 386, 395
   ABC program, 386
Ukraine, 52, 66, 227, 237, 346, 347, 349, 355, 361, 362, 367
Ulam, Stanislaw, 152
Ultisols, 53
UN Development Program (UNDP), 34, 76, 78, 204, 320, 337
UN Population Division (UNPD), 33, 40, 67, 111, 112, 249
undernourishment, 200, 202, 203,208, 242, 318, 405, 409
   causes, 208
United Arab Emirates, 324, 328, 334, 335
United Kingdom, 168n, 242, 244, 260, 415
United States, 2, 3, 18, 32, 36, 37, 43, 54, 55, 59, 60, 61n, 63, 65, 66, 70, 75, 76, 79, 81–84, 90, 101–103, 105, 107, 109, 110, 115, 118, 127, 142, 156, 157, 164, 165, 168n, 169, 181, 182, 201, 250, 253, 254, 258, 260, 263, 272, 274, 292, 302, 303, 306, 324, 342, 411, 413, 414
   comparative advantage, in agriculture, 367
   economic inequality, 372
   GDP per capita growth, 239, 375
   social mobility, 245
Universal Soil Loss Equation (USLE), 135
urbanization
   human fertility, 325
Uribe, Alvaro, 296
Uruguay, 301, 302, 303, 304
Uruguay Round of trade negotiations, 165, 166, 167
US Agency for International Development (USAID), 355
USSR, *see* Former Soviet Union
Uzbekistan, 347, 353, 361, 366, 367

value added, 75, 85n
Venezuela, 225, 295–298, 299, 308, 310, 314, 321
Vietnam, 74, 102, 106, 156, 157, 267, 270, 276, 280, 281, 284, 287

Wales, 24
Washington Consensus, 211, 218
water resources, 77, 80, 122–123, 306, 329, 412
   in Latin America, 306
   in Middle East and North Africa, 78, 319, 329
   in Sub-Saharan Africa, 78
Weinberger, Caspar, 322
women
   economic empowerment of, 4, 231, 276, 381, 410
   participation in labor force, 30, 381
World Bank, 5, 11, 22, 27, 29, 36, 40, 66, 67, 76, 78–81, 101, 102, 113, 122, 161, 164, 166, 167, 182–184, 207, 211, 216, 217, 224, 244, 246, 247, 267, 268, 310, 355, 374, 375
World Food Summit, 204
World Health Organization (WHO), 386, 408, 409, 413
World Trade Organization (WTO), 152, 165, 167–170, 253, 391, 413
   free trade, 253
   liberalization, 169

Yemen, 322, 325, 327, 328, 334, 335, 339

Zaire, *see* Democratic Republic of the Congo
Zambia, 390, 395, 398
Zimbabwe, 375, 379, 387, 395, 399